Lecture Notes in Computer Science 12422

More information about this series at http://www.springer.com/series/7407

Nicholas Olenev · Yuri Evtushenko ·
Michael Khachay · Vlasta Malkova (Eds.)

Optimization and Applications

11th International Conference, OPTIMA 2020
Moscow, Russia, September 28 – October 2, 2020
Proceedings

 Springer

Editors
Nicholas Olenev 🆔
Dorodnicyn Computing Centre
FRC CSC RAS
Moscow, Russia

Michael Khachay 🆔
Krasovsky Institute of Mathematics
and Mechanics
Ekaterinburg, Russia

Yuri Evtushenko 🆔
Dorodnicyn Computing Centre
FRC CSC RAS
Moscow, Russia

Vlasta Malkova 🆔
Dorodnicyn Computing Centre
FRC CSC RAS
Moscow, Russia

ISSN 0302-9743   ISSN 1611-3349 (electronic)
Lecture Notes in Computer Science
ISBN 978-3-030-62866-6   ISBN 978-3-030-62867-3   (eBook)
https://doi.org/10.1007/978-3-030-62867-3

LNCS Sublibrary: SL1 – Theoretical Computer Science and General Issues

This Springer imprint is published by the registered company Springer Nature Switzerland AG
The registered company address is: Gewerbestrasse 11, 6330 Cham, Switzerland

Preface

This book contains the first volume of the refereed proceedings of the 11th International Conference on Optimization and Applications (OPTIMA 2020)[1]. The goal of the conference is to bring together researchers and practitioners working in the field of optimization theory, methods, software, and related areas. Organized annually since 2009, the conference has attracted a significant number of researchers, academics, and specialists in many fields of optimization, operations research, optimal control, game theory, and their numerous applications in practical problems of operations research, data analysis, and software development. The broad scope of OPTIMA has made it an event where researchers involved in different domains of optimization theory and numerical methods, investigating continuous and discrete extremal problems, designing heuristics and algorithms with theoretical bounds, developing optimization software and applying optimization techniques to highly relevant practical problems, can meet together and discuss their approaches and results. We strongly believe that this facilitates cross-fertilization of ideas between scientists elaborating on the modern optimization theory and methods and employing them to valuable practical problems.

This year, the conference was held online, due to the COVID-19 pandemic situation, during September 28 – October 2, 2020. By tradition, the main organizers of the conference were Montenegrin Academy of Sciences and Arts, Montenegro, Dorodnicyn Computing Centre FRC CSC RAS, Russia, Moscow Institute of Physics and Technology, Russia, and the University of Évora, Portugal. This year, the key topics of OPTIMA were grouped into six tracks:

  (i)   Mathematical programming
 (ii)   Combinatorial and discrete optimization
(iii)   Optimal control
 (iv)   Optimization in economics, finance, and social sciences
  (v)   Global optimization
 (vi)   Applications

In the framework of the conference, a welcome session was held dedicated to the anniversary of an Academician of the Montenegro Academy of Sciences and Arts, Milojica Jacimovic, a world-famous scientist in the field of computational mathematics and one of the founders of the conference.

The Program Committee (PC) of the conference brought together over a hundred well-known experts in continuous and discrete optimization, optimal control and game theory, data analysis, mathematical economy, and related areas from leading institutions of 25 countries including Argentina, Australia, Austria, Belarus, Belgium, Croatia, Finland, France, Germany, Greece, India, Israel, Italy, Kazakhstan,

---

[1] http://www.agora.guru.ru/optima-2020.

Montenegro, The Netherlands, Poland, Portugal, Russia, Serbia, Sweden, Taiwan, Ukraine, the UK, and the USA.

In response to the call for papers, OPTIMA 2020 received 108 submissions. Out of 60 full papers considered for reviewing (46 abstracts and short communications were excluded because of formal reasons) only 23 papers were selected by the PC for publication. Thus, the acceptance rate for this volume is about 38%. Each submission was reviewed by at least three PC members or invited reviewers, experts in their fields, in order to supply detailed and helpful comments. In addition, the PC recommended to include 18 papers in the supplementary volume after their presentation and discussion during the conference and subsequent revision with respect to the reviewers' comments.

The conference featured five invited lecturers as well as several plenary and keynote talks. The invited lectures included:

- Prof. Andrei Dmitruk (CEMI RAS and MSU, Russia), "Lagrange Multipliers Rule for a General Extremum Problem with an Infinite Number of Constraints"
- Prof. Nikolai Osmolovskii (Systems Research Institute, Poland), "Quadratic Optimality Conditions for Broken Extremals and Discontinuous Controls"
- Prof. Boris T. Polyak and Ilyas Fatkhullin (Institute of Control Sciences, Russia), "Static feedback in linear control systems as optimization problem"
- Prof. Alexey Tret'yakov (Siedlce University of Natural Sciences and Humanities, Poland), "P–regularity Theory: Applications to Optimization"

We would like to thank all the authors for submitting their papers and the members of the PC and the reviewers for their efforts in providing exhaustive reviews. We would also like to express special gratitude to all the invited lecturers and plenary speakers.

October 2020

Nicholas Olenev
Yuri Evtushenko
Michael Khachay
Vlasta Malkova

Organization

Program Committee Chairs

Milojica Jaćimović	Montenegrin Academy of Sciences and Arts, Montenegro
Yuri G. Evtushenko	Dorodnicyn Computing Centre, FRC CSC RAS, Russia
Igor G. Pospelov	Dorodnicyn Computing Centre, FRC CSC RAS, Russia
Maksat Kalimoldayev	Institute of Information and Computational Technologies, Kazakhstan

Program Committee

Majid Abbasov	St. Petersburg State University, Russia
Samir Adly	University of Limoges, France
Kamil Aida-Zade	Institute of Control Systems of ANAS, Azerbaijan
Alla Albu	Dorodnicyn Computing Centre, FRC CSC RAS, Russia
Alexander P. Afanasiev	Institute for Information Transmission Problems, RAS, Russia
Yedilkhan Amirgaliyev	Süleyman Demirel University, Kazakhstan
Anatoly S. Antipin	Dorodnicyn Computing Centre, FRC CSC RAS, Russia
Sergey Astrakov	Institute of Computational Technologies, Siberian Branch, RAS, Russia
Adil Bagirov	Federation University, Australia
Evripidis Bampis	LIP6 UPMC, France
Oleg Burdakov	Linköping University, Sweden
Olga Battaïa	ISAE-SUPAERO, France
Armen Beklaryan	National Research University Higher School of Economics, Russia
Vladimir Beresnev	Sobolev Institute of Mathematics, Russia
René Van Bevern	Novosibirsk State University, Russia
Sergiy Butenko	Texas A&M University, USA
Vladimir Bushenkov	University of Évora, Portugal
Igor A. Bykadorov	Sobolev Institute of Mathematics, Russia
Alexey Chernov	Moscow Institute of Physics and Technology, Russia
Duc-Cuong Dang	INESC TEC, Portugal
Tatjana Davidovic	Mathematical Institute of Serbian Academy of Sciences and Arts, Serbia
Stephan Dempe	TU Bergakademie Freiberg, Germany

Alexandre Dolgui	IMT Atlantique, LS2N, CNRS, France
Olga Druzhinina	FRC CSC RAS, Russia
Anton Eremeev	Omsk Division of Sobolev Institute of Mathematics, Siberian Branch, RAS, Russia
Adil Erzin	Novosibirsk State University, Russia
Francisco Facchinei	Sapienza University of Rome, Italy
Vladimir Garanzha	Dorodnicyn Computing Centre, FRC CSC RAS, Russia
Alexander V. Gasnikov	Moscow Institute of Physics and Technology, Russia
Manlio Gaudioso	Università della Calabria, Italy
Alexander I. Golikov	Dorodnicyn Computing Centre, FRC CSC RAS, Russia
Alexander Yu. Gornov	Institute for System Dynamics and Control Theory, Siberian Branch, RAS, Russia
Edward Kh. Gimadi	Sobolev Institute of Mathematics, Siberian Branch, RAS, Russia
Andrei Gorchakov	Dorodnicyn Computing Centre, FRC CSC RAS, Russia
Alexander Grigoriev	Maastricht University, The Netherlands
Mikhail Gusev	N.N. Krasovskii Institute of Mathematics and Mechanics, Russia
Vladimir Jaćimović	University of Montenegro, Montenegro
Vyacheslav Kalashnikov	ITESM, Campus Monterrey, Mexico
Valeriy Kalyagin	Higher School of Economics, Russia
Igor E. Kaporin	Dorodnicyn Computing Centre, FRC CSC RAS, Russia
Alexander Kazakov	Matrosov Institute for System Dynamics and Control Theory, Siberian Branch, RAS, Russia
Mikhail Yu. Khachay	Krasovsky Institute of Mathematics and Mechanics, Russia
Oleg V. Khamisov	L. A. Melentiev Energy Systems Institute, Russia
Andrey Kibzun	Moscow Aviation Institute, Russia
Donghyun Kim	Kennesaw State University, USA
Roman Kolpakov	Moscow State University, Russia
Alexander Kononov	Sobolev Institute of Mathematics, Russia
Igor Konnov	Kazan Federal University, Russia
Vladimir Kotov	Belarus State University, Belarus
Vera Kovacevic-Vujcic	University of Belgrade, Serbia
Yury A. Kochetov	Sobolev Institute of Mathematics, Russia
Pavlo A. Krokhmal	University of Arizona, USA
Ilya Kurochkin	Institute for Information Transmission Problems, RAS, Russia
Dmitri E. Kvasov	University of Calabria, Italy
Alexander A. Lazarev	V.A. Trapeznikov Institute of Control Sciences, Russia
Vadim Levit	Ariel University, Israel
Bertrand M. T. Lin	National Chiao Tung University, Taiwan

Alexander V. Lotov	Dorodnicyn Computing Centre, FRC CSC RAS, Russia
Nikolay Lukoyanov	N.N. Krasovskii Institute of Mathematics and Mechanics, Russia
Vittorio Maniezzo	University of Bologna, Italy
Olga Masina	Yelets State University, Russia
Vladimir Mazalov	Institute of Applied Mathematical Research, Karelian Research Center, Russia
Nevena Mijajlović	University of Montenegro, Montenegro
Nenad Mladenovic	Mathematical Institute, Serbian Academy of Sciences and Arts, Serbia
Angelia Nedich	University of Illinois at Urbana-Champaign, USA
Yuri Nesterov	CORE, Université Catholique de Louvain, Belgium
Yuri Nikulin	University of Turku, Finland
Evgeni Nurminski	Far Eastern Federal University, Russia
Nicholas N. Olenev	CEDIMES-Russie, Dorodnicyn Computing Centre, FRC CSC RAS, Russia
Panos Pardalos	University of Florida, USA
Alexander V. Pestcrev	V.A. Trapeznikov Institute of Control Sciences, Russia
Alexander Petunin	Ural Federal University, Russia
Stefan Pickl	Bundeswehr University Munich, Germany
Boris T. Polyak	V.A. Trapeznikov Institute of Control Sciences, Russia
Yury S. Popkov	Institute for Systems Analysis, FRC CSC RAS, Russia
Leonid Popov	IMM UB RAS, Russia
Mikhail A. Posypkin	Dorodnicyn Computing Centre, FRC CSC RAS, Russia
Oleg Prokopyev	University of Pittsburgh, USA
Artem Pyatkin	Novosibirsk State University, Sobolev Institute of Mathematics, Russia
Ioan Bot Radu	University of Vienna, Austria
Soumyendu Raha	Indian Institute of Science, India
Andrei Raigorodskii	Moscow State University, Russia
Larisa Rybak	Belgorod State Technological University, Russia
Leonidas Sakalauskas	Institute of Mathematics and Informatics, Lithuania
Eugene Semenkin	Siberian State Aerospace University, Russia
Yaroslav D. Sergeyev	University of Calabria, Italy
Natalia Shakhlevich	University of Leeds, UK
Alexander A. Shananin	Moscow Institute of Physics and Technology, Russia
Angelo Sifaleras	University of Macedonia, Greece
Mathias Staudigl	Maastricht University, The Netherlands
Petro Stetsyuk	V.M. Glushkov Institute of Cybernetics, Ukraine
Alexander Strekalovskiy	Matrosov Institute for System Dynamics and Control Theory, Siberian Branch, RAS, Russia
Vitaly Strusevich	University of Greenwich, UK
Michel Thera	University of Limoges, France
Tatiana Tchemisova	University of Aveiro, Portugal

Anna Tatarczak	Maria Curie-Skłodowska University, Poland
Alexey A. Tretyakov	Dorodnicyn Computing Centre, FRC CSC RAS, Russia
Stan Uryasev	University of Florida, USA
Vladimir Voloshinov	Institute for Information Transmission Problems RAS, Russia
Frank Werner	Otto von Guericke University Magdeburg, Germany
Oleg Zaikin	Matrosov Institute for System Dynamics and Control Theory, Siberian Branch, RAS, Russia
Vitaly G. Zhadan	Dorodnicyn Computing Centre, FRC CSC RAS, Russia
Anatoly A. Zhigljavsky	Cardiff University, UK
Julius Žilinskas	Vilnius University, Lithuania
Yakov Zinder	University of Technology, Australia
Tatiana V. Zolotova	Financial University under the Government of the Russian Federation, Russia
Vladimir I. Zubov	Dorodnicyn Computing Centre, FRC CSC RAS, Russia
Anna V. Zykina	Omsk State Technical University, Russia

Organizing Committee Chairs

Milojica Jaćimović	Montenegrin Academy of Sciences and Arts, Montenegro
Yuri G. Evtushenko	Dorodnicyn Computing Centre, FRC CSC RAS, Russia
Nicholas N. Olenev	Dorodnicyn Computing Centre, FRC CSC RAS, Russia

Organizing Committee

Gulshat Amirkhanova	Institute of Information and Computational Technologies, Kazakhstan
Natalia Burova	Dorodnicyn Computing Centre, FRC CSC RAS, Russia
Alexander Golikov	Dorodnicyn Computing Centre, FRC CSC RAS, Russia
Alexander Gornov	Institute of System Dynamics and Control Theory, Siberian Branch, RAS, Russia
Vesna Dragović	Montenegrin Academy of Sciences and Arts, Montenegro
Vladimir Jaćimović	University of Montenegro, Montenegro
Mikhail Khachay	Krasovsky Institute of Mathematics and Mechanics, Russia
Yury Kochetov	Sobolev Institute of Mathematics, Russia

Elena A. Koroleva	Dorodnicyn Computing Centre, FRC CSC RAS, Russia
Vlasta Malkova	Dorodnicyn Computing Centre, FRC CSC RAS, Russia
Nevena Mijajlović	University of Montenegro, Montenegro
Oleg Obradovic	University of Montenegro, Montenegro
Mikhail A. Posypkin	Dorodnicyn Computing Centre, FRC CSC RAS, Russia
Kirill B. Teymurazov	Dorodnicyn Computing Centre, FRC CSC RAS, Russia
Yulia Trusova	Dorodnicyn Computing Centre, FRC CSC RAS, Russia
Svetlana Vladimirova	Dorodnicyn Computing Centre, FRC CSC RAS, Russia
Victor Zakharov	FRC CSC RAS, Russia
Elena S. Zasukhina	Dorodnicyn Computing Centre, FRC CSC RAS, Russia
Ivetta Zonn	Dorodnicyn Computing Centre, FRC CSC RAS, Russia
Vladimir Zubov	Dorodnicyn Computing Centre, FRC CSC RAS, Russia

Contents

Optimization of the Values of the Right-Hand Sides of Boundary Conditions with Point and Integral Terms for the ODE System

Kamil Aida-zade[1,2]([✉])[iD] and Vagif Abdullayev[1,3][iD]

[1] Institute of Control Systems of Azerbaijan National Academy of Sciences, Str. B. Vahabzade 9, Baku AZ1141, Azerbaijan
{kamil_aydazade,vaqif_ab}@rambler.ru
[2] Institute of Mathematics and Mechanics of Azerbaijan National Academy of Sciences, Str. B. Vahabzade 9, Baku AZ1141, Azerbaijan
[3] Azerbaijan State Oil and Industry University, Azadlig ave. 20, Baku AZ1010, Azerbaijan

Abstract. In the paper, we investigate the problem of optimal control for a linear system of ordinary differential equations with linear boundary conditions. The boundary conditions include, as terms, the values of the phase variable both at separate intermediate points and their integral values over individual intervals of the independent variable. The values of the right sides of unseparated boundary conditions are optimizing in the problem. In the paper the necessary conditions for the existence and uniqueness of the solution to the boundary value problem, the convexity of the target functional, the necessary optimality conditions for the optimized parameters are obtained. Conditions contain constructive formulas of the gradient components of the functional. The numerical solution of an illustrative problem is considered.

Keywords: Functional gradient · Convexity of functional · Nonlocal conditions · Optimality conditions · Multipoint conditions

1 Introduction

The paper deals with the problem of optimal control of a system of ordinary linear differential equations with nonlocal conditions, containing summands with point and integral values of the phase function.

The study of nonlocal boundary value problems was started in [1–3] and continued in the works of many other researchers [4–13]. The practical importance of investigating these problems and the related control problems is due to the fact that the measured information about the state of the object at any point of the object or at any time instant actually covers the neighborhood of the point or the instant of measurement.

© Springer Nature Switzerland AG 2020
N. Olenev et al. (Eds.): OPTIMA 2020, LNCS 12422, pp. 1–16, 2020.
https://doi.org/10.1007/978-3-030-62867-3_1

It should be noted that various statements of control problems with multipoint and intermediate conditions are considered in many studies [14–16]. Optimality conditions in various forms were obtained for them [17–19] and numerical schemes for solving these problems were proposed; cases of nonlocal (loaded) differential equations with multipoint boundary conditions were investigated in [12, 20].

This research differs from previous studies primarily in that we optimize the values of the right-hand sides of nonlocal boundary conditions. In this paper, we obtain the conditions of existence and uniqueness of the solution to a nonlocal boundary value problem, the convexity of the objective functional of the problem, and the necessary optimality conditions for the optimal control and optimization problem under investigation. Depending on the rank of the matrix of coefficients at the point values of the phase vector function in nonlocal conditions, two techniques are proposed to obtain formulas for the components of the gradient of the functional. One technique uses the idea of the conditional gradient method [21, 22], and the other the Lagrange method [22]. The use of the Lagrange method leads to an increase in both the number of conditions and the dimension of the vector of parameters to be determined. Therefore, depending on the rank of the matrix of conditions, preference is given to the first approach.

The paper gives the results of numerical experiments, using one illustrative problem as an example. The formulas obtained here for the components of the gradient of the objective functional are used for its numerical solution by first-order optimization methods.

2 Formulation of the Problem

We consider an object whose dynamics is described by a linear system of differential equations:

$$\dot{x}(t) = A_1(t)x(t) + A_2(t), \quad t \in [t^1, t^f], \tag{1}$$

with nonlocal (unseparated multipoint and integral) conditions and with optimized values of the right-hand sides:

$$\sum_{i=1}^{l_1} \alpha_i x(\tilde{t}^i) + \sum_{j=1}^{l_2} \int_{\hat{t}^{2j-1}}^{\hat{t}^{2j}} \beta_j(t)x(t)dt = \vartheta. \tag{2}$$

Here: $x(t) \in R^n$ is a phase variable. Given are: matrix functions $A_1(t) \neq const$, $\beta_j(t)$, $j = 1, 2, ..., l_2$ of dimension $(n \times n)$, continuous at $t \in [t^1, t^f]$, and n–dimensional vector function $A_2(t)$; scalar matrices α_i, $i = 1, 2, ..., l_1$ of dimension $(n \times n)$; instants of time $\tilde{t}^i \in [t^1, t^f]$, $i = 1, 2, ..., l_1$, $\tilde{t}^1 < ... < \tilde{t}^{i-1} < \tilde{t}^i < ... < \tilde{t}^{l_1}$, $\hat{t}^1 < ... < \hat{t}^{i-1} < \hat{t}^i < ... < \hat{t}^{2l_2}$, with $\tilde{t}^i \notin [\hat{t}^{2j-1}, \hat{t}^{2j}]$, $i = 1, 2, ..., l_1$, $j = 1, 2, ..., l_2$, and $t^1 = \tilde{t}^1 < \hat{t}^1$, $\hat{t}^{2l_2} < \tilde{t}^{l_1} = t^f$; l_1, l_2 are integers. The optimized parameter of the problem is the vector $\vartheta \in V \subset R^n$, which is normally determined by the impacts of external sources. The set of admissible values V of the optimized vector of parameters ϑ is compact and convex.

The objective functional for finding the vector of parameters ϑ is

$$J(\vartheta) = \int\limits_{t^1}^{t^f} f^0(x(t),\, \vartheta, t)\, dt + \Phi(\bar{x},\, \vartheta) \to \min_{\vartheta \in V}. \tag{3}$$

Here, $f^0(x, \vartheta, t)$, $\Phi(\bar{x}, \vartheta)$ are given functions continuously differentiable in x, ϑ, $\bar{x}$, and the following notation is used:

$$\tilde{t} = (\tilde{t}^1, \tilde{t}^2, ..., \tilde{t}^{l_1}), \quad \hat{t} = (\hat{t}^1, \hat{t}^2, ..., \hat{t}^{2l_2}), \quad \bar{t} = (\tilde{t}^1, \tilde{t}^2, ..., \tilde{t}^{l_1}, \hat{t}^1, \hat{t}^2, ..., \hat{t}^{2l_2}),$$

$$\tilde{x} = (\tilde{x}^1, \tilde{x}^2, ..., \tilde{x}^{l_1})^T = (x(\tilde{t}^1), x(\tilde{t}^2), ..., x(\tilde{t}^{l_1}))^T \in R^{l_1 n},$$

$$\hat{x} = (\hat{x}^1, \hat{x}^2, ..., \hat{x}^{2l_2})^T = (x(\hat{t}^1), x(\hat{t}^2), ..., x(\hat{t}^{2l_2}))^T \in R^{2l_2 n},$$

$$x(\bar{t}) = \bar{x} = (\bar{x}^1, \bar{x}^2, ..., \bar{x}^{l_1 + 2l_2})^T = (x(\tilde{t}^1), x(\tilde{t}^2), ..., x(\tilde{t}^{l_1}), x(\hat{t}^1), ..., x(\hat{t}^{2l_2}))^T,$$

"T"–transpose sign.

Assume that there are n linearly independent conditions in (2). In the case of a smaller number of linearly independent conditions or a smaller number of conditions in general, e.g. n_1, $n_1 < n$, then, accordingly, $\vartheta \in R^{n_1}$. This means that there are $(n - n_1)$ free (non-fixed) initial conditions in the problem. Then, the optimized values of the initial conditions for any $(n - n_1)$ components of the phase vector $x(t)$ can be added to conditions (2), thereby expanding the vector $\vartheta \in R^{n_1}$ to $\vartheta \in R^n$.

3   Obtaining Optimality Conditions

Further we assume that for all parameters ϑ, problem (1), (2) has a solution, and a unique one at that. This requires that condition given in the following theorem be satisfied. In the theorem, the matrix $F(t, \tau)$ is a fundamental matrix of solutions to system (1), i.e. a solution to the Cauchy matrix problem:

$$\dot{F}(t, \tau) = A_1(t)F(t, \tau), \quad t, \tau \in [t^1, t^f], \quad F(t^1, t^1) = E,$$

where E is an n-dimensional identity matrix.

Theorem 1. *Problem (1), (2) for an arbitrary vector of parameters ϑ has a solution, and a unique one at that, if the functions $A_1(t)$ are continuous, $A_2(t)$, $\beta_i(t)$, $i = 1, 2, ..., l_2$, integrable and*

$$rang \left[\sum_{i=1}^{l_1} \alpha_i F(\tilde{t}^i, t^1) + \sum_{j=1}^{l_2} \int\limits_{\hat{t}^{2j-1}}^{\hat{t}^{2j}} \beta_j(t)F(t, t^1)dt \right] = n. \tag{4}$$

Proof. The proof of the theorem follows from a direct substitution of the Cauchy formula with respect to system (1)

$$x(t) = F(t, t^1)x^1 + \int_{t^1}^{t} F(t, \tau) A_2(\tau)d\tau, \tag{5}$$

in conditions (2). After simple transformations and grouping, we obtain an algebraic system with respect to $x^1 = x(t^1)$:

$$Lx^1 = D, \tag{6}$$

$$L = \sum_{i=1}^{l_1} \alpha_i F(\tilde{t}^i, t^1) + \sum_{j=1}^{l_2} \int_{\tilde{t}^{2j-1}}^{\tilde{t}^{2j}} \beta_j(t) F(t, t^1)dt,$$

$$D = \vartheta - \sum_{i=1}^{l_1} \alpha_i \left[\int_{t^1}^{\tilde{t}^i} F(\tilde{t}^i, \tau) A_2(\tau)d\tau \right] - \sum_{j=1}^{l_2} \int_{\tilde{t}^{2j-1}}^{\tilde{t}^{2j}} \beta_j(t) \left[\int_{t^1}^{t} A_2(\tau)d\tau \right] dt.$$

It is known that system of equations (6) has a solution, and a unique one at that, if the matrix L is invertible, i.e. when condition (4) is satisfied. It is clear that $rang\, L$ does not depend on the values of the vector D, and therefore, it does not depend on the vector ϑ. And due to the uniqueness of representation (5) for the solution of the Cauchy problem with respect to system (1), problem (1), (2) also has a solution, and a unique one at that, when condition (4) is satisfied.

The following theorem takes place.

Theorem 2. *Suppose all the conditions imposed on the functions and parameters in problem (1)–(3) are satisfied, and the functions $f^0(x, \vartheta, t)$ and $\Phi(\bar{x}, \vartheta)$ are convex in $x, \bar{x}, \vartheta$. Then the functional $J(\vartheta)$ is convex. If, in addition, one of the functions $f^0(x, \vartheta, t)$ and $\Phi(\bar{x}, \vartheta)$ is strongly convex, then the functional of the problem is also strongly convex.*

The proof of the theorem follows from a direct check of the convexity condition of the integer functional under the conditions of convexity of the functions $f^0(x, \vartheta, t)$ and $\Phi(\bar{x}, \vartheta)$.

Let us investigate the differentiability of functional (3) and obtain formulas for the components of its gradient with respect to the optimized parameters $\vartheta \in R^n$.

The derivatives $\partial f^0/\partial x$, $\partial f^0/\partial \vartheta$, $\partial \Phi/\partial \bar{x}^i$ will be understood as rows of the corresponding dimensions. For arbitrary functions $f(t)$, we use the notation:

$$f(t_\pm) = f(t \pm 0) = \lim_{\varepsilon \to +0} f(t \pm \varepsilon), \quad \Delta f(t) = f(t_+) - f(t_-),$$

and we assume that $f(t_+^f) = f(t_-^1) = 0$.

$\chi_{[\hat{t}^{2j-1},\ \hat{t}^{2j}]}(t)$, $j = 1, 2, ..., l_2$ is a characteristic function:

$$\chi_{[\hat{t}^{2j-1},\ \hat{t}^{2j}]}(t) = \begin{cases} 0, & t \notin [\hat{t}^{2j-1}, \hat{t}^{2j}], \\ 1, & t \in [\hat{t}^{2j-1}, \hat{t}^{2j}], \end{cases} \quad j = 1, 2, ..., l_2.$$

Suppose the rank of the augmented matrix $\alpha = [\alpha_1, \alpha_2, ..., \alpha_{l_1}]$ of dimension $n \times l_1 n$ is $\bar{n}$. Then

$$rang\, \alpha = \bar{n} \le n. \tag{7}$$

In the case of $\bar{n} < n$, conditions (2), due to their linear combination, can be reduced to such a form that the last $(n - \bar{n})$ rows of the matrix $\alpha = [\alpha_1, \alpha_2, ..., \alpha_{l_1}]$ will be zero, the integral summands in conditions (2) also undergoing a linear combination. But it is important that these transformations do not violate condition (4) for the existence and uniqueness of the solution of boundary value problem (1), (2) for arbitrary ϑ. To avoid introducing new notation, suppose the matrices α_i, $i = 1, 2, ..., l_1$ have dimension $\bar{n} \times n$, and the rank of their augmented matrix is $\bar{n}$. Then we divide constraints (2) into two parts, writing the first $\bar{n}$ constraints in the following form:

$$\sum_{i=1}^{l_1} \alpha_i x(\hat{t}^i) + \sum_{j=1}^{l_2} \int_{\hat{t}^{2j-1}}^{\hat{t}^{2j}} \beta_j^1(t) x(t) dt = \vartheta^{(1)}, \tag{8}$$

and in the last $n - \bar{n}$ constraints, there will be no point values of the function $x(t)$:

$$\sum_{j=1}^{l_2} \int_{\hat{t}^{2j-1}}^{\hat{t}^{2j}} \beta_j^2(t) x(t) dt = \vartheta^{(2)}. \tag{9}$$

Here, the matrices α_i, $\beta_j^1(t)$, $i = 1, 2, ..., l_1$, $j = 1, 2, ..., l_2$ have dimension $\bar{n} \times n$, and the matrices $\beta_j^2(t)$, $j = 1, 2, ..., l_2$ have dimension $(n - \bar{n}) \times n$, the vectors $\vartheta^{(1)} \in R^{\bar{n}}$, $\vartheta^{(2)} \in R^{n-\bar{n}}$, $\vartheta = (\vartheta^{(1)}, \vartheta^{(2)}) \in R^n$.

Therefore, it is possible to extract from the augmented matrix α the matrix (minor) $\widehat{\alpha}$ of rank $\bar{n}$ formed by $\bar{n}$ columns of the matrix α.

Suppose k_i is the number of the column of the matrix α_{s_i}, $1 \le s_i \le l_1$, $i = 1, 2, ..., \bar{n}$, included as the i-th column in the matrix $\widehat{\alpha}$, i.e. the i-th column of the matrix $\widehat{\alpha}$ is the k_i-th column of the matrix α_{s_i}.

Suppose $\widehat{x} = (\widehat{x}^1, \widehat{x}^2, ..., \widehat{x}^{\bar{n}})^T = (x_{k_1}(\tilde{t}^{s_1}), x_{k_2}(\tilde{t}^{s_2}), ..., x_{k_{\bar{n}}}(\tilde{t}^{s_{\bar{n}}}))^T = (\tilde{x}_{k_1}^{s_1}, \tilde{x}_{k_2}^{s_2}, ..., \tilde{x}_{k_{\bar{n}}}^{s_{\bar{n}}})^T$ is an $\bar{n}$-dimensional vector consisting of the components of the vector $x(\tilde{t})$ formed from the values of the k_j-th coordinates of the n-dimensional vector $x(t)$ corresponding to the matrix $\widehat{\alpha}$ at time instants $\tilde{t}^{s_j}$, $1 \le s_j \le l_1$, $j = 1, 2, ..., \bar{n}$.

Suppose $\breve{\alpha}$ and $\breve{x}$ are residual $(\bar{n} \times (l_1 n - \bar{n}))$-dimensional matrix and residual $(l_1 n - \bar{n})$-dimensional vector, respectively, obtained by removing from the matrix α and from the vector $\tilde{x}$ $\bar{n}$ columns included in the matrix $\widehat{\alpha}$ and $\bar{n}$ components included in the vector $\widehat{x}$, respectively.

Suppose the i-th column of the matrix $\breve{\alpha}$ is the g_i-the column of the matrix α_{q_i}, $1 \le g_i \le n$, $1 \le q_i \le l_1$, $i = 1, 2, ..., (l_1 n - \bar{n})$.

$$\breve{x} = (x_{g_1}(\tilde{t}^{q_1}), x_{g_2}(\tilde{t}^{q_2}), ..., x_{g_{(l_1 n - \bar{n})}}(\tilde{t}^{q(l_1 n - \bar{n})}))^T = (\tilde{x}_{g_1}^{q_1}, \tilde{x}_{g_2}^{q_2}, ..., \tilde{x}_{g_{(l_1 n - \bar{n})}}^{q(l_1 n - \bar{n})})^T.$$

Obviously, $(g_i, q_i) \ne (s_j, k_j)$, $j = 1, 2, ..., \bar{n}$, $i = 1, 2, ..., (l_1 n - \bar{n})$.

For simplicity of formula notation, the $\bar{n}$-dimensional square matrix $\widehat{\alpha}^{-1}$ is denoted by C with elements c_{ij}, and the $(\bar{n} \times (l_1 n - \bar{n}))$ matrix $(-\widehat{\alpha}^{-1}\breve{\alpha})$ is denoted by B with elements b_{ij}.

Next, we consider separately the cases when $\bar{n} = n$ and $\bar{n} < n$.

In the case of $\bar{n} = n$, the following theorem takes place.

Theorem 3. *In the conditions imposed on the functions and parameters involved in problem (1)–(3), functional (3) is differentiated with respect to the parameters ϑ of the right-hand sides of nonlocal boundary conditions. The gradient of the functional of the problem for $\operatorname{rang} \alpha = \operatorname{rang} \widehat{\alpha} = n$ is determined by the formulas*

$$\frac{\partial J}{\partial \vartheta_k} = \sum_{i=1}^{n} \left[\frac{\partial \Phi(\bar{x}, \vartheta)}{\partial \tilde{x}_{k_i}^{s_i}} + \Delta \psi_{k_i}(\tilde{t}^{s_i}) \right] c_{ik} + \frac{\partial \Phi(\bar{x}, \vartheta)}{\partial \vartheta_k} + \int_{t^1}^{t^f} \frac{\partial f^0(x, \vartheta, t)}{\partial \vartheta_k} dt, \quad (10)$$

where $k = 1, 2, ..., n$, the vector function $\psi(t)$, which is continuously differentiable over the interval $[t^1, t^f]$ except the points $\tilde{t}^i$, $\hat{t}^j$, $i = 2, 3, .., l_1 - 1$, $j = 1, 2, .., 2l_2$, is the solution to the adjoint problem:

$$\dot{\psi}(t) = -A_1^T(t)\psi(t) + \left(\frac{\partial f^0(x, \vartheta, t)}{\partial x} \right)^T + \sum_{j=1}^{l_2} \left(\chi(\hat{t}^{2j}) - \chi(\hat{t}^{2j-1}) \right) \beta_j^T(t) \left(\widehat{\alpha}^{-1} \right)^T$$

$$\times \sum_{i=1}^{n} \left(\frac{\partial \Phi(\bar{x}, \vartheta)}{\partial \tilde{x}_{k_i}^{s_i}} + \psi_{k_i}(\tilde{t}_-^{s_i}) - \psi_{k_i}(\tilde{t}_+^{s_i}) \right), \quad (11)$$

$$\psi_{g_\nu}(\tilde{t}_+^{q_\nu}) = \psi_{g_\nu}(\tilde{t}_-^{q_\nu}) + \sum_{i=1}^{n} b_{i\nu} \left(\frac{\partial \Phi(\bar{x}, \vartheta)}{\partial \tilde{x}_{k_i}^{s_i}} + \psi_{k_i}(\tilde{t}_-^{s_i}) - \psi_{k_i}(\tilde{t}_+^{s_i}) \right) + \frac{\partial \Phi(\bar{x}, \vartheta)}{\partial \tilde{x}_{g_\nu}^{q_\nu}}, \quad (12)$$

$$\psi_i(\hat{t}_+^j) = \psi_i(\hat{t}_-^j) + \frac{\partial \Phi(\bar{x}, \vartheta)}{\partial \hat{x}_i^j}, \quad i = 1, 2, ..., n, \quad j = 1, 2, ..., 2l_2, \quad (13)$$

where $\nu = 1, 2, ..., l_1 n$.

Proof. Suppose $x(t) \in R^n$ is the solution to boundary value problem (1), (2) for an admissible vector of parameters $\vartheta \in V$, and $x^1(t) = x(t) + \Delta x(t)$ is the solution to problem (1), (2) corresponding to the increment admissible vector $\vartheta^1 = \vartheta + \Delta\vartheta \in V$:

$$\dot{x}^1(t) = A_1(t)x^1(t) + A_2(t), \quad t \in [t^1, t^f], \quad (14)$$

$$\sum_{i=1}^{l_1} \alpha_i x^1(\tilde{t}^i) + \sum_{j=1}^{l_2} \int_{\hat{t}^{2j-1}}^{\hat{t}^{2j}} \beta_j(t) x^1(t) dt = \vartheta^1. \tag{15}$$

It follows from (1), (2) and (14), (15) that the following takes place:

$$\Delta \dot{x}(t) = A_1(t) \Delta x(t), \ t \in [t^1, t^f], \tag{16}$$

$$\sum_{i=1}^{l_1} \alpha_i \Delta x(\tilde{t}^i) + \sum_{j=1}^{l_2} \int_{\hat{t}^{2j-1}}^{\hat{t}^{2j}} \beta_j(t) \Delta x(t) dt = \Delta \vartheta. \tag{17}$$

Then for the increment of functional (3), we have

$$\Delta J(\vartheta) = J(\vartheta^1) - J(\vartheta) = \int_{t^1}^{t^f} \left[\frac{\partial f^0(x, \vartheta, t)}{\partial x} \Delta x(t) + \frac{\partial f^0(x, \vartheta, t)}{\partial \vartheta} \Delta \vartheta \right] dt$$

$$+ \sum_{i=1}^{l_1} \frac{\partial \Phi(\bar{x}, \vartheta)}{\partial \tilde{x}^i} \Delta x(\tilde{t}^i) + \sum_{j=1}^{2l_2} \frac{\partial \Phi(\bar{x}, \vartheta)}{\partial \hat{x}^j} \Delta x(\hat{t}^j) + \frac{\partial \Phi(\bar{x}, \vartheta)}{\partial \vartheta} \Delta \vartheta + R. \tag{18}$$

Here $R = o\left(\|\Delta x(t)\|_{C^{1,n}[t^1, t^f]}, \|\Delta \vartheta\|_{R^n}\right)$ is a residual member. Under the assumptions made for the data of problems (1), (2), using the known technique [22], we can obtain an estimate of the form $\|\Delta x(t)\|_{C^{1,n}[t^1, t^f]} \leq c_1 \|\Delta \vartheta\|_{R^n}$ where $c_1 > 0$ is independent of $x(t)$. From here, taking into account formula (18), we have the differentiability of the functional $J(\vartheta)$ with respect to ϑ.

Let us transpose the right-hand side members of (16) to the left and multiply scalarwise both sides of the resulting equality by the as yet arbitrary n-dimensional vector function $\psi(t)$, which is continuously differentiable over the intervals $(\bar{t}^i, \bar{t}^{i+1})$, $i = 1, 2, ..., (l_1 + 2l_2 - 1)$. Integrating the obtained equality by parts and adding the result to (18), after simple transformations, we will have:

$$\Delta J(\vartheta) = \sum_{i=1}^{(l_1+2l_2-1)} \int_{\bar{t}^i}^{\bar{t}^{i+1}} \left[-\dot{\psi}^T(t) - \psi^T(t) A_1(t) + \frac{\partial f^0(x, \vartheta, t)}{\partial x} \right] \Delta x(t) dt$$

$$+ \left[\int_{t^1}^{t^f} \frac{\partial f^0(x, \vartheta, t)}{\partial \vartheta} dt + \frac{\partial \Phi(\bar{x}, \vartheta)}{\partial \vartheta} \right] \Delta \vartheta + \left\{ \sum_{i=1}^{l_1} \frac{\partial \Phi(\bar{x}, \vartheta)}{\partial \tilde{x}^i} \Delta x(\tilde{t}^i) \right.$$

$$+ \sum_{j=1}^{2l_2} \frac{\partial \Phi(\bar{x}, \vartheta)}{\partial \hat{x}^j} \Delta x(\hat{t}^j) + \psi^T(t^f) \Delta x(t^f) - \psi^T(t^1) \Delta x(t^1)$$

$$+ \sum_{i=2}^{l_1-1} \left[\psi(\tilde{t}^i_-) - \psi(\tilde{t}^i_+) \right]^T \Delta x(\tilde{t}^i) + \sum_{j=1}^{2l_2} \left[\psi(\hat{t}^j_-) - \psi(\hat{t}^j_+) \right]^T \Delta x(\hat{t}^j) \right\} + R. \tag{19}$$

Let us deal with the summands inside the curly brackets.

In (19), we use conditions (17) to obtain the dependence of any n components of the (nl_1)–dimensional vector

$$\Delta x(\tilde{t}) = \Delta \tilde{x} = \left(\Delta x_1(\tilde{t}^1), \Delta x_2(\tilde{t}^1), ..., \Delta x_n(\tilde{t}^1), ..., \Delta x_i(\tilde{t}^j), ..., \Delta x_n(\tilde{t}^{l_1}) \right),$$

through the other $n(l_1 - 1)$ components.

Then relation (17) can be written as:

$$\widehat{\alpha} \, \Delta\widehat{x} + \breve{\alpha}\Delta\breve{x} + \sum_{j=1}^{l_2} \int_{\hat{t}^{2j-1}}^{\hat{t}^{2j}} \beta_j(t)\Delta x(t)dt = \Delta\vartheta.$$

Hence, taking into account (7), we have

$$\Delta\widehat{x} = \widehat{\alpha}^{-1}\Delta\vartheta - \widehat{\alpha}^{-1}\breve{\alpha}\Delta\breve{x} - \sum_{j=1}^{l_2} \int_{\hat{t}^{2j-1}}^{\hat{t}^{2j}} \widehat{\alpha}^{-1}\beta_j(t)\Delta x(t)dt. \tag{20}$$

Further, for simplicity of presentation of technical details, alongside matrix operations, we will use component-wise formula notation. Given the agreed notation: $C = -\widehat{\alpha}^{-1}$, $B = -\widehat{\alpha}^{-1}\breve{\alpha}$, (20) in a component-wise form will look as follows:

$$\Delta\widehat{x}_i = \Delta x_{k_i}(\tilde{t}^{s_i}) = \sum_{k=1}^{n} c_{ik}\Delta\vartheta_k + \sum_{\nu=1}^{l_1 n} b_{i\nu}\Delta x_{g_\nu}(\tilde{t}^{q_\nu})$$

$$-\sum_{j=1}^{l_2}\sum_{k=1}^{n}\int_{\hat{t}^{2j-1}}^{\hat{t}^{2j}} \widehat{\alpha}^{-1}\beta_{ik}^j(t)\Delta x_k(t)dt, \; i = 1, 2, ..., n, \; 1 \le g_\nu \le n. \tag{21}$$

The last 4–7 summands in (19) are written as follows:

$$\psi^T(t^f)\Delta x(t^f) = \sum_{j=1}^{n}\psi_j(t^f)\Delta x_j(t^f), \quad \psi^T(t^1)\Delta x(t^1) = \sum_{j=1}^{n}\psi_j(t^1)\Delta x_j(t^1). \tag{22}$$

From (19), taking into account (21)–(22), we can obtain:

$$\Delta J(\vartheta) = \sum_{i=1}^{(l_1+2l_2-1)} \int_{\tilde{t}^i}^{\tilde{t}^{i+1}} \left[-\dot{\psi}^T(t) - \psi^T(t)A_1(t) + \frac{\partial f^0(x, \vartheta, t)}{\partial x} \right.$$

$$\left. -\sum_{i=1}^{n}\left(\frac{\partial\Phi(\tilde{x}, \vartheta)}{\partial\tilde{x}_{k_i}^{s_i}} + \Delta\psi_{k_i}(\tilde{t}^{s_i}) \right) \sum_{j=1}^{l_2}\left(\chi(\hat{t}^{2j}) - \chi(\hat{t}^{2j-1}) \right) \widehat{\alpha}^{-1}\beta_j(t) \right] \Delta x(t)dt$$

$$+ \sum_{k=1}^{n} \left\{ \sum_{i=1}^{n} \left(\frac{\partial \Phi(\tilde{x}, \vartheta)}{\partial \tilde{x}_{ki}^{s_i}} + \Delta \psi_{k_i}(\tilde{t}^{s_i}) \right) c_{ik} + \frac{\partial \Phi(\tilde{x}, \vartheta)}{\partial \vartheta_k} + \int_{t^1}^{t^f} \frac{\partial f^0(x, \vartheta, t)}{\partial \vartheta_k} dt \right\} \Delta \vartheta_k$$

$$+ \sum_{\nu=1}^{l_1 n} \left[\sum_{i=1}^{n} b_{i\nu} \left(\frac{\partial \Phi(\tilde{x}, \vartheta)}{\partial \tilde{x}_{ki}^{s_i}} + \Delta \psi_{k_i}(\tilde{t}^{s_i}) \right) + \left(\frac{\partial \Phi(\tilde{x}, \vartheta)}{\partial \tilde{x}_{g_\nu}^{q_\nu}} + \Delta \psi_{g_\nu}(\tilde{t}^{q_\nu}) \right) \right] \Delta x_{g_\nu}(\tilde{t}^{q_\nu})$$

$$+ \sum_{j=1}^{2l_2} \sum_{i=1}^{n} \left[\frac{\partial \Phi(\tilde{x}, \vartheta)}{\partial \hat{x}_i^j} + \Delta \psi_i(\hat{t}^j) \right] \Delta x_i(\hat{t}^j) + R. \tag{23}$$

Since the vector functions $\psi(t)$ are arbitrary, let us require that the expressions in the first and last two square brackets (23) be 0. From the first requirement, we obtain adjoint system of differential equations (11), and from the other two requirements, we obtain the following expressions:

$$\sum_{i=1}^{n} b_{i\nu} \left[\frac{\partial \Phi(\tilde{x}, \vartheta)}{\partial \tilde{x}_{ki}^{s_i}} + \Delta \psi_{k_i}(\tilde{t}^{s_i}) \right] + \left[\frac{\partial \Phi(\tilde{x}, \vartheta)}{\partial \tilde{x}_{g_\nu}^{q_\nu}} + \Delta \psi_{g_\nu}(\tilde{t}^{q_\nu}) \right] = 0, \quad \nu = 1, 2, ..., l_1 n,$$

$$\frac{\partial \Phi(\tilde{x}, \vartheta)}{\partial \hat{x}_i^j} + \Delta \psi_i(\hat{t}^j) = 0, \quad i = 1, 2, ..., n, \quad j = 1, 2, ..., 2l_2.$$

Hence conditions (12), (13).

Then, the desired components of the gradient of the functional with respect to ϑ will be determined from (23) as the linear parts of the increment of the functional for $\Delta \vartheta$ from formulas (10).

From Theorem 3, we can obtain formulas for simpler special cases when the ranks of any of the matrices α_i, $i = 1, 2, ..., l_1$ are n and then they can be taken as the matrix $\hat{\alpha}$.

For instance, when $rang \alpha_1 = n$, and therefore α_1^{-1} exists, for the components of the gradient of the functional with respect to ϑ, we can obtain the formula:

$$\nabla J(\vartheta) = -(\alpha_1^{-1})^T \left[\psi(t^1) - \frac{\partial \Phi^T(\tilde{x}, \vartheta)}{\partial \tilde{x}^1} \right] + \frac{\partial \Phi^T(\tilde{x}, \vartheta)}{\partial \vartheta} + \int_{t^1}^{t^f} \frac{\partial f^0(x, \vartheta, t)}{\partial \vartheta} dt, \tag{24}$$

where the adjoint boundary value problem has the form

$$\dot{\psi}(t) = -A_1^T(t) \psi(t) + \left(\frac{\partial f^0(x, \vartheta, t)}{\partial x} \right)^T$$

$$+ \sum_{j=1}^{l_2} \left[\chi(\hat{t}^{2j}) - \chi(\hat{t}^{2j-1}) \right] \beta_j^T(t) \left(\alpha_1^{-1} \right)^T \left(\psi(t^1) - \frac{\partial \Phi^T(\tilde{x}, \vartheta)}{\partial \tilde{x}^1} \right), \tag{25}$$

$$\alpha_{l_1}^T (\alpha_1^{-1})^T \psi(t^1) + \psi(t^f) = -\frac{\partial \Phi^T(\tilde{x}, \vartheta)}{\partial \tilde{x}^{l_1}} + \alpha_{l_1}^T (\alpha_1^{-1})^T \frac{\partial \Phi^T(\tilde{x}, \vartheta)}{\partial \tilde{x}^1}, \tag{26}$$

$$\psi(\tilde{t}_+^i) = \psi(\tilde{t}_-^i) + \alpha_i^T (\alpha_1^{-1})^T \psi(t^1) + \frac{\partial \Phi^T(\tilde{x}, \vartheta)}{\partial \tilde{x}^i}$$

$$-\alpha_i^T (\alpha_1^{-1})^T \frac{\partial \Phi^T (\bar{x}, \vartheta)}{\partial \tilde{x}^1}, \quad i = 2, 3, ..., l_1 - 1, \tag{27}$$

$$\psi(\hat{t}_+^j) = \psi(\hat{t}_-^j) + \frac{\partial \Phi^T (\bar{x}, \vartheta)}{\partial \hat{x}^j}, \quad j = 1, 2, ..., 2l_2. \tag{28}$$

Let us now consider the case when $rang\alpha = \bar{n} < n$.

Theorem 4. *Suppose all the conditions imposed on the data of the problem are satisfied, and rang $\alpha = \bar{n} < n$. Then integer functional (3) is differentiable with respect to the parameters $\vartheta = (\vartheta_1, \vartheta_2)$, and the components of its gradient are determined by the following formulas:*

$$\frac{\partial J}{\partial \vartheta_k^{(1)}} = \sum_{i=1}^{\bar{n}} \left[\frac{\partial \Phi (\bar{x}, \vartheta)}{\partial \tilde{x}_{k_i}^{s_i}} + \Delta \psi_{k_i} (\tilde{t}^{s_i}) \right] c_{ik}$$

$$+ \frac{\partial \Phi (\bar{x}, \vartheta)}{\partial \vartheta_k^{(1)}} + \int_{t^1}^{t^f} \frac{\partial f^0 (x, \vartheta, t)}{\partial \vartheta_k^{(1)}} dt, \quad k = 1, 2, ..., \bar{n}, \tag{29}$$

$$\frac{\partial J}{\partial \vartheta_k^{(2)}} = -\lambda + \frac{\partial \Phi (\bar{x}, \vartheta)}{\partial \vartheta_k^{(2)}} + \int_{t^1}^{t^f} \frac{\partial f^0 (x, \vartheta, t)}{\partial \vartheta_k^{(2)}} dt, \quad k = 1, 2, ..., (n - \bar{n}), \tag{30}$$

where $\psi(t) \in R^n, \lambda \in R^{\bar{n}}$ satisfy the conditions of the adjoint problem:

$$\dot{\psi}(t) = -A_1^T (t) \psi(t) + \left(\frac{\partial f^0 (x, \vartheta, t)}{\partial x} \right)^T$$

$$+ \sum_{j=1}^{l_2} \left(\chi(\hat{t}^{2j}) - \chi(\hat{t}^{2j-1}) \right) (\beta_j^1 (t))^T (\bar{\alpha}^{-1})^T \sum_{i=1}^{n} \left(\frac{\partial \Phi (\bar{x}, \vartheta)}{\partial \tilde{x}_{k_i}^{s_i}} + \psi_{k_i} (\hat{t}_-^{s_i}) - \psi_{k_i} (\hat{t}_+^{s_i}) \right)$$

$$- \sum_{j=1}^{l_2} \left(\chi(\hat{t}^{2j}) - \chi(\hat{t}^{2j-1}) \right) (\beta_j^2 (t))^T \lambda, \tag{31}$$

$$\psi_{g_\nu} (\tilde{t}_+^{q_\nu}) = \psi_{g_\nu} (\tilde{t}_-^{q_\nu}) + \sum_{i=1}^{\bar{n}} b_{i\nu} \left(\frac{\partial \Phi (\bar{x}, \vartheta)}{\partial \tilde{x}_{k_i}^{s_i}} + \psi_{k_i} (\tilde{t}_-^{s_i}) - \psi_{k_i} (\tilde{t}_+^{s_i}) \right)$$

$$+ \frac{\partial \Phi (\bar{x}, \vartheta)}{\partial \tilde{x}_{g_\nu}^{q_\nu}}, \nu = 1, .., (l_1 n - \bar{n}), \tag{32}$$

$$\psi_i (\hat{t}_+^j) = \psi_i (\hat{t}_-^j) + \frac{\partial \Phi (\bar{x}, \vartheta)}{\partial \hat{x}_i^j}, \quad i = 1, 2, ..., n, \ j = 1, 2, ..., 2l_2. \tag{33}$$

Proof. To take into account conditions (9), we use the method of Lagrange multipliers. Then, to increment the functional, we have:

$$\Delta J(\vartheta) = \int_{t^1}^{t^f} \psi^T (t) \left[\Delta \dot{x}(t) - A_1(t) \Delta x(t) \right] dt$$

$$+ \lambda^T \left[\sum_{j=1}^{l_2} \int_{\hat{t}^{2j-1}}^{\hat{t}^{2j}} \beta_j^2(t) \Delta x(t) dt - \Delta \vartheta^{(2)} \right], \tag{34}$$

where the as yet arbitrary $\lambda - (n - \bar{n})$-dimensional vector of Lagrange multipliers, $\psi(t)$ – conjugate variables.

To take into account conditions (8), we select an invertible $\bar{n}$-dimensional square matrix $\widehat{\alpha}$ from the augmented $(\bar{n} \times l\,n)$ matrix $\alpha = (\alpha_1, \alpha_2, ..., \alpha_{l_1})$. This will allow expressing from the analogue of relations (17) any $\bar{n}$ components of the (nl_1)-dimensional vector $\Delta x(\bar{t})$ through $(nl_1 - \bar{n})$ the remaining components and obtain a formula similar to (20).

Further, the course of the proof of the theorem is similar to the proof of Theorem 3, with the only difference in the dimensions of the matrices $\widehat{\alpha}$, $\breve{\alpha}$ and vectors $\Delta \widehat{x}$, $\Delta \breve{x}$ and, of course, in the added second summand in (34).

Note a special case when $\bar{n} = 0$. Obtaining formulas for the components of the gradient with respect to ϑ in this case is simplified, because the Lagrange method will be used for all conditions (2). This will, of course, lead to an increase in the dimension of the problem due to the increase in the dimension of the vector λ to n, i.e. $\lambda \in R^n$. Using the calculations presented in the proof of Theorem 4 with respect to conditions (9), the components of the gradient of the functional can be obtained in the following form:

$$\nabla J(\vartheta) = \int_{t^1}^{t^f} \frac{\partial f^0(x, \vartheta, t)}{\partial \vartheta} dt + \frac{\partial \Phi(\bar{x}, \vartheta)}{\partial \vartheta} - \lambda. \tag{35}$$

$\psi(t)$ and λ are the solution of the following adjoint boundary value problem with $2n$ conditions:

$$\dot{\psi}(t) = -A_1^T(t)\psi(t) + \sum_{j=1}^{l_2} \left[\chi(\hat{t}^{2j}) - \chi(\hat{t}^{2j-1}) \right] (\beta_j(t))^T \lambda + \left(\frac{\partial f^0(x, \vartheta, t)}{\partial x} \right)^T, \tag{36}$$

$$\psi(t^1) = \frac{\partial \Phi^T(\bar{x}, \vartheta)}{\partial \bar{x}^1} + \alpha_1^T \cdot \lambda, \tag{37}$$

$$\psi(t^f) = -\frac{\partial \Phi(\bar{x}, \vartheta)}{\partial \bar{x}^{l_1}} - \alpha_{l_1}^T \cdot \lambda, \tag{38}$$

and jump conditions:

$$\psi(\bar{t}_+^i) = \psi(\bar{t}_-^i) + \frac{\partial \Phi(\bar{x}, \vartheta)}{\partial \bar{x}^i} + \alpha_i^T \cdot \lambda, \quad i = 2, 3, ..., l_1 - 1, \tag{39}$$

$$\psi(\hat{t}_+^j) = \psi(\hat{t}_-^j) + \frac{\partial \Phi(\bar{x}, \vartheta)}{\partial \hat{x}^j}, \quad j = 1, 2, ..., 2l_2. \tag{40}$$

Formulas (36)–(38) differ from the formulas given in Theorems 3 and 4 in the dimension of the parameters and the number of conditions in the adjoint problem. In the formulas given in Theorem 3 for the case of $rang\,\alpha = \bar{n} = n$, the number of boundary conditions for the conjugate variable was n, and there were no unknown parameters. In the case of $\bar{n} < n$, formulas (30), (31) and (35)–(38) include the $(n - \bar{n})$-dimensional vector of Lagrange multipliers λ, for the determination of which there are also additional $(n-\bar{n})$ boundary conditions totaling $(2n - \bar{n})$.

Theorem 5. *In order for the pair $(\vartheta^*, x^*(t))$ to be a solution to problem (1)– (3), it is necessary and sufficient that the following take place for the arbitrary admissible vector of parameters $\vartheta \in V$:*

$$(\nabla J(\vartheta^*), \vartheta - \vartheta^*) \geq 0,$$

where $\nabla J(\vartheta^*)$ is determined depending on the rank of the augmented matrix $\alpha = [\alpha_1, \alpha_2, ..., \alpha_{l_1}]$ by one of formulas (10), (24), (29), (30) or (35).

The proof of the theorem follows from the convexity of the admissible domain V, the convexity and differentiability of the objective functional $J(\vartheta)$ ([21,22]).

4   The Scheme of the Numerical Solution of the Problem and The Results of Computer Experiments

For a numerical solution of system of differential equations with multipoint conditions both for the direct problem (1), (2) and the adjoint problem for the given vector of parameters ϑ, we use the method of condition shift proposed by the authors in [8,9,23].

Problem 1. *We present the results of numerical experiments obtained by solving the following problem described for $t \in [0,1]$ by a system of differential equations:*

$$\begin{cases} \dot{x}_1(t) = 2x_1(t) + x_2(t) - 6\cos(8t) - 22\sin(8t) - 3t^2 + 4t + 6, \\ \dot{x}_2(t) = tx_1(t) + x_2(t) - (24 - 2t)\cos(8t) - 3\sin(8t) - 2t^3 + t^2 - 1, \end{cases} \tag{41}$$

with nonlocal conditions

$$\begin{cases} x_1(0) + x_1(0.5) + x_2(1) + \int\limits_{0.6}^{0.8} (x_1(t) + 2x_2(t))dt = \vartheta_1, \\ \int\limits_{0.2}^{0.4} (x_1(t) - x_2(t))dt = \vartheta_2. \end{cases} \tag{42}$$

There are the following constraints on optimized parameters:

$$-3 \leq \vartheta_1 \leq 3, \quad -2 \leq \vartheta_2 \leq 2, \quad -5 \leq u(t) \leq 5.$$

The objective functional has the form:

$$J(\vartheta) = \int\limits_0^1 [x_1(t) + x_2(t) - t^2 - 2]^2 dt + \delta \cdot [(\vartheta_1 + 0.24)^2 + (\vartheta_2 + 1.17)^2]$$

$$+ (x_2(0.5) + 1.52)^2 + (x_1(1) + 0.29)^2 + (x_2(1) - 2.97)^2 \Big] \to \min. \quad (43)$$

Adjoint problem (31) has the form:

$$\begin{cases} \dot{\psi}_1(t) = -2\psi_1(t) - t\psi_2(t) + \lambda \left[\chi(0.4) - \chi(0.2) \right] + \\ \qquad + \psi_1(0) \left[\chi(0.8) - \chi(0.6) \right] + 2(x_1(t) + x_2(t) - t^2 - 2), \\ \dot{\psi}_2(t) = -\psi_1(t) - \psi_2(t) - \lambda \left[\chi(0.4) - \chi(0.2) \right] + \\ \qquad + 2\psi_1(0) \left[\chi(0.8) - \chi(0.6) \right] + 2(x_1(t) + x_2(t) - t^2 - 2). \end{cases} \quad (44)$$

The augmented matrix

$$\alpha = \begin{pmatrix} 1\ 0\ 1\ 0\ 0\ 1 \\ 0\ 0\ 0\ 0\ 0\ 0 \end{pmatrix}$$

has rank 1. This corresponds to the case in Theorem 4 considered in the previous paragraphs. We build the matrix $\hat{\alpha}$ from the first column and the first row of the matrix α_1: $\hat{\alpha} = (1)$, $\breve{\alpha} = (0\ 1\ 0\ 0\ 1)$.
Then for the matrices B and C, we have:

$$B = -\hat{\alpha}^{-1}\breve{\alpha} = (0\ -1\ 0\ 0\ -1), \quad C = \hat{\alpha}^{-1} = (1).$$

For the elements of the vector $\frac{\partial \Phi(\bar{x}, \vartheta)}{\partial \bar{x}^i}$, $i = 1, 2, 3$, we have:

$$\frac{\partial \Phi(\bar{x}, \vartheta)}{\partial \bar{x}_{k_1}^{s_1}} = \frac{\partial \Phi(\bar{x}, \vartheta)}{\partial \bar{x}_1^1} = 0, \quad \frac{\partial \Phi(\bar{x}, \vartheta)}{\partial \bar{x}_{g_2}^{q_1}} = \frac{\partial \Phi(\bar{x}, \vartheta)}{\partial \bar{x}_2^1} = 0, \quad \frac{\partial \Phi(\bar{x}, \vartheta)}{\partial \bar{x}_{g_2}^{q_2}} = \frac{\partial \Phi(\bar{x}, \vartheta)}{\partial \bar{x}_1^2} = 0,$$

$$\frac{\partial \Phi(\bar{x}, \vartheta)}{\partial \bar{x}_{g_3}^{q_3}} = \frac{\partial \Phi(\bar{x}, \vartheta)}{\partial \bar{x}_2^2} = 2(x_2(0.5) + 1.52), \quad \frac{\partial \Phi(\bar{x}, \vartheta)}{\partial \bar{x}_{g_4}^{q_4}} = \frac{\partial \Phi(\bar{x}, \vartheta)}{\partial \bar{x}_1^3} = 2(x_1(1) + 0.29),$$

$$\frac{\partial \Phi(\bar{x}, \vartheta)}{\partial x_{g_5}^{q_5}} = \frac{\partial \Phi(\bar{x}, \vartheta)}{\partial \bar{x}_2^3} = 2(x_2(1) - 2.97), \quad \frac{\partial \Phi(x, \vartheta)}{\partial \bar{x}_i^j} = 0, \quad i = 1, 2, \quad j = 1, 2, 3, 4.$$

Then conditions (32) and (33) take the form:

$$\begin{cases} \psi_1(0) + \psi_2(1) = -2(x_2(1) - 2.97), \\ \psi_2(0) = 0, \quad \psi_1(1) = -2(x_1(1) + 0.29), \\ \psi_1(0.5^+) = \psi_1(0.5^-) + \psi_1(0), \\ \psi_2(0.5^+) = \psi_2(0.5^-) + 2(x_2(0.5) + 1.52), \\ \psi_i(\hat{t}_+^j) = \psi_i(\hat{t}_-^j), \quad i = 1, 2, \quad j = 1, 2, 3, 4. \end{cases} \quad (45)$$

The components of the gradient of the functional with respect to the vector ϑ, according to formulas (29), (30), are determined as follows:

$$\frac{\partial J(\vartheta)}{\partial \vartheta_1} = -\psi_1(0) + 2(\vartheta_1 + 0.24), \quad \frac{\partial J(\vartheta)}{\partial \vartheta_2} = -\lambda + 2(\vartheta_2 + 1.17). \quad (46)$$

The iterative procedure of the gradient projection method [21] was carried out with accuracy with respect to the functional $\varepsilon = 10^{-5}$ from different starting points $\vartheta^{(0)}$. The auxiliary Cauchy problems used by the method of condition

shift ([8]) to solve both direct problem (41)–(42) and adjoint problem (44)–
(45) were solved by the fourth-order Runge-Kutta method with different steps
$h = 0.01$, $0.02, 0.05$.

Figure 1 shows the graphs of solutions to the direct (a) and adjoint (b) boundary value problems for the number of time interval partitions $N = 200$ obtained
at the twentieth iteration of optimization. The value of the functional at the
starting point $\vartheta_1^{(0)} = -0.125$; $\vartheta_2^{(0)} = 1.5$; was $J(\vartheta^{(0)}) = 397.30552$, and the
value of $\lambda = -240.1465$. At the twentieth iteration, we obtained the value of
the functional $J(\vartheta^{(20)}) = 1.4 \cdot 10^{-8}$, and the parameter values were as follows:
$\vartheta_1^{(20)} = -0.2403$, $\vartheta_2^{(20)} = -1.1719$, $\lambda^{(20)} = 0.0002$.

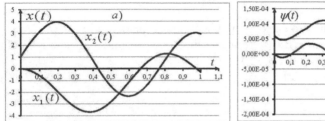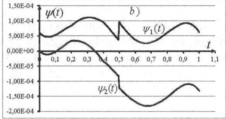

Fig. 1. Graphs of the obtained solutions of the direct (a) and adjoint (b) boundary
value problems.

5   Conclusion

In the paper, we investigate the linear optimal control problem for a dynamic
object, in which the values of the right-hand sides of linear nonlocal boundary
conditions are optimized. The boundary conditions include, as terms, the values
of the phase variable at intermediate points and the integral values of the phase
variable over several intervals. The conditions for the existence and uniqueness
of the solution to the boundary value problem with unseparated boundary conditions and convexity of the objective functional to the problem are investigated.
The necessary optimality conditions are formulated by using the technique of the
Lagrange method and definition of the conditional gradient. The obtained results
can be used to investigate control problems described by nonlinear systems of
differential equations with linear unseparated boundary conditions, including the
point and integral values of phase variables and optimized right-hand sides.

References

1. Nicoletti, O.: Sulle condizioni iniziali che determiniano gli integrali della diffenziali
   ordinazie. Att della R. Acc. Sc, Torino (1897)
2. Tamarkin, Y.D.: On some general problems in the theory of ordinary differential
   equations and on series expansions of arbitrary functions. Petrograd (1917)

3. De la Vallée-Poussin, Ch.J.: Sur l'équation différentielle linéare du second ordre. Détermination d'une integrale par deux valeurs assignées. Extension aux équations d'orde n. J. Math. Pures Appl. **8**(9) (1929)
4. Kiguradze, I.T.: Boundary value problems for system of ordinary differential equations. Itogi Nauki Tekh. Sovrem. Probl. Mat. Nov. Dostizheniya **30**, 3–103 (1987)
5. Nakhushev, A.M.: Loaded Equations and Applications. Nauka, Moscow (2012)
6. Dzhumabaev, D.S., Imanchiev, A.E.: Boundary value problems for system of ordinary differential equations. Mat. J. **5**(15), 30–38 (2005)
7. Assanova, A.T., Imanchiyev, A.E., Kadirbayeva, ZhM: Solvability of nonlocal problems for systems of Sobolev-type differential equations with a multipoint condition. Izv. Vyssh. Uchebn. Zaved. Mat. **12**, 3–15 (2019)
8. Aida-zade, K.R., Abdullaev, V.M.: On the solution of boundary value problems with nonseparated multipoint and integral conditions. Differ. Equ. **49**(9), 1114–1125 (2013). https://doi.org/10.1134/S0012266113090061
9. Abdullaev, V.M., Aida-Zade, K.R.: Numerical method of solution to loaded nonlocal boundary value problems for ordinary differential equations. Comput. Math. Math. Phys. **54**(7), 1096–1109 (2014). https://doi.org/10.1134/S0965542514070021
10. Assanova, A.T.: Solvability of a nonlocal problem for a hyperbolic equation with integral conditions. Electron. J. Differ. Equ. **170**, 1–12 (2017)
11. Aida-zade, K.R., Abdullayev, V.M.: Optimizing placement of the control points at synthesis of the heating process control. Autom. Remote Control **78**(9), 1585–1599 (2017). https://doi.org/10.1134/S0005117917090041
12. Abdullayev, V.M., Aida-zade, K.R.: Numerical solution of the problem of determining the number and locations of state observation points in feedback control of a heating process. Comput. Math. Math. Phys. **58**(1), 78–89 (2018). https://doi.org/10.1134/S0965542518010025
13. Aida-zade, K.R., Hashimov, V.A.: Optimization of measurement points positioning in a border control synthesis problem for the process of heating a rod. Autom. Remote Control **79**(9), 1643–1660 (2018). https://doi.org/10.1134/S0005117918090096
14. Abdullayev, V.M.: Numerical solution to optimal control problems with multipoint and integral conditions. Proc. Inst. Math. Mech. **44**(2), 171–186 (2018)
15. Devadze, D., Beridze, V.: Optimality conditions and solution algorithms of optimal control problems for nonlocal boundary-value problems. J. Math. Sci. **218**(6), 731–736 (2016). https://doi.org/10.1007/s10958-016-3057-x
16. Zubova, S.P., Raetskaya, E.V.: Algorithm to solve linear multipoint problems of control by the method of cascade decomposition. Autom. Remote Control **78**(7), 1189–1202 (2017). https://doi.org/10.1134/S0005117917070025
17. Abdullayev, V.M., Aida-zade, K.R.: Optimization of loading places and load response functions for stationary systems. Comput. Math. Math. Phys. **57**(4), 634–644 (2017). https://doi.org/10.1134/S0965542517040029
18. Aschepkov, L.T.: Optimal control of system with intermediate conditions. J. Appl. Math. Mech. **45**(2), 215–222 (1981)
19. Vasil'eva, O.O., Mizukami, K.: Dynamical processes described by boundary problem: necessary optimality conditions and methods of solution. J. Comput. Syst. Sci. Int. (A J. Optim. Control) **1**, 95–100 (2000)
20. Abdullayev, V.M., Aida-zade, K.R.: Approach to the numerical solution of optimal control problems for loaded differential equations with nonlocal conditions. Comput. Math. Math. Phys. **59**(5), 696–707 (2019). https://doi.org/10.1134/S0965542519050026

21. Polyak, B.T.: Introduction to Optimization. Lenand, Moscow (2019)
22. Vasil'ev, F.P.: Methods of Optimization. Faktorial Press, Moscow (2002)
23. Aida-Zade, Kamil, Abdullayev, Vagif: Numerical method for solving the parametric identification problem for loaded differential equations. Bull. Iran. Math. Soc. **45**(6), 1725–1742 (2019). https://doi.org/10.1007/s41980-019-00225-3
24. Moszynski, K.: A method of solving the boundary value problem for a system of linear ordinary differential equation. Algorytmy. Varshava. **11**(3), 25–43 (1964)
25. Abramov, A.A.: A variation of the 'dispersion' method. USSR Comput. Math. Math. Phys. **1**(3), 368–371 (1961)

Saddle-Point Method in Terminal Control with Sections in Phase Constraints

Anatoly Antipin[1] and Elena Khoroshilova[2]

[1] Dorodnicyn Computing Centre, FRC CSC RAS,
Vavilov Street 40, 119333 Moscow, Russia
asantip@yandex.ru

[2] CMC Faculty, Lomonosov MSU, Leninskiye Gory, 119991 Moscow, Russia
khorelena@gmail.com

Abstract. A new approach to solving terminal control problems with phase constraints, based on saddle-point sufficient optimality conditions, is considered. The basis of the approach is Lagrangian formalism and duality theory. We study linear controlled dynamics in the presence of phase constraints. The cross section of phase constraints at certain points in time leads to the appearance of new intermediate finite-dimensional convex programming problems. In fact, the optimal control problem, defined over the entire time interval, is split into a number of independent intermediate subproblems, each of which is defined in its own subsegment. Combining the solutions of these subproblems together, we can obtain solutions5 to the original problem on the entire time interval. To this end, a gradient flow is launched to solve all intermediate problems at the same time. The convergence of computing technology to the solution of the optimal control problem in all variables is proved.

Keywords: Optimal control · Lagrange function · Duality · Saddle point · Iterative solution methods · Convergence

1 Introduction

Dynamic problems of terminal control under state constraints are among the most complex in optimal control theory. For quite a long time, from the moment of their occurrence and the first attempts at practical implementation in technical fields, these problems have been studied by experts from different angles. Much attention is traditionally paid to the development of computational methods for solving this class of problems in a wide area of applications [9,13]. At the same time, directions are being developed related to further generalizations and the development of the Pontryagin maximum principle [7,10], as well as with the extension of the classes of problem statements [8,12,14]. The questions of the existence, stability, optimality of solutions are studied in [11] and others.

Supported by the Russian Science Foundation (Research Project 18-01-00312).

In our opinion, one of the most important areas of the theory of solving optimal control problems is the study of various approaches to the development of evidence-based methods for solving terminal control problems. The theory of evidence-based methods is currently an important tool in various application areas of mathematical modeling tools. In this theory, emphasis is placed on the ideas of proof, validity, and guarantee of the result. The latter assumes that the developed computing technology (computing process) generates a sequence of iterations that has a limit point on a bounded set, and this point is guaranteed to be a solution to the original problem with a given accuracy.

In this paper, we consider the problem of terminal control with phase constraints and their cross sections at discretization points. Intermediate spaces are associated with sampling points, the dimension of which is equal to the dimension of the phase trajectory vector. The sections of the phase trajectory in the spaces of sections form a polyhedral set. On this set, we pose the problem of minimizing a convex objective function. At each sampling point, we obtain some finite-dimensional problem. To iteratively proceed to the next phase trajectory, it is enough to take a gradient-type step in the section space for each intermediate problem. These steps together on all intermediate problems form a saddle-point gradient flow. This computational flow with an increase in the number of iterations leads us to the solution of the problem.

2 Statement of Terminal Control Problem with Continuous Phase Constraints

We consider a linear dynamic controlled system defined on a given time interval $[t_0, t_f]$, with a fixed left end and a moving right end under phase constraints on the trajectory. The dynamics of the controlled phase trajectory $x(t)$ is described by a linear system of differential equations with an implicit condition at the right end of the time interval. A terminal condition is defined as a solution to a linear programming problem that is not known in advance. In this case, it is necessary to choose a control so that the phase trajectory satisfies the phase constraints, and its right end coincides with the solution of the boundary value problem. The control problem is considered in a Hilbert function space.

Formally, everything said in the case of continuous phase constraints can be represented as a problem: find the optimal control $u(t) = u^*(t)$ and the corresponding trajectory $x(t) = x^*(t)$, $t \in [t_0, t_f]$, that satisfy the system

$$\frac{d}{dt}x(t) = D(t)x(t) + B(t)u(t), \ t_0 \le t \le t_f, \ x(t_0) = x_0, \ x(t_f) = x_f^*,$$
$$G(t)x(t) \le g(t), \ x(t) \in \mathrm{R}^n \ \forall t \in [t_0, t_f],$$
$$u(t) \in \mathrm{U} = \{u(t) \in \mathrm{L}_2^r[t_0, t_f] \mid u(t) \in [u^-, u^+] \ \forall t \in [t_0, t_f]\},$$
$$x_f^* \in \mathrm{Argmin}\{\langle \varphi_f, x_f \rangle \mid G_f x_f \le g_f, \ x_f \in \mathrm{R}^n\}, \tag{1}$$

where $D(t), B(t), G(t)$ are continuous matrices of size $n \times n$, $n \times r$, $m \times n$ respectively; $g(t)$ is a given continuous vector function; $G_f = G(t_f)$, $g_f = g(t_f)$,

$x_f = x(t_f)$ are the values at the right-hand end of the time interval; φ_f is the given vector (normal to the linear objective functional), $x(t_0) = x_0$ is the given initial condition. The inclusion $x(t) \in \mathrm{R}^n$ means that the vector $x(t)$ for each t belongs to the finite-dimensional space R^n. The controls $u(t)$ for each $t \in [t_0, t_f]$ belong to the set U, which is a convex compact set from R^r. Problem (1) is considered as an analogue of the linear programming problem formulated in a functional Hilbert space.

To solve the differential system in (1), it is necessary to use the initial condition x_0 and some control $u(t) \in$ U. For each admissible $u(t)$, in the framework of the classical theorems of existence and uniqueness, we obtain a unique phase trajectory $x(t)$. The right end of the optimal trajectory must coincide with the finite-dimensional solution of the boundary value problem, i. e. $x^*(t_f) = x_f^*$. An asterisk means that $x^*(t)$ is the optimal solution; in particular, x_f^* is a solution to the boundary value optimization problem. The control must be selected so that phase constraints are additionally fulfilled. The left end x_0 of the trajectory is fixed and is not an object of optimization.

The formulated problem with phase constraints from the point of view of developing evidence-based computational methods is one of the difficult problems. Traditionally, optimal control problems (without a boundary value problem) are studied in the framework of the Hamiltonian formalism, the peak of which is the maximum principle. This principle is a necessary condition for optimality and is the dominant tool for the study of dynamic controlled problems. However, the maximum principle does not allow constructing methods that are guaranteed to give solutions with a predetermined accuracy. In the case of convex problems of type (1), it seems more reasonable to conduct a study in the framework of the Lagrangian formalism. Moreover, the class of convex problems in optimal control is wide enough, and almost any smooth problem can be approximated by a convex, quadratic, or linear problem.

Problem (1) without phase restrictions was investigated by the authors in [1–5]. In the linear-convex case, relying on the saddle-point inequalities of the Lagrange function, the authors proved the convergence of extragradient and extraproximal methods to solving the terminal control problem in all solution components: weak convergence in controls, strong convergence in phase and conjugate trajectories, and also in terminal variables of intermediate (boundary value) problems. This turned out to be possible due to the fact that the saddle-point inequalities for the Lagrange function in the case under consideration represent sufficient optimality conditions. These conditions, in contrast to the necessary conditions of the maximum principle, allow us to develop an evidence-based (without heuristic) theory of methods for solving optimal control problems, which was demonstrated in [1–5].

3   Phase Constraints Sections and Finite-Dimensional Intermediate Problems Generated by Them

In statement (1), we presented the phase constraints $G(t)x(t) \leq g(t)$, $t \in [t_0, t_f]$, of continuous type. An approach will be described below, where instead of continuous phase constraints their sections $G_s x_s \leq g_s$ are considered at certain instants of time t_s on a discrete grid

$$\Gamma = \{t_0, t_1, ..., t_{s-1}, t_s, t_{s+1}, ..., t_f\}.$$

At these moments of time, finite-dimensional cross-section problems are formed, and between these moments (on the discretization segments $[t_{s-1}, t_s]$, $s = \overline{1, f}$) intermediate terminal control problems arise. Thus, the original problem formulated on the entire segment $[t_0, t_f]$ is decomposed into a set of independent problems, each of which is defined on its own sub-segment. The obtained intermediate problems no longer have phase constraints, since the phase constraints on the sub-segments have passed to the boundary value problems at the ends of these sub-segments. This approach does not require the existence of a functional Slater condition. In fact, the existence of finite-dimensional saddle points in the intermediate spaces R^n (that are generated by sections of phase constraints at given moments t_s) is sufficient. Each section has its own boundary-value problem, and then through all these solutions (like a thread through a coal ear), the desired phase trajectory is drawn over the entire time interval. In finite-dimensional section spaces, the Slater condition for convex problems is always satisfied by definition.

Except for discrete phase constraints, the rest of problem (1) remains continuous. The combination of the trajectories and other components of the problem over all time sub-segments results in the solution of the original problem over the entire time interval $[t_0, t_f]$. The approach based on sections can be interpreted as a method of decomposing a complex problem into a number of simple ones.

Thus, on each of the segments $[t_{s-1}, t_s]$, a specific segment $x_s(t)$ of the phase trajectory of differential equation (1) is defined. At the common point of the adjacent segments $[t_{s-1}, t_s]$ and $[t_s, t_{s+1}]$ the values $x_s(t_s)$ and $x_{s+1}(t_s)$ coincide in construction: $x_s(t_s) = x_{s+1}(t_s)$, i.e. on each segment of the partition, the right end of the trajectory coincides with the starting point of the trajectory in the next segment.

As a result of discretization based on (1), the following statement of the multi-problem is obtained:

$$\frac{d}{dt} x_s(t) = D(t)x_s(t) + B(t)u_s(t), \ t \in [t_{s-1}, t_s],$$

$$x_s(t_{s-1}) = x^*_{s-1}(t_{s-1}), \ x_s(t_s) = x^*_s, \ u_s(t) \in \mathrm{U},$$

$$x^*_1 \in \mathrm{Argmin}\{\langle \varphi_1, x_1 \rangle \mid G_1 x_1 \leq g_1, \ x_1 \in \mathrm{R}^n\}, \ x^*_1 \in X_1,$$

$$x^*_2 \in \mathrm{Argmin}\{\langle \varphi_2, x_2 \rangle \mid G_2 x_2 \leq g_2, \ x_2 \in \mathrm{R}^n\}, \ x^*_2 \in X_2,$$

$$\cdots\cdots\cdots\cdots\cdots\cdots\cdots\cdots\cdots\cdots\cdots\cdots\cdots\cdots\cdots\cdots\cdots\cdots$$

$$x^*_f \in \mathrm{Argmin}\{\langle \varphi_f, x_f \rangle \mid G_f x_f \leq g_f \ x_f \in \mathrm{R}^n\}, \ x^*_f \in X_f. \tag{2}$$

Here $x_s(t_s)$ is the value of the function $x_s(t)$ at the right end of segment $[t_{s-1}, t_s]$, x^*_s is the solution of sth intermediate linear programming problem; φ_s is the normal to the objective function; X_s is an intermediate reachability set; $G_s = G(t_s)$, $g_s = g(t_s)$, $s = \overline{1, f}$. If we combine all parts of the trajectories $x_s(t)$ then we get the full trajectory on the entire segment $x(t)$, $t \in [t_0, t_f]$. In other words, we "broke" the original problem (1) into f independent problems of the same kind.

For greater clarity, imagine system (2) in an expanded form. Discretization of Γ generates time intervals $[t_{s-1}, t_s]$, on which functions $x_s(t)$ are defined for all $s = \overline{1, f}$. Each of these functions is the restriction of the phase trajectory $x(t)$ to the segment $[t_{s-1}, t_s]$. In this model, for each sth time interval $[t_{s-1}, t_s]$, the sth controlled trajectory $x_s(t)$ and the sth intermediate problem are defined:

$$\frac{d}{dt}x_1(t) = D(t)x_1(t) + B(t)u_1(t), \ t \in [t_0, t_1],$$

$$x_1(t_0) = x_0, \ x_1(t_1) = x_1^*, \ u_1(t) \in U,$$

$$x_1^* \in \operatorname{Argmin}\{\langle \varphi_1, x_1 \rangle \mid G_1 x_1 \le g_1, \ x_1 \in \mathbb{R}^n\}, \ x_1^* \in X_1, x_1(t_1) = x_1,$$

$$\cdots\cdots\cdots\cdots\cdots\cdots\cdots$$

$$\frac{d}{dt}x_s(t) = D(t)x_s(t) + B(t)u_s(t), \ t \in [t_{s-1}, t_s],$$

$$x_s(t_{s-1}) = x_{s-1}^*, \ x_s(t_s) = x_s^*, \ u_s(t) \in U,$$

$$x_s^* \in \operatorname{Argmin}\{\langle \varphi_s, x_s \rangle \mid G_s x_s \le g_s, \ x_s \in \mathbb{R}^n\}, \ x_s^* \in X_s, x_s(t_s) = x_s, \quad (3)$$

$$\cdots\cdots\cdots\cdots\cdots\cdots\cdots$$

$$\frac{d}{dt}x_f(t) = D(t)x_f(t) + B(t)u_f(t), \ t \in [t_{f-1}, t_f],$$

$$x_f(t_{f-1}) = x_{f-1}^*, \ x_f(t_f) = x_f^*, \ u_f(t) \in U,$$

$$x_f^* \in \operatorname{Argmin}\{\langle \varphi_f, x_f \rangle \mid G_f x_f \le g_f, \ x_f \in \mathbb{R}^n\}, \ x_f^* \in X_f, x_f(t_f) = x_f.$$

So, within the framework of the proposed approach, the initial problem with phase constraints (1) is split into a finite set of independent intermediate terminal control problems without phase constraints. Each of these problems can be solved independently, starting with the first problem. Then, conducting a phase trajectory through solutions of intermediate problems, we find a solution to the terminal control problem over the entire segment $[t_0, t_f]$. To solve any of the subproblems of system (3), the authors developed methods in [1,2].

4   Problem Statement in Vector-Matrix Form

For greater clarity, we present the system (2) or (3) in a more compact vector-matrix form:

dynamics

$$\begin{pmatrix} \frac{dx_1}{dt} \\ \frac{dx_2}{dt} \\ \vdots \\ \frac{dx_f}{dt} \end{pmatrix} = \begin{pmatrix} D_1 & 0 & \cdots & 0 \\ 0 & D_2 & \cdots & 0 \\ \vdots & \vdots & \ddots & \vdots \\ 0 & 0 & \cdots & D_f \end{pmatrix} \begin{pmatrix} x_1 \\ x_2 \\ \vdots \\ x_f \end{pmatrix} + \begin{pmatrix} B_1 & 0 & \cdots & 0 \\ 0 & B_2 & \cdots & 0 \\ \vdots & \vdots & \ddots & \vdots \\ 0 & 0 & \cdots & B_f \end{pmatrix} \begin{pmatrix} u_1 \\ u_2 \\ \vdots \\ u_f \end{pmatrix}$$

where $x(t_0) = x_0$, $x_s(t_s) = x_s^*$, $x_f(t_f) = x_f^*$, $u_s(t) \in U$, and intermediate problems

$$
\begin{pmatrix} x_1^* \\ x_2^* \\ \vdots \\ x_f^* \end{pmatrix} \in \operatorname{Argmin} \left\{ (\varphi_1 \ \varphi_2 \ \cdots \ \varphi_f) \begin{pmatrix} x_1 \\ x_2 \\ \vdots \\ x_f \end{pmatrix} \middle| \begin{pmatrix} G_1 & 0 & \cdots & 0 \\ 0 & G_2 & \cdots & 0 \\ \vdots & \vdots & \ddots & \vdots \\ 0 & 0 & \cdots & G_f \end{pmatrix} \begin{pmatrix} x_1 \\ x_2 \\ \vdots \\ x_f \end{pmatrix} \leq \begin{pmatrix} g_1 \\ g_2 \\ \vdots \\ g_f \end{pmatrix} \right\}
$$

(4)

Recall once again that each function $x(t)$ generates a vector with components $(x(t_1), ..., x(t_s), ..., x(t_f))$, and the number of components is equal to the number of sampling points of the segment $[t_0, t_f]$. Each component of this vector, in turn, is a vector of size n. Thus, we have a space of dimension $\mathrm{R}^{n \times f}$. In this space, the diagonal matrix $G(t_s)$, $s = \overline{1, f}$, is defined, each component of which is submatrix $G_s(t_s)$ from (4), whose dimension is $n \times n$. We described the matrix functional constraint of the inequality type at the right-hand side, which is given by vector $g = (g_1, g_2, ..., g_f)$. The linear objective function that completes the formulation of the finite-dimensional linear programming problem in (4) is a scalar product of vectors φ and x.

Thus, in macro format, we can represent problem (4) in the form

$$
\begin{cases} \dfrac{d}{dt} x(t) = D(t)x(t) + B(t)u(t), \ t_0 \leq t \leq t_f, \ x(t_0) = x_0, \ x(t_f) = x_f^*, \\ \qquad x^* \in \operatorname{Argmin}\{\langle \varphi_f, x \rangle \mid Gx \leq g, \ x \in \mathrm{R}^n\}, u(t) \in \mathrm{U}. \end{cases}
$$

(5)

Note that the macro system (5) obtained as a result of scalarization of intermediate problems (3) (or (4)) almost completely coincides with the terminal control problem with the boundary value problem on the right-hand end suggested and explored in [1,2]. Therefore, the method for solving problem (5) and the proof of its convergence as a whole will repeat the logic of reasoning.

As a solution to differential system (5), we mean any pair $(x(t), u(t)) \in L_2^n[t_0, t_f] \times \mathrm{U}$ that satisfies the condition

$$
x(t) = x(t_0) + \int_{t_0}^t (D(\tau)x(\tau) + B(\tau)u(\tau))d\tau, \quad t_0 \leq t \leq t_f.
$$

(6)

The trajectory $x(t)$ in (6) is an absolutely continuous function. The class of absolutely continuous functions is a linear variety everywhere dense in $L_2^n[t_0, t_f]$. In the future, this class will be denoted as $AC^n[t_0, t_f] \subset L_2^n[t_0, t_f]$. For any pair of functions $(x(t), u(t)) \in AC^n[t_0, t_f] \times \mathrm{U}$, the Newton-Leibniz formula and, accordingly, the integration-by-parts formula are satisfied.

5   Saddle-Point Sufficient Optimality Conditions. Dual Approach

Drawing the corresponding analogies with the theory of linear programming, we write out the primal and dual Lagrange functions for the problem (5). To do this, we scalarize system (5) and introduce a linear convolution known as the Lagrange function

$$\mathcal{L}(p, \psi(t); x, x(t), u(t)) = \langle \varphi, x \rangle + \langle p, Gx - g \rangle$$
$$+ \int_{t_0}^{t_f} \langle \psi(t), D(t)x(t) + B(t)u(t) - \frac{d}{dt}x(t) \rangle dt,$$

defined for all $p \in R_+^m$, $\psi(t) \in \Psi_2^n[t_0, t_f]$, $x \in R^n$, $(x(t), u(t)) \in AC^n[t_0, t_f] \times U$. Here x is a finite-dimensional vector composed of the values of trajectory $x(t)$ at the sampling points; $\Psi_2^n[t_0, t_f]$ is a linear manifold of absolutely continuous functions from an adjoint space. The variety $\Psi_2^n[t_0, t_f]$ is everywhere dense in $L_2^n[t_0, t_f]$.

The saddle point $(p^*, \psi^*(t); x^*, x^*(t), u^*(t))$ of the Lagrange function is formed by primal $(x^*, x^*(t), u^*(t))$ and dual $(p^*, \psi^*(t))$ solutions of problem (5) and, by definition, satisfies the system of inequalities

$$\langle \varphi, x^* \rangle + \langle p, Gx^* - g \rangle + \int_{t_0}^{t_f} \langle \psi(t), D(t)x^*(t) + B(t)u^*(t) - \frac{d}{dt}x^*(t) \rangle dt$$

$$\leq \langle \varphi, x^* \rangle + \langle p^*, Gx^* - g \rangle + \int_{t_0}^{t_f} \langle \psi^*(t), D(t)x^*(t) + B(t)u^*(t) - \frac{d}{dt}x^*(t) \rangle dt$$

$$\leq \langle \varphi, x \rangle + \langle p^*, Gx - g \rangle + \int_{t_0}^{t_f} \langle \psi^*(t), D(t)x(t) + B(t)u(t) - \frac{d}{dt}x(t) \rangle dt$$

for all $p \in R_+^m$, $\psi(t) \in \Psi_2^n[t_0, t_f]$, $x \in R^n$, $(x(t), u(t)) \in AC^n[t_0, t_f] \times U$.

If the original problem (5) has a primal and dual solution, then this pair is a saddle point of the Lagrange function. Here, as in the finite-dimensional case, the dual solution is formed by the coordinates of the normal to the supporting plane at the minimum point.

The converse is also true: the saddle point of the Lagrange function consists of the primal and dual solutions to original problem (5).

Using formulas to go over to conjugate linear operators

$$\langle \psi, Dx \rangle = \langle D^T \psi, x \rangle, \ \langle \psi, Bu \rangle = \langle B^T \psi, u \rangle$$

and the integrating-by-parts formula on segment $[t_0, t_f]$

$$\langle \psi(t_f), x(t_f) \rangle - \langle \psi(t_0), x(t_0) \rangle = \int_{t_0}^{t_f} \langle \frac{d}{dt}\psi(t), x(t) \rangle dt + \int_{t_0}^{t_f} \langle \psi(t), \frac{d}{dt}x(t) \rangle dt,$$

we write out the dual Lagrange function and saddle-point system in the conjugate form:

$$\mathcal{L}^T(p, \psi(t); x, x(t), u(t)) = \langle \varphi + G^T p - \psi_f, x \rangle - \langle g, p \rangle + \langle \psi_0, x_0 \rangle$$
$$+ \int_{t_0}^{t_f} \langle D^T(t)\psi(t) + \frac{d}{dt}\psi(t), x(t) \rangle dt + \int_{t_0}^{t_f} \langle B^T(t)\psi(t), u(t) \rangle dt$$

for all $p \in R_+^m$, $\psi(t) \in \Psi_2^n[t_0, t_f]$, $x \in R^n$, $(x(t), u(t)) \in AC^n[t_0, t_f] \times U$, $x_0 = x(t_0)$, $\psi_0 = \psi(t_0)$, $\psi_f = \psi(t_f)$.

The dual saddle-point system has the form

$$\langle \varphi + G^T p - \psi_f, x^* \rangle + \langle -g, p \rangle + \langle \psi_0, x_0^* \rangle$$

$$+ \int_{t_0}^{t_f} \langle D^T(t)\psi(t) + \tfrac{d}{dt}\psi(t), x^*(t) \rangle dt + \int_{t_0}^{t_f} \langle B^T(t)\psi(t), u^*(t) \rangle dt \leq$$

$$\leq \langle \varphi + G^T p^* - \psi_f^*, x^* \rangle + \langle -g, p^* \rangle + \langle \psi_0^*, x_0^* \rangle$$

$$+ \int_{t_0}^{t_f} \langle D^T(t)\psi^*(t) + \tfrac{d}{dt}\psi^*(t), x^*(t) \rangle dt + \int_{t_0}^{t_f} \langle B^T(t)\psi^*(t), u^*(t) \rangle dt \leq$$

$$\leq \langle \varphi + G^T p^* - \psi_f^*, x \rangle + \langle -g, p^* \rangle + \langle \psi_0^*, x_0 \rangle$$

$$+ \int_{t_0}^{t_f} \langle D^T(t)\psi^*(t) + \tfrac{d}{dt}\psi^*(t), x(t) \rangle dt + \int_{t_0}^{t_f} \langle B^T(t)\psi^*(t), u(t) \rangle dt$$

for all $p \in R_+^m$, $\psi(t) \in \Psi_2^n[t_0, t_1]$, $x \in R^n$, $(x(t), u(t)) \in AC^n[t_0, t_f] \times U$.

Both Lagrangians (primal and dual) have the same saddle point $(p^*, \psi^*(t); x^*, x^*(t), u^*(t))$, which satisfies the saddle-point conjugate system.

From the analysis of the saddle-point inequalities, we can write out mutually dual problems:

the primal problem:

$$x^* \in \text{Argmin}\{\langle \varphi, x \rangle \mid Gx \leq g, \ x \in R^n,$$

$$\tfrac{d}{dt}x(t) = D(t)x(t) + B(t)u(t), \quad x(t_0) = x_0, \ u(t) \in U\};$$

the dual problem:

$$(p^*, \psi^*(t)) \in \text{Argmax}\{\langle -g, p \rangle + \langle \psi_0, x_0^* \rangle + \int_{t_0}^{t_f} \langle \psi(t), B(t)u^*(t) \rangle dt \ \Big|$$

$$D^T(t)\psi(t) + \tfrac{d}{dt}\psi(t) = 0, \ \psi_f = \varphi + G^T p,$$

$$p \in R_+^m, \ \psi(t) \in \Psi_2^n[t_0, t_f]\},$$

$$\int_{t_0}^{t_f} \langle B^T(t)\psi^*(t), u^*(t) - u(t) \rangle dt \leq 0, \ u(t) \in U.$$

6   Method for Solving. Convergence Technique

Replacing the variational inequalities in the above system with the corresponding equations with the projection operator, we can write the differential system in operator form. Then, based on this system, we write out a saddle-point method of extragradient type to calculate the saddle point of the Lagrange function. The two components of the saddle point are the primal and dual solutions to problem (5).

The formulas of this iterative method are as follows:

1) *predictive half-step*

$$\tfrac{d}{dt}x^k(t) = D(t)x^k(t) + B(t)u^k(t), \ x^k(t_0) = x_0,$$

$$\bar{p}^k = \pi_+(p^k + \alpha(Gx^k - g)),$$

$$\tfrac{d}{dt}\psi^k(t) + D^T(t)\psi^k(t) = 0, \ \psi^k = \varphi + G^T p^k,$$

$$\bar{u}^k(t) = \pi_U(u^k(t) - \alpha B^T(t)\psi^k(t));$$

2) *basic half-step*

$$\tfrac{d}{dt}\bar{x}^k(t) = D(t)\bar{x}^k(t) + B(t)\bar{u}^k(t), \ \bar{x}^k(t_0) = x_0,$$

$$p^{k+1} = \pi_+(p^k + \alpha(G\bar{x}^k - g)),$$

$$\tfrac{d}{dt}\bar{\psi}^k(t) + D^T(t)\bar{\psi}^k(t) = 0, \ \bar{\psi}^k = \varphi + G^T\bar{p}^k,$$

$$u^{k+1}(t) = \pi_U(u^k(t) - \alpha B^T(t)\bar{\psi}^k(t)), \ k = 0, 1, 2, ...$$

Here, at each half-step, two differential equations are solved and an iterative step along the controls is carried out. Below, a theorem on the convergence of the method to the solution is formulated.

Theorem 1. *If the set of solutions* $(p^*, \psi^*(t); x^*, x^*(t), u^*(t))$ *for problem (5) is not empty, then sequence* $\{(p^k, \psi^k(t); x^k, x^k(t), u^k(t))\}$ *generated by the method with the step length* $\alpha \le \alpha_0$ *contains subsequence* $\{(p^{k_i}, \psi^{k_i}(t); x^{k_i}, x^{k_i}(t), u^{k_i}(t))\}$, *which converges to the solution of the problem, including: weak convergence in controls, strong convergence in trajectories, conjugate trajectories, and also in terminal variables.*

The proof of the theorem is carried out in the same way as in [6]. The computational process presented in this paper implements the idea of evidence-based computing. It allows us to receive guaranteed solutions to the problem with a given accuracy, consistent with the accuracy of the initial information.

Conclusions. In this paper, we study a terminal control problem with a finite-dimensional boundary value problem at the right-hand end of the time interval and phase constraints distributed over a finite given number of points of this interval. The problem has a convex structure, which makes it possible, within the duality theory, using the saddle-point properties of the Lagrangian, to develop a theory of saddle-point methods for solving terminal control problems. The approach proposed here makes it possible to deal with a complex case with intermediate phase constraints on the controlled phase trajectories. The convergence of the computation process for all components of the solution has been proved: namely, weak convergence in controls and strong convergence in phase and dual trajectories and in terminal variables.

References

1. Antipin, A.S., Khoroshilova, E.V.: Linear programming and dynamics. Ural Math. J. **1**(1), 3–19 (2015)
2. Antipin, A.S., Khoroshilova, E.V.: Saddle-point approach to solving problem of optimal control with fixed ends. J. Global Optim. **65**(1), 3–17 (2016)
3. Antipin, A.S., Khoroshilova, E.V.: On methods of terminal control with boundary value problems: Lagrange approach. In: Goldengorin, B. (ed.) Optimization and Applications in Control and Data Sciences, pp. 17–49. Springer, New York (2016)
4. Antipin, A.S., Khoroshilova, E.V.: Lagrangian as a tool for solving linear optimal control problems with state constraints. In: Proceedings of the International Conference on Optimal Control and Differential Games Dedicated to L.S. Pontryagin on the Occasion of His 110th Birthday, pp. 23–26 (2018)
5. Antipin, A., Khoroshilova, E.: Controlled dynamic model with boundary-value problem of minimizing a sensitivity function. Optim. Lett. **13**(3), 451–473 (2017). https://doi.org/10.1007/s11590-017-1216-8
6. Antipin, A.S., Khoroshilova, E.V.: Dynamics, phase constraints, and linear programming. Comput. Math. Math. Phys. **60**(2), 184–202 (2020)
7. Dmitruk, A.V.: Maximum principle for the general optimal control problem with phase and regular mixed constraints. Comput. Math. Model **4**, 364–377 (1993)
8. Dykhta, V., Samsonyuk, O.: Some applications of Hamilton-Jacobi inequalities for classical and impulsive optimal control problems. Eur. J. Control **17**(1), 55–69 (2011)
9. Gornov, A.Y., Tyatyushkin, A.I., Finkelstein, E.A.: Numerical methods for solving terminal optimal control problems. Comput. Math. Math. Phys. **56**(2), 221–234 (2016)
10. Hartl, R.F., Sethi, S.P., Vickson, R.G.: A survey of the maximum principles for optimal control problems with state constraints. SIAM Rev. **37**(2), 181–218 (1995)
11. Mayne, D.Q., Rawlings, J.B., Rao, C.V., Scokaert, P.O.M.: Survey constrained model predictive control: stability and optimality. Automatica (J. IFAC) **36**(6), 789–814 (2001)
12. Pales, Z., Zeidan, V.: Optimal control problems with set-valued control and state constraints. SIAM J. Optim. **14**(2), 334–358 (2003)
13. Pytlak, R.: Numerical Methods for Optimal Control Problems with State Constraints. Lecture Notes in Mathematics, vol. 1707, p. 214p. Springer, Berlin, Heidelberg (1999)
14. Rao, A.V.: A Survey of Numerical Methods for Optimal Control. Advances in the Astronautical Sciences. Preprint AAS 09-334 (2009)

Pricing in Dynamic Marketing: The Cases of Piece-Wise Constant Sale and Retail Discounts

Igor Bykadorov[1,2,3(✉)]

[1] Sobolev Institute of Mathematics, 4 Koptyug Ave., 630090 Novosibirsk, Russia
bykadorov.igor@mail.ru
[2] Novosibirsk State University, 1 Pirogova St., 630090 Novosibirsk, Russia
[3] Novosibirsk State University of Economics and Management, Kamenskaja street 56, 630099 Novosibirsk, Russia

Abstract. We consider a stylized distribution channel, where a manufacturer sells a single kind of good to a retailer. In classical setting, the profit of manufacturer is quadratic w.r.t. wholesales discount, while the profit of retailer is quadratic w.r.t. retail discount (pass-through). Thus, the wholesale prices and the retail prices are continuous. These results are elegant mathematically but not adequate economically. Therefore, we assume that wholesale discount and retail discounts are piece-wise constant. We show the strict concavity of retailer's profits w.r.t. retail discount levels. As for the manufacturer's profits w.r.t. wholesale discount levels, we show that strict concavity can be guaranteed only in the case when retail discount is constant and sufficiently large.

Keywords: Retailer · Piece-wise constant prices · Retailer · Wholesale discount · Retail discount · Concavity

1 Introduction

Typically, economic agents stimulate production and sales through communications (advertising, promotion, etc.), as well as various types of influence on pricing. Moreover, in the structure of "producer - retailer - consumer", various types of discounts are often used. One of the first work in this direction can be considered the paper [1], see also [2]. Among many works on this subject, let us note [3–8].

In the presented paper, we study dynamic marketing model, cf. [9–12]. At every moment $t \in [t_1, t_2]$, the manufacturer stimulates retailer by wholesale discount as the manufacturer's control $\alpha(t)$, while the retailer stimulates sales by retail discount as the retailer's control $\beta(t)$. We assume that the controls $\alpha(t)$ and $\beta(t)$ are piece-wise constant; moreover, time switches of discount levels are fixed and known[1]. This way, the optimal control problems reduce to the

---

[1] It seems realistic, cf. [12].

© Springer Nature Switzerland AG 2020
N. Olenev et al. (Eds.): OPTIMA 2020, LNCS 12422, pp. 27–39, 2020.
https://doi.org/10.1007/978-3-030-62867-3_3

mathematical programming problems where the profit of the manufacturer is quadratic with respect to wholesale discount level(s), while the profit of the retailer is quadratic with respect to pass-through level(s).

The first question is the concavity property of the profits. This allows getting the optimal behavior strategies of the manufacturer and the retailer.

The main result of [12] is that the retailer's profit is strictly concave w.r.t. pass-through levels. The proof takes into account very strongly the structure of the Hessian matrix of the retailer's profit, it turns out to be inapplicable to the study of the concavity of the manufacturer's profit.

The presented paper is devoted to the development of [12]. First, we show that Hessian matrix of the retailer's profit is strict diagonal dominance matrix. This allows not only to obtain a simpler proof of the main result of [12], but also to prove the strict concavity of the manufacturer's profit in the case when the retail discount is constant and sufficiently large. As for the case when the retail discount is piece-wise constant, we get a sufficient condition (a'la "retail discount levels should not differ much from each other"), guaranteeing strict concavity of the manufacturer's profit. Finally, examples are given when the manufacturer's profit is not concave, which indicates the "unimprovability" of the result.

The paper is organized as follows. In Sect. 2.1 we set the basic model as in [9–12]. In Sect. 2.2 we consider explicitly the case when wholesale discount and pass-through are piece-wise constant. Here we repeat the results of [12] about the form of cumulative sales (Proposition 1) and about the strictly concavity of retailer's profit (Proposition 2). Moreover, we get that Hessian matrix of the retailer's profit is strict diagonal dominance matrix (Lemma 1) and that the retailer's profit is strictly concave for constant path-trough (Proposition 3). Finally, we get a sufficient condition for strict concavity of manufacturer's profit for piece-wise constant path-trough (Lemma 2) and examples when the manufacturer's profit is not concave. Section 3 contains the proofs of Lemmas and Propositions, the discussions of examples. Section 4 concludes.

2   Model

Let us remember the model, see [12].

2.1   Basic Model

Let us consider a vertical distribution channel. On the market, there are manufacturer ("firm"), retailer and consumer.

The firm produces and sells a single product during the time period $[t_1, t_2]$. Let p be the unit price in a situation where the firm sells the product directly to the consumer, bypassing the retailer, $p > 0$. To increase its profits, the firm uses the services of a retailer. To encourage the retailer to sell the commodity, the firm provides it with wholesale discount

$$\alpha(t) \in [A_1, A_2] \subset [0, 1].$$

Thus, the wholesale price of the goods is

$$p_w(t) = (1 - \alpha(t))p.$$

In turn, the retailer directs retail discount ("pass-through"), i.e., a part

$$\beta(t) \in [B_1, B_2] \subset [0, 1]$$

of the discount $\alpha(t)$ to reduce the market price of the commodity. Therefore, the retail price of the commodity is equal to

$$(1 - \beta(t)\alpha(t))p.$$

Then the retailer's profit per unit from the sale is the difference between retail price and wholesale price, i.e.,

$$\alpha(t)(1 - \beta(t))p.$$

Let $x(t)$ be the accumulated sales during the period $[t_1, t]$ while c_0 be a unit production cost.

At the end of the selling period, the total profit of the firm is

$$\Pi_m = \int_{t_1}^{t_2} (p_w(t) - c_0)\, \dot{x}(t)dt = \int_{t_1}^{t_2} (q - \alpha(t)p)\, \dot{x}(t)dt,$$

where $q = p - c_0$. The total profit of the retailer is

$$\Pi_r = p \int_{t_1}^{t_2} \dot{x}(t)\alpha(t)(1 - \beta(t))dt.$$

We assume that accumulated sales $x(t)$ and the motivation of the retailer $M(t)$, satisfy the differential equations

$$\dot{M}(t) = \gamma\dot{x}(t) + \varepsilon\left(\alpha(t) - \overline{\alpha}\right),$$

$$\dot{x}(t) = -\theta x(t) + \delta M(t) + \eta\alpha(t)\beta(t),$$

where $\gamma > 0$, $\varepsilon > 0$, $\theta > 0$, $\delta >$, $\eta > 0$; see [12] for details[2].

Let $\overline{M} > 0$ be the initial motivation of the retailer.

Thus, the *Manufacturer-Retailer Problem* is

$$\Pi_m \longrightarrow \max_\alpha$$
$$\Pi_r \longrightarrow \max_\beta$$
$$\dot{x}(t) = -\theta x(t) + \delta M(t) + \eta\alpha(t)\beta(t),$$
$$\dot{M}(t) = \gamma\dot{x}(t) + \varepsilon\left(\alpha(t) - \overline{\alpha}\right),$$
$$x(t_1) = 0,\ M(t_1) = \overline{M},$$
$$\alpha(t) \in [A_1, A_2] \subset [0, 1],$$
$$\beta(t) \in [B_1, B_2] \subset [0, 1].$$

---

[2] Parameter $\overline{\alpha} \in [A_1, A_2]$ takes into account the fact that the retailer has some expectations about the wholesale discount: the motivation is reduced if the retailer is dissatisfied with the wholesale discount, i.e., if $\alpha(t) < \overline{\alpha}$; on the contrary, the motivation increases if $\alpha(t) > \overline{\alpha}$.

2.2   The Case: Wholesale Discount and Pass-Through Are Piece-Wise Constant

Let for some $t_1 = \tau_0 < \tau_1 < \ldots < \tau_n < \tau_{n+1} = t_2$

$$\alpha(t) = \begin{cases} \alpha_1, & t \in (\tau_0, \tau_1) \\ \alpha_2, & t \in (\tau_1, \tau_2) \\ \ldots \\ \alpha_{n+1}, & t \in (\tau_n, \tau_{n+1}) \end{cases} \qquad \beta(t) = \begin{cases} \beta_1, & t \in (\tau_0, \tau_1) \\ \beta_2, & t \in (\tau_1, \tau_2) \\ \ldots \\ \beta_{n+1}, & t \in (\tau_n, \tau_{n+1}) \end{cases}$$

i.e., $\alpha(t) = \alpha_i$, $\beta(t) = \beta_i$, $t \in (\tau_{i-1}, \tau_i)$, $i \in \{1, \ldots n+1\}$. Then, due to continuity of space variables,

$$x(t) = \begin{cases} x_1(t), & t \in [\tau_0, \tau_1] \\ x_2(t), & t \in [\tau_1, \tau_2] \\ \ldots \\ x_{n+1}(t), & t \in [\tau_n, \tau_{n+1}] \end{cases} \qquad M(t) = \begin{cases} M_1(t), & t \in [\tau_1, \tau_1] \\ M_2(t), & t \in [\tau_1, \tau_2] \\ \ldots \\ M_{n+1}(t), & t \in [\tau_n, \tau_{n+1}] \end{cases}$$

i.e.,

$$x(t) = x_i(t) , \ M(t) = M_i(t) , \ t \in [\tau_{i-1}, \tau_i], \ i \in \{1, \ldots n+1\},$$

where $x_i(t)$ and $M_i(t)$ are the solutions of the systems[3]

$$\begin{aligned} &\dot{x}_i(t) = -\theta x_i(t) + \delta M_i(t) + \eta \alpha_i \beta_i, \\ &\dot{M}_i(t) = \gamma \dot{x}_i(t) + \varepsilon (\alpha_i - \overline{\alpha}), \\ &x_i(\tau_i) = x_{i-1}(\tau_i), \\ &M_i(\tau_i) = M_{i-1}(\tau_i), \\ &t \in [\tau_{i-1}, \tau_i], i \in \{1, \ldots n+1\}. \end{aligned}$$

We get

$$\Pi_r = p \cdot \sum_{i=1}^{n+1} (1 - \beta_i) \alpha_i (x_i(\tau_i) - x_i(\tau_{i-1})), \tag{1}$$

$$\Pi_m = p \cdot \sum_{i=1}^{n} (\alpha_{i+1} - \alpha_i) x_i(\tau_i) + (q - \alpha_{n+1}p) x(t_2). \tag{2}$$

Therefore, we need the expressions for $x_i(\tau_i)$. Let $a = \theta - \gamma\delta$. Under rather natural condition (the concavity of cumulative sales for constant wholesale price, see details in [9, 11]), we assume

$$a > 0. \tag{3}$$

Let[4]

$$K(t) = \tfrac{\delta \overline{M}}{a} \cdot \left(1 - e^{a(t_1 - t)}\right) + \tfrac{\overline{\alpha}\delta\varepsilon}{a^2} \cdot \left(1 - e^{a(t_1 - t)} + a(t_1 - t)\right),$$
$$H_i(t) = \tfrac{\eta}{a} \cdot \left(1 - e^{a(\tau_{i-1} - t)}\right), \ t \geq \tau_{i-1}, \ i \in \{1, \ldots n+1\},$$
$$L_i(t) = -\tfrac{\delta\varepsilon}{a^2} \cdot \left(1 - e^{a(\tau_{i-1} - t)} + a(\tau_{i-1} - t)\right), \ t \geq \tau_{i-1}, \ i \in \{1, \ldots n+1\}.$$

---

[3] Note that $x_1(\tau_0) = 0$ while $M_1(\tau_0) = \overline{M}$.
[4] Due to (3), these formulas are well defined.

Proposition 1. *(See [12].) For $t \in [\tau_{i-1}, \tau_i]$, $i \in \{1, \ldots n + 1\}$*

$$x_i(t) = K(t) + (H_i(t)\beta_i + L_i(t))\alpha_i +$$

$$\sum_{j=1}^{i-1}((H_j(t) - H_{j+1}(t))\beta_j + L_j(t) - L_{j+1}(t))\alpha_j.$$

To optimize the profits, we need first to study their concavity. The main result of [12] is

Proposition 2. *(See [12].) The retailer's profit Π_r is strictly concave with respect to pass-through levels β_i, $i \in \{1, \ldots n + 1\}$.*

The proof in [12] takes into account very strongly the structure of the Hessian matrix of the retailer's profit, it turns out to be inapplicable to the study of the concavity of the manufacturer's profit.

The presented paper is devoted to the development of [12]. First, we show

Lemma 1. *Hessian matrix of the retailer's profit is strict diagonal dominance matrix.*

Proof. See Sect. 3.1.

This Lemma allows not only to obtain a simpler proof of Proposition 2, but also to prove the strict concavity of the manufacturer's profit in the case when the retail discount is constant. More precisely, the following Proposition holds.

Proposition 3. *Let $\beta_i = \beta$, $i \in \{1, \ldots n + 1\}$ and*

$$\beta \geq \frac{\delta\varepsilon}{a\eta}. \tag{4}$$

Then the retailer's profit Π_m is strictly concave with respect to wholesale discount levels α_i, $i \in \{1, \ldots n + 1\}$.

Proof. See Sect. 3.2.

As for the case when the retail discount is piece-wise constant, we can get a sufficient condition (a'la "retail discount levels should not differ much from each other"), guaranteeing strict concavity of the manufacturer's profit. For simplicity, let us consider the case $n = 1$.

Lemma 2. *Let $n = 1$ and*

$$\beta_1 \in \left[\frac{\delta\varepsilon}{a\eta}, 2\beta_2 - \frac{\delta\varepsilon}{a\eta}\right].$$

Then the retailer's profit Π_m is strictly concave with respect to wholesale discount levels α_1, α_2.

Proof. See Sect. 3.3.

Let us presents examples when the producer's profit is not concave, which indicates the "unimprovability" of the result. (See Sect. 3.4 for details.)

Example 1. Let $n = 1, t_1 = 0, \tau_1 = 7, t_2 = 10, \beta_1 = 0.88, \beta_2 = 0.1, \delta = 0.2, \varepsilon = 0.1, \eta = 2, \theta = 1, \gamma = 0.6$. Then $a = \theta - \gamma\delta = 0.88$ and $\det \Pi''_m < 0$.

Example 2. Let $n = 1, t_1 = 0, \tau_1 = 3, t_2 = 6, \beta_1 = 0.9, \beta_2 = 0.1, \delta = 0.2, \varepsilon = 0.1, \eta = 2, \theta = 1, \gamma = 0.6$. Then $a = \theta - \gamma\delta = 0.88$ and $\det \Pi''_m < 0$.

Example 2 shows that even if τ_1 is the middle of $[t_1, t_2]$, profit Π_m can be non-concave.

3 Proofs. Examples

3.1 Proof of Lemma 1

Due to (1), we get

$$\frac{\partial^2 \Pi_r}{\partial \beta_i \partial \beta_j}$$

$$= \begin{cases} -2p\alpha_i \cdot \left(\dfrac{\partial x_i(\tau_i)}{\partial \beta_i} - \dfrac{\partial x_i(\tau_{i-1})}{\partial \beta_i} \right), & i = j \in \{1, \ldots, n+1\} \\ -p\alpha_j \cdot \left(\dfrac{\partial x_j(\tau_j)}{\partial \beta_i} - \dfrac{\partial x_j(\tau_{j-1})}{\partial \beta_i} \right), & i \in \{1, \ldots, n\}, j \in \{i+1, \ldots, n+1\} \\ -p\alpha_i \cdot \left(\dfrac{\partial x_i(\tau_i)}{\partial \beta_j} - \dfrac{\partial x_i(\tau_{i-1})}{\partial \beta_j} \right), & j \in \{1, \ldots, n\}, i \in \{j+1, \ldots, n+1\} \end{cases}$$

i.e.,

$$\frac{\partial^2 \Pi_r}{\partial \beta_i \partial \beta_j}$$

$$= \begin{cases} -2(\alpha_i)^2 \cdot \dfrac{p\eta}{a} \cdot s_i, & i = j \in \{1, \ldots, n+1\} \\ \alpha_i\alpha_j \cdot \dfrac{p\eta}{a} \cdot s_i s_j \cdot \displaystyle\prod_{k=i+1}^{j-1} (1 - s_k), & i \in \{1, \ldots, n\}, j \in \{i+1, \ldots, n+1\} \\ \alpha_i\alpha_j \cdot \dfrac{p\eta}{a} \cdot s_i s_j \cdot \displaystyle\prod_{k=j+1}^{i-1} (1 - s_k), & j \in \{1, \ldots, n\}, i \in \{j+1, \ldots, n+1\} \end{cases}$$

where

$$s_i = 1 - e^{T_i} = 1 - e^{a(\tau_{i-1} - \tau_i)} \in (0, 1), \ i \in \{1, \ldots, n+1\},$$
$$T_i = a(\tau_{i-1} - \tau_i) < 0, \ i \in \{1, \ldots, n+1\}.$$

We get

$$\det \Pi''_r = \left(\frac{p\eta}{a} \right)^{n+1} \cdot \prod_{i=1}^{n+1} (\alpha_i)^2 \det A,$$

where the eliments of matrix A are

$$
A_{ij} = \begin{cases} -2 \cdot s_i, & i = j \in \{1, \ldots, n+1\} \\ s_i s_j \cdot \prod\limits_{k=i+1}^{j-1} (1 - s_k), & i \in \{1, \ldots, n\}, j \in \{i+1, \ldots, n+1\} \\ s_i s_j \cdot \prod\limits_{k=j+1}^{i-1} (1 - s_k), & j \in \{1, \ldots, n\}, i \in \{j+1, \ldots, n+1\} \end{cases}
$$

Further,

$$
\sum_{j=1}^{n+1} A_{ij} = \sum_{j=1}^{i-1} A_{ij} + A_{ii} + \sum_{j=i+1}^{n+1} A_{ij}
$$

$$
= s_i \cdot \left(\sum_{j=1}^{i-1} \left(s_j \cdot \prod_{k=j+1}^{i-1} (1 - s_k) \right) - 2 + \sum_{j=i+1}^{n+1} \left(s_j \cdot \prod_{k=i+1}^{j-1} (1 - s_k) \right) \right)
$$

$$
= s_i \cdot \left(\sum_{j=1}^{i-1} \left((1 - r_j) \cdot \prod_{k=j+1}^{i-1} r_k \right) - 2 + \sum_{j=i+1}^{n+1} \left((1 - r_j) \cdot \prod_{k=i+1}^{j-1} r_k \right) \right),
$$

where

$$
r_i = 1 - s_i \in (0, 1), \quad i \in \{1, \ldots, n+1\}.
$$

Let us show that

$$
\sum_{j=1}^{i-1} \left((1 - r_j) \cdot \prod_{k=j+1}^{i-1} r_k \right) = 1 - \prod_{k=1}^{i-1} r_k, \quad i \in \{2, \ldots, n+1\}, \tag{5}
$$

and

$$
\sum_{j=i+1}^{n+1} \left((1 - r_j) \cdot \prod_{k=i+1}^{j-1} r_k \right) = 1 - \prod_{k=i+1}^{n+1} r_k, \quad i \in \{1, \ldots n\}. \tag{6}
$$

First we show (5) by the Induction on i.
For $i = 2$,

$$
\sum_{j=1}^{1} \left((1 - r_j) \cdot \prod_{k=j+1}^{1} r_k \right) = (1 - r_j) \cdot \prod_{k=2}^{1} r_k = 1 - r_1.
$$

For $i = 3$,

$$
\sum_{j=1}^{2} \left((1 - r_j) \cdot \prod_{k=j+1}^{2} r_k \right) = (1 - r_1) \cdot r_2 + (1 - r_2) \cdot \prod_{k=3}^{2} r_k
$$
$$
= (1 - r_1) \cdot r_2 + 1 - r_2 = 1 - r_1 r_2.
$$

Let (5) holds for $i = f$, i.e., let

$$\sum_{j=1}^{f-1} \left((1 - r_j) \cdot \prod_{k=j+1}^{f-1} r_k \right) = 1 - \prod_{k=1}^{f-1} r_k .$$

Then we get for $i = f + 1$

$$\sum_{j=1}^{f} \left((1 - r_j) \cdot \prod_{k=j+1}^{f} r_k \right) = \sum_{j=1}^{f-1} \left((1 - r_j) \cdot \prod_{k=j+1}^{f} r_k \right) + (1 - r_f) \cdot \prod_{k=F+1}^{f} r_k$$

$$= \sum_{j=1}^{f-1} \left((1 - r_j) \cdot \prod_{k=j+1}^{f-1} r_k \right) \cdot r_f + 1 - r_f = \left(1 - \prod_{k=1}^{f-1} r_k \right) \cdot r_f + 1 - r_f = 1 - \prod_{k=1}^{f} r_k .$$

So (5) holds.

As to (6), let us apply the induction on n.

For $n = 1$, we have $i = 1$ and

$$\sum_{j=2}^{2} \left((1 - r_2) \cdot \prod_{k=2}^{1} r_k \right) = (1 - r_2) \cdot \prod_{k=2}^{1} r_k = 1 - r_2 .$$

For $n = 2$,

$$i = 1 : \qquad \sum_{j=i+1}^{3} \left((1 - r_j) \cdot \prod_{k=i+1}^{j-1} r_k \right) = \sum_{j=2}^{3} \left((1 - r_j) \cdot \prod_{k=2}^{j-1} r_k \right)$$

$$= (1 - r_2) \cdot \prod_{k=2}^{1} r_k + (1 - r_3) \cdot \prod_{k=2}^{2} r_k = 1 - r_2 + (1 - r_3) \cdot r_2 = 1 - r_2 r_3 ,$$

$$i = 2 : \qquad \sum_{j=i+1}^{3} \left((1 - r_j) \cdot \prod_{k=i+1}^{j-1} r_k \right) = \sum_{j=3}^{3} \left((1 - r_j) \cdot \prod_{k=3}^{j-1} r_k \right)$$

$$= (1 - r_3) \cdot \prod_{k=3}^{2} r_k = 1 - r_3 .$$

Let (6) holds for $n = F - 1$, i.e., let

$$\sum_{j=i+1}^{F} \left((1 - r_j) \cdot \prod_{k=i+1}^{j-1} r_k \right) = 1 - \prod_{k=i+1}^{F} r_k , \qquad i \in \{1, \ldots F - 1\} .$$

Then we get for $n = F$

$$\sum_{j=i+1}^{F+1} \left((1 - r_j) \cdot \prod_{k=i+1}^{j-1} r_k \right)$$

$$= \sum_{j=i+1}^{F} \left((1 - r_j) \cdot \prod_{k=i+1}^{j-1} r_k \right) + (1 - r_{F+1}) \cdot \prod_{k=i+1}^{F} r_k$$

$$= 1 - \prod_{k=i+1}^{F} r_k + (1 - r_{F+1}) \cdot \prod_{k=i+1}^{F} r_k = 1 - \prod_{k=i+1}^{F+1} r_k, \qquad i \in \{1, \ldots F\}.$$

So (6) holds.

Due to (5) and (6),

$$\sum_{j=1}^{n+1} A_{ij} = \left(\sum_{j=1}^{i-1} \left((1 - r_j) \cdot \prod_{k=j+1}^{i-1} r_k \right) - 2 + \sum_{j=i+1}^{n+1} \left((1 - r_j) \cdot \prod_{k=i+1}^{j-1} r_k \right) \right) s_i$$

$$= s_i \cdot \left(1 - \prod_{k=1}^{i-1} r_k - 2 + 1 - \prod_{k=i+1}^{n+1} r_k \right) = -s_i \cdot \left(\prod_{k=1}^{i-1} r_k + \prod_{k=i+1}^{n+1} r_k \right) < 0.$$

Therefore, A is strictly diagonally dominant.

3.2 Proof of Proposition 3

Due to (2), we get

$$\frac{\partial^2 \Pi_m}{\partial \beta_i \partial \beta_j}$$

$$= \begin{cases} \dfrac{2p}{a^2} \cdot (\delta \varepsilon \cdot T_i - b_i s_i), & i = j \in \{1, \ldots, n+1\} \\[2ex] \dfrac{pb_i}{a^2} \cdot s_i s_j \cdot \displaystyle\prod_{k=i+1}^{j-1} (1 - s_k), & i \in \{1, \ldots, n\}, j \in \{i+1, \ldots, n+1\} \\[2ex] \dfrac{pb_j}{a^2} \cdot s_i s_j \cdot \displaystyle\prod_{k=i+1}^{j-1} (1 - s_k), & j \in \{1, \ldots, n\}, i \in \{j+1, \ldots, n+1\} \end{cases}$$

where

$$b_i = a\eta\beta_i - \delta\varepsilon, \quad i \in \{1, \ldots, n+1\}.$$

Let

$$\beta_i = \beta, \ i \in \{1, \ldots, n+1\}.$$

Then due to (4)

$$b_i = b = a\eta\beta - \delta\varepsilon \geq 0, \ i \in \{1, \ldots, n+1\},$$

and

$$\frac{\partial^2 \Pi_m}{\partial \beta_i \partial \beta_j}$$

$$= \begin{cases} \dfrac{2p}{a^2} \cdot (\delta\varepsilon \cdot T_i - b \cdot s_i), & i = j \in \{1, \ldots, N\} \\[2mm] \dfrac{pb}{a^2} \cdot s_i s_j \cdot \displaystyle\prod_{k=i+1}^{j-1} (1 - s_k), & i \in \{1, \ldots, N-1\}, j \in \{i+1, \ldots, N\} \\[4mm] \dfrac{pb}{a^2} \cdot s_i s_j \cdot \displaystyle\prod_{k=i+1}^{j-1} (1 - s_k), & j \in \{1, \ldots, N-1\}, i \in \{j+1, \ldots, N\} \end{cases}$$

We get

$$\det \Pi_m'' = \left(\frac{p}{a^2}\right)^{n+1} \cdot \det B,$$

where the elements of matrix B are

$$B_{ij} = \begin{cases} 2 \cdot (\delta\varepsilon \cdot T_i - b s_i), & i = j \in \{1, \ldots, n+1\} \\[2mm] b \cdot s_i s_j \cdot \displaystyle\prod_{k=i+1}^{j-1} (1 - s_k), & i \in \{1, \ldots, n\}, j \in \{i+1, \ldots, n+1\} \\[4mm] b \cdot s_i s_j \cdot \displaystyle\prod_{k=i+1}^{j-1} (1 - s_k), & j \in \{1, \ldots, n\}, i \in \{j+1, \ldots, n+1\} \end{cases}$$

We get

$$\sum_{j=1}^{n+1} B_{ij} = \sum_{j=1}^{i-1} B_{ij} + B_{ii} + \sum_{j=i+1}^{n+1} B_{ij}$$

$$= 2 \cdot \delta\varepsilon \cdot T_i + b \cdot s_i \cdot \left(\sum_{j=1}^{i-1} s_j \cdot \prod_{k=i+1}^{j-1} (1 - s_k) - 2 + \sum_{j=i+1}^{n+1} s_j \cdot \prod_{k=i+1}^{j-1} (1 - s_k) \right)$$

$$= 2 \cdot \delta\varepsilon \cdot T_i + b \cdot s_i \cdot \left(\sum_{j=1}^{i-1} (1 - r_j) \cdot \prod_{k=i+1}^{j-1} r_k - 2 + \sum_{j=i+1}^{n+1} (1 - r_j) \cdot \prod_{k=i+1}^{j-1} r_k \right)$$

(due to (5) and (6))

$$= 2 \cdot \delta\varepsilon \cdot T_i - b \cdot s_i \cdot \left(\prod_{k=1}^{i-1} r_k + \prod_{k=i+1}^{n+1} r_k \right) < 0.$$

Therefore, matrix B is strictly diagonally dominant.

3.3   Proof of Lemma 2

Let $n = 1$, then

$$\frac{\partial^2 \Pi_m}{\partial \alpha_1^2} = -\frac{2p}{a^2} \cdot (b_1 \cdot s_1 - \delta\varepsilon \cdot T_1) < 0$$

$$\frac{\partial^2 \Pi_m}{\partial \alpha_2^2} = -\frac{2p}{a^2} \cdot (b_2 \cdot s_2 - \delta\varepsilon \cdot T_2) < 0$$

$$\frac{\partial^2 \Pi_m}{\partial \alpha_1 \partial \alpha_2} = \frac{pb_1}{a^2} \cdot s_1 s_2$$

We get

$$\frac{\partial^2 \Pi_m}{\partial \alpha_1^2} + \frac{\partial^2 \Pi_m}{\partial \alpha_1 \partial \alpha_2}$$

$$= \frac{p}{a^2} \cdot ((-2 + s_2) \cdot b_1 s_1 + 2 \cdot \delta\varepsilon \cdot T_1) < 0 \quad \forall b_1 > 0$$

$$\frac{\partial^2 \Pi_m}{\partial \alpha_2^2} + \frac{\partial^2 \Pi_m}{\partial \alpha_1 \partial \alpha_2}$$

$$= \frac{p}{a^2} \cdot ((-2 \cdot b_2 + b_1 \cdot s_1) \cdot s_2 + 2 \cdot \delta\varepsilon \cdot T_2) < 0 \quad \forall 2 \cdot b_2 \geq b_1 \geq 0$$

Hence, if $2 \cdot b_2 \geq b_1 \geq 0$ then matrix Π_m'' is strictly diagonally dominant.

3.4   Examples

Example 1. Let $n = 1$,

$$t_1 = 0, \tau_1 = 7, t_2 = 10, \beta_1 = 0.88, \beta_2 = 0.1, \delta = 0.2, \varepsilon = 0.1, \eta = 2, \ \theta = 1, \gamma = 0.6.$$

Then

$$a = \theta - \gamma\delta = 0.88, b_1 = 1.5288, b_2 = 0.156,$$
$$T_1 = -6.16, T_2 = -2.64, s_1 \approx 0.997, s_2 \approx 0.928$$

and

$$4 \cdot (b_1 s_1 - \delta\varepsilon T_1)(b_2 s_2 - \delta\varepsilon T_2) - (b_1 s_1 s_2)^2 \approx -0.625 < 0.$$

Hence $\det \Pi_m'' < 0$.

Example 2. Let $n = 1$,

$$t_1 = 0, \tau_1 = 3, t_2 = 6, \beta_1 = 0.9, \beta_2 = 0.1, \delta = 0.2, \varepsilon = 0.1, \eta = 2, \ \theta = 1, \gamma = 0.6.$$

Then

$$a = \theta - \gamma\delta = 0.88, b_1 = 1.564, b_2 = 0.156,$$
$$T_1 = T_2 = T = -2.64, \ s_1 = s_2 = s \approx 0.928$$

and

$$4 \cdot (b_2 s - \delta\varepsilon T) \cdot (b_1 s - \delta\varepsilon T) - (b_1 s^2)^2 \approx -0.6988 < 0.$$

Hence $\det \Pi_m'' < 0$.

4   Conclusion

In this paper, we study a stylized vertical control distribution channel in the structure "manufacturer-retailer-consumer" and develop the results of [9–12].

More precisely, we consider the situation when the wholesale discount and pass-through are piece-wise constant. The switching times are assumed to be known and fixed. This case seems to be economically adequate. The arising optimization problems contain quadratic objective functions (manufacturer's profit and retailer's profit) with respect to wholesale discount and pass-through level(s).

We get an exhaustive answer to the question of the concavity of the manufacturer's profit with respect to wholesale discount levels levels and of the retailer's profit with respect to pass-through levels.

As for the topics of further research, we plan to study the equilibrium (Nash and Stackelberg) in the structure "manufacturer-retailer-consumer".

Moreover, it seems interesting to study the interaction of several manufacturers and several retailers.

Acknowledgments. The study was carried out within the framework of the state contract of the Sobolev Institute of Mathematics (project no. 0314-2019-0018) and within the framework of Laboratory of Empirical Industrial Organization, Economic Department of Novosibirsk State University. The work was supported in part by the Russian Foundation for Basic Research, projects 18-010-00728 and 19-010-00910 and by the Russian Ministry of Science and Education under the 5–100 Excellence Programme.

References

1. Nerlove, M., Arrow, K.J.: Optimal advertising policy under dynamic conditions. Economica **29**(144), 129–142 (1962)
2. Vidale, M.L., Wolfe, H.B.: An Operations-Research Study of Sales Response to Advertising. Mathematical Models in Marketing. Lecture Notes in Economics and Mathematical Systems (Operations Research), vol. 132, pp. 223–225. Springer, Berlin, Heidelberg (1976). https://doi.org/10.1007/978-3-642-51565-1_72
3. Bykadorov, I.A., Ellero, A., Moretti, E.: Minimization of communication expenditure for seasonal products. RAIRO Oper. Res. **36**(2), 109–127 (2002)
4. Mosca, S., Viscolani, B.: Optimal goodwill path to introduce a new product. J. Optim. Theory Appl. **123**(1), 149–162 (2004)
5. Giri, B.C., Bardhan, S.: Coordinating a two-echelon supply chain with price and inventory level dependent demand, time dependent holding cost, and partial backlogging. Int. J. Math. Oper. Res. **8**(4), 406–423 (2016)
6. Printezis, A., Burnetas, A.: The effect of discounts on optimal pricing under limited capacity. Int. J. Oper. Res. **10**(2), 160–179 (2011)
7. Lu, L., Gou, Q., Tang, W., Zhang, J.: Joint pricing and advertising strategy with reference price effect. Int. J. Prod. Res. **54**(17), 5250–5270 (2016)
8. Zhang, S., Zhang, J., Shen, J., Tang, W.: A joint dynamic pricing and production model with asymmetric reference price effect. J. Ind. Manage. Optim. **15**(2), 667–688 (2019)

9. Bykadorov, I., Ellero, A., Moretti, E., Vianello, S.: The role of retailer's performance in optimal wholesale price discount policies. Eur. J. Oper. Res. **194**(2), 538–550 (2009)
10. Bykadorov, I.A., Ellero, A., Moretti, E.: Trade discount policies in the differential games framework. Int. J. Biomed. Soft Comput. Hum. Sci. **18**(1), 15–20 (2013)
11. Bykadorov, I.: Dynamic marketing model: optimization of retailer's role. Commun. Comput. Inf. Sci. **974**, 399–414 (2019)
12. Bykadorov, I.: Dynamic marketing model: the case of piece-wise constant pricing. Commun. Comput. Inf. Sci. **1145**, 150–163 (2020)

The Generalized Algorithms of Global Parametric Optimization and Stochastization for Dynamical Models of Interconnected Populations

Anastasia Demidova[1] , Olga Druzhinina[2,3(✉)] , Milojica Jacimovic[4] ,
Olga Masina[5] , Nevena Mijajlovic[6] , Nicholas Olenev[2] ,
and Alexey Petrov[5]

[1] Peoples' Friendship University of Russia (RUDN University), Moscow, Russia
`ademidova@sci.pfu.edu.ru`
[2] Federal Research Center "Computer Science and Control" of RAS, Moscow, Russia
`{ovdruzh,nolenev}@mail.ru`
[3] V.A. Trapeznikov Institute of Control Sciences of RAS, Moscow, Russia
[4] Montenegrin Academy of Sciences and Arts, Podgorica, Montenegro
`milojica@jacimovic.me`
[5] Bunin Yelets State University, Yelets, Russia
`olga121@inbox.ru, xeal91@yandex.ru`
[6] University of Montenegro, Podgorica, Montenegro
`nevenamijajlovic@hotmail.com`

Abstract. We consider the issue of synthesis and analysis of multidimensional controlled model with consideration of predator-prey interaction and taking into account migration flows, and propose new formulations of corresponding optimal control problems. In search for optimal trajectories, we develop a generalized algorithm of global parametric optimization, which is based on the development of algorithm for generating control function and on modifications of classical numerical methods for solving differential equations. We present the results of the search for optimal trajectories and control functions generation. Additionally, we propose an algorithm for the transition to stochastic controlled models based on the development of a method for constructing self-consistent stochastic models. The tool software as a part of a software package for modeling controlled dynamical systems has been developed. We have also carried out a computer study of the constructed models. The results can be used in the problems of modeling dynamics processes, taking into account the requirements of control and optimization.

Keywords: Multidimensional nonlinear models · Dynamics of interconnected populations · Global parametric optimization methods · Differential equations · Stochastic models · Optimal control · Software package

© Springer Nature Switzerland AG 2020
N. Olenev et al. (Eds.): OPTIMA 2020, LNCS 12422, pp. 40–54, 2020.
https://doi.org/10.1007/978-3-030-62867-3_4

1   Introduction

When solving the problems of studying the dynamics of interconnected communities, one of the important directions is the construction and study of new multidimensional mathematical models which take into account various types of interspecific and interspecific interactions as well as control effects. In this framework, the problems of optimal control in model of population dynamics are of theoretical and practical interest.

Problems of constructing multidimensional populations models are considered, for examples in [1–4], while models of interconnected communities dynamics taking into account competition and migration flows were studied in [5–12] among others. Migration and population models are widely used in analyzing and predicting the dynamics of real populations and communities, as well as in solving problems of their optimal exploitation. These problems include, in particular, the problems of studying the dynamics of population models on an unlimited trophic resource located in communicating areals, studying the dynamics of the population, and optimization of trade. The results obtained in the course of solving these problems can be used in modeling environmental, demographic, and socio-economic systems. The effects of migration flows in population models are considered in [8–12] and in other papers, with a number of papers considering both deterministic and stochastic dynamic models. For stochastic modeling of various types of dynamic systems, a method for constructing self-consistent one-step models is proposed [13] and a specialized software package is developed [14,15]. This software package allows to perform computer research of models based on the implementation of algorithms for numerical calculation of stochastic differential equations, as well as algorithms for generating trajectories of multidimensional Wiener processes and multipoint distributions. However, until now, the application of this software package was not considered for the problems associated with the modeling of stochastic controlled systems.

It should be noted that, when studying models of dynamics of interconnected communities, the use of applied mathematical packages and General-purpose programming languages is relevant [16–18]. Analysis of models, even in three-and four-dimensional cases, is associated with such a factor as the cumbersomeness of intermediate calculations, in particular, when searching for stationary States. Computer research allows not only to obtain the results of numerical experiments in the framework of stationary states analysis, search for trajectories and estimation of model parameters, but also to identify qualitative new effects caused by the model structure and external influences. An important role is also played by the ability to perform a comparative analysis of models that differ in the types of relationships of phase variables during computational experiments.

Classes of three-dimensional and four-dimensional unmanaged models with competition and migration flows are considered in [10]. This article provides a comparative analysis of the qualitative properties of four-dimensional models, taking into account changes in migration rates, as well as coefficients of intraspecific and interspecific interaction. In [11], four-dimensional nonlinear models of population dynamics of interconnected communities are proposed and studied, taking into account migration and competition, as well as migration, competition and mutualism. In [12], it is proposed to construct multidimensional models taking into account competition and mutualism, as well as taking into account migration flows.

The type of community interaction called "predator-prey" interaction is discussed in numerous papers (see, for example, [2–4,19,20]). A significant part of the results are obtained for two-dimensional and three-dimensional cases, while for higher-dimensional models, there are fewer of them. The two-dimensional "predator-prey" model, taking into account the intraspecific competition of victims, is considered in [2]. For a model with intraspecific competition of victims, the conditions for the existence of a global attractor, as well as conditions for the asymptotic stability of a positive equilibrium state corresponding to the equilibrium coexistence of predators and victims, are obtained. The two-dimensional "predator-prey" model, taking into account the intraspecific competition of predators, is considered in [3]. The two-dimensional "predator-prey" model, taking into account the intraspecific competition of predators and victims, is studied in [4]. A three-dimensional model with one victim and two predators, taking into account the competition of victims and the saturation of predators, is studied in [19]. It is shown that at certain values of parameters, the saturation of predators can lead to the possibility of their coexistence when consuming one type of preys, but only in an oscillatory mode. The three-dimensional model "predator – two populations of preys" with consideration of interspecific competition of victims is considered in [3]. It is shown that for certain parameter values, the presence of a predator in a community can ensure the coexistence of competing populations, which is impossible in the absence of a predator.

The control problems for population dynamics models are set and studied in [21,22] and in the papers of other researchers. Some aspects of optimal control of distributed population models are studied in [21]. The optimality criterion for autoreproduction systems in the framework of the analysis of evolutionarily stable behavior is presented in [22]. In [11,12], optimal control problems are proposed for certain classes of population-migration models with competition under phase and mixed constraints.

Due to the complex structure of controlled population models with migration flows and various types of interspecies interactions, the creation of algorithms and design of programs for global parametric optimization is an urgent problem. Features of global parametric optimization problems include the high dimensionality of the search space, the complex landscape, and the high computational complexity of target functions. Nature-inspired algorithms are quite effective for solving these problems [23,24]. Algorithms for single-criteria global optimization of trajectories of dynamic systems with switches for a software package for modeling switched systems are proposed in [25]. Algorithms for searching for optimal parameters for some types of switchable models, taking into account the action of non-stationary forces, are considered in [26]. This article provides a comparative analysis of single-criteria global optimization methods and discusses their application to find coefficients of parametric control functions.

In this paper, we consider nonlinear models of population dynamics of inter-connected communities, taking into account the interaction "predator-prey" and taking into account migration flows. Optimal control problems are proposed for these models with migration flows. The problem of optimal control under phase constraints for a three-dimensional population model is considered, taking into account the features of the trophic chain and taking into account migration flows. Numerical optimization methods and generalized symbolic calculation algorithms are used to solve the optimal control problem. A computer study of the trajectories of the controlled model is performed. A controlled stochastic model with migration flows by the predator-prey interaction is constructed. To construct this model, we use the method of constructing self-consistent stochastic models and a generalized stochastization algorithm taking into account the controlled case. The properties of models in the deterministic and stochastic cases are characterized. Specialized software systems are used as tools for studying models and solving optimal control problems. Software packages are designed for conducting numerical experiments based on the implementation of algorithms for constructing motion paths, algorithms for parametric optimization and generating control functions, as well as for numerical solution of differential equations systems using modified Runge–Kutta methods.

In Sect. 2, we formulate the optimal control problems for the three-dimensional controlled model with consideration of predator-prey interaction and taking into account migration flows. In Sect. 3 we develop a generalized algorithm of global parametric optimization. We present the results of the search for optimal trajectories and generation of control functions. In Sect. 4, we propose an algorithm for the transition to stochastic controlled models based on the development of a method for constructing self-consistent stochastic models. We use a software package for modeling controlled dynamic systems.

2   The Optimal Control Problems for Deterministic Models of Interacting Communities with Trophic Chains and Migration Flows

A multidimensional model of m communities which takes into account trophic chains and migration flows is given by a system of ordinary differential equations of the following type

$$
\begin{aligned}
\dot{x}_i &= x_i(a_i - p_i x_i) + \beta_{i+1} x_{i+1} - \gamma_i x_i - q_i x_i x_{i+2}, & i &= 3k - 2, \\
\dot{x}_i &= x_i(a_i - p_i x_i) - \beta_i x_i + \gamma_{i-1} x_{i-1}, & i &= 3k - 1, \\
\dot{x}_i &= x_i(a_i - p_i x_i) + r_i x_i x_{i-2}, & i &= 3k.
\end{aligned}
\tag{1}
$$

Here k is the number of the three-element set (prey, prey with shelter, predator), $k = 1, \ldots, m$. The quantity $m \geq 1$ is the number of such sets. Then the total number of considered equations in (1) is $3m$. In this model it is assumed that the number of predator populations is equal to the number of prey populations, and each prey population has the ability to use a refuge. The following assumptions and designations are accepted in (1). The first m equations describe such dynamics of prey populations of densities x_i, and in these equation we taken into account the conditions of interaction of preys with predators ($i = 1, 4, \ldots$), and the conditions outside of interaction with predators in the presence of shelter ($i = 2, 5, \ldots$). For $i = 3, 6, \ldots, 3m$, the equations describe the dynamics of predator populations. Further, a_i are the coefficients of natural growth, p_i are intraspecific competition coefficients, β and γ are the migration coefficients, q_i and r_i are the coefficients of interspecific interactions. The dimension of this model $3m$ satisfies condition $m \geq 1$.

A special case of the model (1) is a model described by a system of the form

$$\dot{x}_1 = a_1 x_1 - p_1 x_1^2 - q x_1 x_3 + \beta x_2 - \gamma x_1,$$
$$\dot{x}_2 = a_2 x_2 - p_2 x_2^2 + \gamma x_1 - \beta x_2, \tag{2}$$
$$\dot{x}_3 = a_3 x_3 - p_3 x_3^2 + r x_1 x_3,$$

where x_1 and x_3 are the densities of prey and predator population in the first areal, x_2 is the prey population density in the second areal, $p_i (i = 1, 2, 3)$ are the intraspecific competition coefficients, q and r are the interaction coefficients of predator and prey, $a_i (i = 1, 2, 3)$ are the natural growth coefficients, β, γ are the migration coefficients of the species between two areas, while the second is a refuge. In the absence of migration, model (2) corresponds to the classic predator-prey model [27,28]. When considering multidimensional models, difficulties arise when calculating symbolic parameters, in particular interest, when finding equilibrium states and constructing phase portraits. In this regard, it is advisable to perform a series of computer experiments, during which the most representative sets of numerical parameter values are considered. Computer research allows us to conduct a comparative analysis of the models properties in the deterministic and stochastic cases. Computer research with consideration of control actions is of interest. Next, we consider optimal control problems in the models of the interacting communities dynamics, taking into account predator-prey interactions in the presence of migration flows.

We formulate the optimal control problems for the three-dimensional model of a "predator-prey" with migration flows. The dynamics of the controlled model is determined by a system of differential equations

$$\dot{x}_1 = a_1 x_1 - p_1 x_1^2 - q x_1 x_3 + \beta x_2 - \gamma x_1 - u_1 x_1,$$
$$\dot{x}_2 = a_2 x_2 - p_2 x_2^2 + \gamma x_1 - \beta x_2 - u_2 x_2, \tag{3}$$
$$\dot{x}_3 = a_3 x_3 - p_3 x_3^2 + r x_1 x_3 - u_3 x_3,$$

where $u_i = u_i(t)$ are control functions.

We set the constraints for the model (3) in the form

$$x_1(0) = x_{10}, \ x_2(0) = x_{20}, \ x_3(0) = x_{30}, \ x_1(T) = x_{11},$$
$$x_2(T) = x_{21}, \ x_3(T) = x_{31}, t \in [0, T], \tag{4}$$

$$0 \leq u_1 \leq u_{11}, \ 0 \leq u_2 \leq u_{21}, \ 0 \leq u_3 \leq u_{31}, \ t \in [0, T]. \tag{5}$$

In relation to the problem (3)–(5), the functional to be maximized is written in the form

$$J(u) = \int_0^T \sum_{i=1}^3 (l_i x_i - c_i) u_i(t) dt. \tag{6}$$

The quality control criterion (6) corresponds to the maximum profit from the use of populations, and l_i is the cost of the i-th population, c_i is the cost of technical equipment corresponding to the i-th population.

The optimal control problem C_1 for the model (3) can be formulated as follows.

(C_1) To find the maximum of functional (6) under conditions (4), (5). It is also of interest to study the following type of restrictions imposed on $u_i(t)$:

$$0 \leq u_1(t) + u_2(t) + u_3(t) \leq M, \ u_i(t) \geq 0, \ i = 1, 2, 3, \ t \in [0, T]. \tag{7}$$

The optimal control problem C_2 for the model (3) is formulated as follows.

(C_2) To find the maximum of functional (6) under conditions (4), (7).

In problems of dynamics of interacting communities, the conditions of non-negativity of phase variables are used. Moreover, restrictions on the growth of the i-th species can be imposed, which leads to mixed restrictions. Given these features, along with problems C_1 and C_2, the optimal control problems with phase and mixed constraints are of interest. However, due to the difficulties of the analytical study of multidimensional dynamic models and the peculiarities of the control quality criterion, methods of numerical optimization are often used. Next, we consider the application of numerical optimization methods as applied to optimal control problems in predator-prey models with migration flows.

3   The Results of Computer Experiments

We consider the algorithm for solving the problem C_1 using global parametric optimization algorithms [23,24]. We propose an approximation of the control function $u(t)$ in the form of polynomials of the n-th degree. This approach to numerical optimization is based on the spline approximation method. Issues of using this method are discussed, for example, in [29,30]. The solution of this problem can be represented in the form of algorithm I, consisting of the following steps.

Step 1. The construction of model trajectories.
Step 2. The evaluation of the control quality criterion.
Step 3. The adjustment of polynomial coefficients using the optimization algorithm. If the stopping condition is not reached, we return to step 1.

Algorithm I is a generalized global parametric optimization algorithm for generating control functions while modeling the dynamics of interacting communities. Within the framework of the generalized algorithm for constructing the control function in a symbolic form, there are wide possibilities for using such methods as a symbol tree method and artificial neural networks.

To solve the problem C_1 we developed a program in the Python programming language using the differential evolution algorithm from the Scipy library. The choice of the differential evolution algorithm in this problem is determined by sufficiently effective indicators with respect to the error and calculation time in comparison to other algorithms, in particular, to the Powell algorithm. A detailed comparative analysis of the differential evolution algorithm compared to other optimization algorithms is given in [31]. The maximizing problem of functional (6) can be reduced to the problem of minimizing the following control quality criterion:

$$\|(\delta, e^{-J})\| \to \min,$$

where δ is the absolute deviation of the trajectories from x_{11}, x_{21}, x_{31} taking into account the constraints (3) in the optimal control problem. The notation e^{-J} is accepted for the inverse exponent corresponding to functional (6).

In our numerical experiment, we will consider a optimal control problem on a class of control functions of the form

$$u_i(t) = \|R_i S\|, \quad R_i = (r_{i0}, r_{i1}, \ldots, r_{in}), \quad S = (t^0, t^1, \ldots, t^n)^{\mathsf{T}}, \tag{8}$$

where $R_i S = r_{i0} t^0 + r_{i1} t^1 + \cdots + r_{in} t^n$ are polynomials of the degree n and $\| \cdot \|$ is the Cartesian norm of the corresponding vector, R_i are the parametric coefficients.

For model (3), in the framework of solving C_1, a series of computer experiments are carried out. The experimental results and a comparison of approximating dependences $x_1(0) = 1, x_2(0) = 0.5, x_3(0) = 1, x_{i1} = 0.2, l_i = 10, c_1 = 1, c_2 = 0.5, c_3 = 1, p_1 = p_2 = p_3 = q = r = 1$ are presented in Table 1.

Table 1. The values of functional (6) for various parameters R.

Values	Error	Value of functional (6)	Calculation time, sec.
$n = 0$, $R_1 = 1.113 \times 10^{-5}$, $R_2 = -1.338$, $R_3 = -1.027$	0.0201	56.332	3.8
$n = 1$, $R_1 = (0.1115, -0.031)$, $R_2 = (-1.034, 0.2816)$, $R_3 = (0.7892, 0.0451)$	0.0091	67.781	23.4
$n = 2$, $R_1 = (-0.2052, 0.0909, -0.0092)$, $R_2 = (2.3093, -1.2374, 0.1250)$, $R_3 = (2.895, -1.558, 0.1532)$	0.0061	73.458	402.1
$n = 3$, increased integration step $R_1 = (-2.7781, 1.1816, -0.1576, 0.0068)$, $R_2 = (-2.3412, 2.1955, -0.4520, 0.0279)$, $R_3 = (-1.6646, 2.2734, -0.5220, 0.0338)$	0.0170	58.779	5210

We observe the following effects arising from the implementation of the algorithm for the optimal control problem in the predator-prey model with migration flows. With an increase in n from 0 to 2, the value of criterion (6) increases and the calculation error decreases. However, for $n > 2$, the value of the criterion begins to decrease. This fact may be associated with an increase in the integration step. The specified increase in this case is necessary to reduce the calculation time.

Figure 1 shows the trajectories of the system (3) for $n = 0$, which corresponds to the case $u_i = const$. Here and further along the abscissa axis, time is indicated, along the ordinate axis, population density for the system (3).

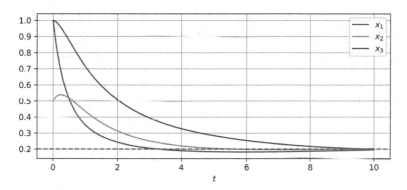

Fig. 1. The trajectories of the system (3) for $n = 0$, $R_1 = 1.11338543 \times 10^{-5}$, $R_2 = -1.33895442$, $R_3 = -1.02759506$.

It can be noted that the trajectories have a character close to monotonically decreasing. We trace an insignificant dependence of the predator population density on the preys population density.

Figure 2 shows the trajectories of the model (3) for $n = 1$. The use of linear control functions significantly increases the value of the criterion (6), while the error decreases (see Table 1). However, in this case, such an efficiency indicator of the algorithm as calculation time worsens. This indicator increases several times.

Figure 3 presents the results of constructing the trajectories of the model (3) for $n = 2$. For this case, we reveal the following effect: the density of the predator population fluctuates and depends on the population density of the preys, and we note a significant increase of the value of the functional (6). In connection with these, it should be noted that there is a direct dependence of the value of the motion quality criterion on the degree of the controlling polynomial. Figure 4 presents the results of constructing the trajectories of system (3) for $n = 3$. According to the results, the indicators of the criterion (6) and absolute error worsened (see Table 1). The negative effect can be explained by an increase in the integration step when performing calculations (for n from 0 to 2 we use 100 steps, for $n = 3$ we have to limit ourselves to the number of steps equal to 22). Moreover,

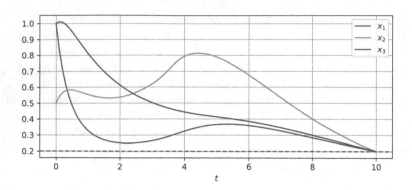

Fig. 2. The trajectories of the system (3) for $n = 1$, $R_1 = (0.11147736, -0.0305515)$, $R_2 = (-1.03398361, 0.28162707)$, $R_3 = (0.78872214, 0.04509503)$.

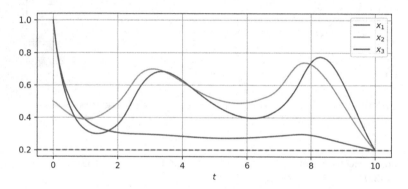

Fig. 3. The trajectories of the system (3) for $n = 2$, $R_1 = (-0.205182, 0.090929, -0.009207)$, $R_2 = (2.309313, -1.237407, 0.125034)$, $R_3 = (2.894907, -1.557567, 0.153203)$.

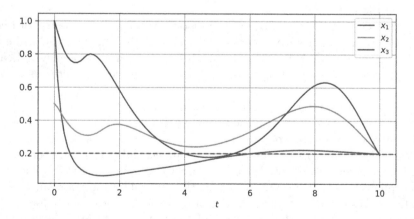

Fig. 4. The trajectories of the system (3) for $n = 3$, $R_1 = (-2.778136, 1.181603, -0.157622, 0.006812)$, $R_2 = (-2.341297, 2.195542, -0.452087, 0.027912)$, $R_3 = (-1.664612, 2.273436, -0.522019, 0.033816)$.

despite the increase in the integration step by several times, the calculation time for one experiment is more than 1.5 h. In the future, we plan to conduct a series of computational experiments with a smaller integration step. Based on the results shown in Table 1, we can conclude that the effectiveness of control functions increases with increasing degree of polynomial. However, it should be noted that an increase of value n significantly increases the computational complexity of algorithm I. The largest values of functional (6) for the model (3) correspond to oscillating interdependent trajectories of "predator-prey".

The presented results based on the application of the differential evolution method belong to step 3 of the generalized global parametric optimization algorithm (algorithm I).

The obtained results can be used to search of the control functions using symbolic regression and artificial neural networks. In particular, it is possible to use the results obtained in the population dynamics with switching control generalized models constructing.

4   The Construction of a Stochastic Predator-Prey Model with Migration Flows

Furthermore in this paper we stochastize the model (3) using the method of constructing self-consistent stochastic models [8]. From transition from the deterministic to the stochastic case and to study the character of accidental effects, a software package for stochastization of one-step processes is developed in a computer algebra system [9, 10, 14, 16]. In this paper, we performed such a modification of this software package that allows us to take into account the controlled actions in the model. For an uncontrolled case, the software package is used in research of the dynamics models of interacting communities, taking into account competition and mutualism [16, 25, 26].

To implement the described software package, the computer system SymPy [11] is used, which is a powerful library of symbolic calculations for the Python language. In addition, the output data obtained using the SymPy library can be transferred for numerical calculations using both the NumPy [12] and SciPy [13] libraries.

For transition from the model (3) to the corresponding stochastic model, we write the interaction scheme, which has the following form:

$$
\begin{aligned}
&X_i \xrightarrow{a_i} 2X_i, && i = 1, 2, 3; \\
&X_i + X_i \xrightarrow{p_i} X_i, && i = 1, 2, 3; \\
&X_1 + X_3 \xrightarrow{q} X_3, \\
&X_1 + X_3 \xrightarrow{r} 2X_3, \\
&X_1 \xrightarrow{\gamma} X_2, && X_2 \xrightarrow{\beta} X_1, \\
&X_i \xrightarrow{u_i} 0, && i = 1, 2, 3.
\end{aligned}
\tag{9}
$$

In this interaction scheme (9), the first line corresponds to the natural reproduction of species in the absence of other factors, the 2-nd line symbolizes intraspecific competition, and the 3-rd and 4-th describe the "predator-prey" relationships between populations x_1 and x_3. The fifth line is a description of a species migration process from one areal to another. The last line is responsible for control. Further, for the obtained interaction scheme using the developed software package, we obtained expressions for the coefficients of the Fokker–Planck equation. The Fokker – Planck equation is a partial differential equation describing the time evolution of the population probability density function. The specified equation in the three-dimensional case, we write as follows:

$$\partial_t P(x,t) = -\sum_{i=1}^{3} [A_i(x)P(x,t)] + \frac{1}{2}\sum_{i,j=1}^{3} \partial_{x_i}\partial_{x_j}[B_{ij}P(x,t)]. \qquad (10)$$

where, for a controlled predator-prey model with migration flows, the drift vector and diffusion matrix, respectively, have the form

$$A(x) = \begin{pmatrix} a_1x_1 - p_1x_1^2 - qx_1x_3 + \beta x_2 - \gamma x_1 - u_1x_1 \\ a_2x_2 - p_2x_2^2 - \beta x_2 + \gamma x_1 - u_2x_2 \\ a_3x_3 - p_3x_3^2 - rx_1x_3 - u_3x_3 \end{pmatrix},$$

$$B(x) = \begin{pmatrix} B_{11} & -\beta x_2 - \gamma x_1 & 0 \\ -\beta x_2 - \gamma x_1 & B_{22} & 0 \\ 0 & 0 & B_{33} \end{pmatrix},$$

where $x = (x_1, x_2, x_3)$ is the system phase vector, $B_{11} = a_1x_1 + p_1x_1^2 + qx_1x_3 + \beta x_2 + \gamma x_1 + u_1x_1$, $B_{22} = a_2x_2 + p_2x_2^2 + \beta x_2 + \gamma x_1 + u_2x_2$, $B_{33} = a_3x_3 + p_3x_3^2 + rx_1x_3 + u_3x_3$.

The generated set of coefficients is transferred to another module of the stochastization software package for one-step processes in order to search for a numerical solution of the obtained stochastic differential equation.

For the numerical experiment of the obtained stochastic model, the same parameters are chosen as for the numerical analysis of the deterministic model (3). The results of the numerical solving of the stochastic differential equation are presented in graphs (Figs. 5, 6, 7 and 8).

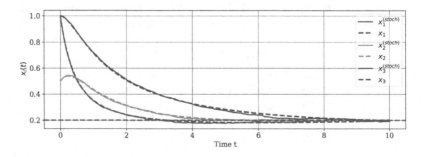

Fig. 5. Comparison of trajectories in deterministic and stochastic cases ($n = 0$).

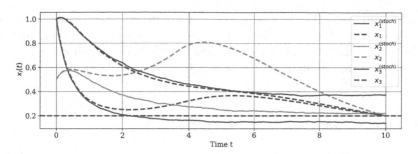

Fig. 6. Comparison of trajectories in deterministic and stochastic cases ($n = 1$).

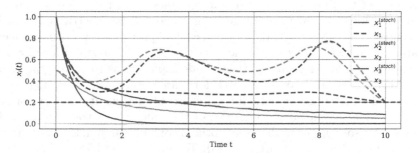

Fig. 7. Comparison of trajectories in deterministic and stochastic cases ($n = 2$).

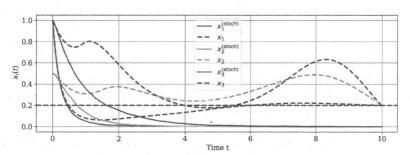

Fig. 8. Comparison of trajectories in deterministic and stochastic cases ($n = 3$).

A comparative analysis of the deterministic and stochastic behavior of the system described by the system of the equations (3) showed that in the first case, namely, for $n = 0$, which corresponds to the case $u_i = const$, the introduction of stochastics weakly affects the behavior of the system. The solutions remain close to the boundary conditions $x_{1i} = 0.2$ specified for the model (3).

In the second, third and fourth cases, namely, for $n = 1$, $n = 2$ and $n = 3$, the transition to the stochastic case significantly changes the behavior of the system. In contrast to deterministic systems, the trajectories of stochastic models solutions for a given set of parameters are monotonous character and go to the stationary mode. Thus, in this case, alternative methods must be used to obtain optimal solutions to the stochastic model. In particular, it is advisable to use the feedback control $u_i(t, x)$. The consideration and implementation of methods for obtaining optimal solutions to stochastic differential equations describing the predator-prey interaction with regard to migration flows is a problem of promising research.

5   Conclusions

In this paper, we have proposed a new approach to the synthesis and analysis of controlled models of the interacting communities dynamics, taking into account the features of trophic chains and migration flows. As a part of this approach, we have developed a generalized algorithm for global parametric optimization using control generation methods. Optimal control problems for models with consideration of predator-prey interactions in the presence of migration flows are considered. Computer research of these models allowed us to obtain the results of numerical experiments on the search for trajectories and generating control functions. The case of representability of control functions in the form of positive polynomials is studied. Peculiarities of the differential evolution algorithm application for solving the problems of optimal control search in the systems with trophic chain and migration are revealed. To solve the problems of population models optimal control, it is proposed to use numerical optimization methods and methods of symbolic calculations.

The implementation of the generalized stochastization algorithm and the analysis of the stochastic model taking into account the predator-prey interaction in the presence of migration flows demonstrated the effectiveness of constructing self-consistent stochastic models method for the controlled case. For a number of parameters sets, it is possible to conduct a series of computer experiments to construct stochastic model trajectories taking into account control actions. A comparative analysis of the studied deterministic and stochastic models is carried out. The choice of control functions for the stochastic model is determined by the structure of the constructed Fokker–Planck equation and is related to the results of numerical experiments conducted with the deterministic model based on global parametric optimization methods. The study of the stochastic model, as opposed to the deterministic one, allows us to take into account the probabilistic nature of the processes of birth and death. In addition, the transition to stochastisation makes it possible to evaluate the effects of the external environment that can cause random fluctuations in the model parameters. For the models studied in this paper, it is shown that in the deterministic case, the proposed global optimization algorithm demonstrates adequate results when constructing optimal trajectories, and for the stochastic case, additional research is required.

The tool software created in the framework of this work can serve as the basis for the new modules (control and numerical optimization modules) for the software package for analyzing self-consistent stochastic models and for the modeling dynamic systems with switching software package. The use of the developed tool software, symbolic calculations and generalized classical and non-classical numerical methods has demonstrated sufficient efficiency for computer research of multidimensional nonlinear models with trophic chain and migration.

The prospects for further research are the synthesis and computer study of partially controlled migration models and the expansion of the numerical methods range for global parametric optimization in the study of models with migration flows. In addition, one of the promising areas is the extension of the previously obtained results for multidimensional uncontrolled models to the controlled case, as well as the development of methods for finding optimal controls using artificial neural networks. It should be noted that the development of methods and tools for analyzing multidimensional models, taking into account various features of trophic chains in the presence of migration flows, as well as the relationship of competition and mutualism, are of interest for further research.

References

1. Murray, J.D.: Mathematical Biology: I. An Introduction. Springer-Verlag, New York (2002). https://doi.org/10.1007/b98868
2. Svirezhev, Y.M., Logofet, D.O.: Stability of Biological Communities. Nauka, Moscow (1978)
3. Bazykin, A.D.: Nonlinear Dynamics of Interacting Populations. Institute of Computer Research, Moscow-Izhevsk (2003)
4. Bratus, A.S., Novozhilov, A.S., Platonov, A.P.: Dynamical Systems and Models of Biology. Draft, Moscow (2011)
5. Lu, Z., Takeuchi, Y.: Global asymptotic behavior in single-species discrete diffusion systems. J. Math. Biol. **32**(1), 67–77 (1993)
6. Zhang, X.-A., Chen, L.: The linear and nonlinear diffusion of the competitive Lotka–Volterra model. Nonlinear Anal. **66**, 2767–2776 (2007)
7. Chen, X., Daus, E.S., Jüngel, A.: Global existence analysis of cross-diffusion population systems for multiple species. Arch. Ration. Mech. Anal. **227**(2), 715–747 (2018)
8. Sinitsyn, I.N., Druzhinina, O.V., Masina, O.N.: Analytical modeling and stability analysis of nonlinear broadband migration flow. Nonlinear world **16**(3), 3–16 (2018)
9. Demidova, A.V., Druzhinina, O., Jacimovic, M., Masina, O.: Construction and analysis of nondeterministic models of population dynamics. In: Vishnevskiy, V.M., Samouylov, K.E., Kozyrev, D.V. (eds.) DCCN 2016. CCIS, vol. 678, pp. 498–510. Springer, Cham (2016). https://doi.org/10.1007/978-3-319-51917-3_43
10. Demidova, A.V., Druzhinina, O.V., Masina, O.N., Tarova, E.D.: Computer research of nonlinear stochastic models with migration flows. CEUR Workshop Proceedings, vol. 2407, pp. 26–37 (2019)
11. Druzhinina, O.V., Masina, O.N., Tarova, E.D.: Analysis and synthesis of nonlinear dynamic models taking into account migration flows and control actions. Nonlinear World **17**(4), 24–37 (2019)
12. Demidova, A., Druzhinina, O., Jaćimović, M., Masina, O., Mijajlovic, N.: Problems of synthesis, analysis and optimization of parameters for multidimensional mathematical models of interconnected populations dynamics. In: Jaćimović, M., Khachay, M., Malkova, V., Posypkin, M. (eds.) OPTIMA 2019. CCIS, vol. 1145, pp. 56–71. Springer, Cham (2020). https://doi.org/10.1007/978-3-030-38603-0_5

13. Demidova, A. V., Gevorkyan, M. N., Egorov, A. D., Kulyabov, D. S., Korolkova, A. V., Sevastyanov, L. A.: Influence of stochastization on one-step models. RUDN J. Math. Inf. Sci. Phys. (1), 71–85 (2014)
14. Gevorkyan, M.N., Velieva, T.R., Korolkova, A.V., Kulyabov, D.S., Sevastyanov, L.A.: Stochastic Runge–Kutta software package for stochastic differential equations. In: Zamojski, W., Mazurkiewicz, J., Sugier, J., Walkowiak, T., Kacprzyk, J. (eds.) Dependability Engineering and Complex Systems. AISC, vol. 470, pp. 169–179. Springer, Cham (2016). https://doi.org/10.1007/978-3-319-39639-2_15
15. Gevorkyan, M.N., Demidova, A.V., Velieva, T.R., Korol'kova, A.V., Kulyabov, D.S., Sevast'yanov, L.A.: Implementing a method for stochastization of one-step processes in a computer algebra system. Program. Comput. Softw. **44**, 86–93 (2018)
16. Oliphant, T.E.: Python for scientific computing. Comput. Sci. Eng. **9**, 10–20 (2007)
17. Lamy, R.: Instant SymPy Starter. Packt Publishing, Birmingham (2013)
18. Oliphant, T.E.: Guide to NumPy, 2nd edn. CreateSpace Independent Publishing Platform, Scotts Valley (2015)
19. Kirlinger, G.: Permanence of some ecological systems with several predator and one preys species. J. Math. Biol. **26**, 217–232 (1988)
20. Kirlinger, G.: Two predators feeding on two preys species: a result on permanence. Math. Biosci. **96**(1), 1–32 (1989)
21. Moskalenko, A.I.: Methods of Nonlinear Mappings in Optimal Control. Theory and Applications to Models of Natural Systems. Nauka, Novosibirsk (1983)
22. Kuzenkov, O.A., Kuzenkova, G.V.: Optimal control of self-reproduction systems. J. Comput. Syst. Sci. Int. **51**, 500–511 (2012)
23. Karpenko, A.P.: Modern Search Engine Optimization Algorithms. Algorithms Inspired by Nature, 2nd edn. N.E. Bauman MSTU, Moscow (2016)
24. Sakharov, M., Karpenko, A.: Meta-optimization of mind evolutionary computation algorithm using design of experiments. In: Abraham, A., Kovalev, S., Tarassov, V., Snasel, V., Sukhanov, A. (eds.) IITI'18 2018. AISC, vol. 874, pp. 473–482. Springer, Cham (2019). https://doi.org/10.1007/978-3-030-01818-4_47
25. Petrov, A.A.: The structure of the software package for modeling technical systems under conditions of switching operating modes. Electromagn. Waves Electro. Syst. **23**(4), 61–64 (2018)
26. Druzhinina, O., Masina, O., Petrov, A.: The synthesis of the switching systems optimal parameters search algorithms. In: Evtushenko, Y., Jaćimović, M., Khachay, M., Kochetov, Y., Malkova, V., Posypkin, M. (eds.) OPTIMA 2018. CCIS, vol. 974, pp. 306–320. Springer, Cham (2019). https://doi.org/10.1007/978-3-030-10934-9_22
27. Lotka, A.: Elements of Physical Ecology. Williams and Wilkins, Baltimora (1925)
28. Volterra, V.: Mathematical Theory of the Struggle for Existence [Russian Translation]. Nauka, Moscow (1976)
29. Dimitrienko, Y.I., Drogolyub, A.N., Gubareva, E.A.: Spline approximation-based optimization of multi-component disperse reinforced composites. Sci. Educ. Bauman MSTU **2**, 216–233 (2015)
30. Laube P., Franz M., Umlauf G.: Deep learning parametrization for B-spline curve approximation. In: 2018 International Conference on 3D Vision (3DV), pp. 691–699 (2018)
31. Padhye N., Mittal P., Deb K.: Differential evolution: performances and analyses. In: 2013 IEEE Congress on Evolutionary Computation, CEC, pp. 1960–1967 (2013)

On Solving a Generalized Constrained Longest Common Subsequence Problem

Marko Djukanovic[1], Christoph Berger[1,2(✉)], Günther R. Raidl[1(✉)], and Christian Blum[2(✉)]

[1] Institute of Logic and Computation, TU Wien, Vienna, Austria
{djukanovic,raidl}@ac.tuwien.ac.at, cberger03@gmail.com
[2] Artificial Intelligence Research Institute (IIIA-CSIC), Campus UAB, Bellaterra, Spain
christian.blum@iiia.csic.es

Abstract. Given a set of two input strings and a pattern string, the constrained longest common subsequence problem deals with finding a longest string that is a subsequence of both input strings and that contains the given pattern string as a subsequence. This problem has various applications, especially in computational biology. In this work we consider the $\mathcal{NP}$–hard case of the problem in which more than two input strings are given. First, we adapt an existing Λ^* search from two input strings to an arbitrary number m of input strings ($m \geq 2$). With the aim of tackling large problem instances approximately, we additionally propose a greedy heuristic and a beam search. All three algorithms are compared to an existing approximation algorithm from the literature. Beam search turns out to be the best heuristic approach, matching almost all optimal solutions obtained by A* search for rather small instances.

Keywords: Longest common subsequences · Constrained subsequences · Beam search · A* search

1 Introduction

Strings are commonly used to represent DNA and RNA in computational biology, and it is often necessary to obtain a measure of similarity for two or more input strings. One of the most well-known measures is calculated by the so-called *longest common subsequence* (LS) problem. Given a number of input strings, this problem asks to find a longest string that is a subsequence of all input strings. Hereby, a *subsequence* t of a string s is obtained by deleting zero or more characters from s. Apart from computational biology, the LCS problem finds also application in video segmentation [3] and text processing [13], just to name a few.

During the last three decades, several variants of the LCS problem have arisen from practice. One of these variants is the *constrained longest common subsequence* (CLCS) problem [14], which can be stated as follows. Given m input

© Springer Nature Switzerland AG 2020
N. Olenev et al. (Eds.): OPTIMA 2020, LNCS 12422, pp. 55–70, 2020.
https://doi.org/10.1007/978-3-030-62867-3_5

strings and a pattern string P, we seek for a longest common subsequence of the input strings that has P as a subsequence. This problem presents a useful measure of similarity when additional information concerning common structure of the input strings is known beforehand. The most studied CLCS variant is the one with only two input strings (2–CLCS); see, for example, [2,5,14]. In addition to these works from the literature, we recently proposed an A* search for the 2–CLCS problem [6] and showed that this algorithm is approximately one order of magnitude faster than other exact approaches.

In the following we consider the general variant of the CLCS problem with $m \geq 2$ input strings $S = \{s_1, \ldots, s_m\}$, henceforth denoted by m–CLCS. Note that the m–CLCS problem is $\mathcal{NP}$–hard [1]. An application of this general variant is motivated from computational biology when it is necessary to find the commonality for not just two but an arbitrary number of DNA molecules under the consideration of a specific known structure.

To the best of our knowledge, the approximation algorithm by Gotthilf et al. [9] is the only existing algorithm for solving the general m–CLCS problem so far. We first extend the general search framework and the A* search from [6] to solve the more general m–CLCS problem. For the application to large-scale instances we additionally propose two heuristic techniques: (i) a greedy heuristic that is efficient in producing reasonably good solutions within a short runtime, and (ii) a beam search (BS) which produces high-quality solutions at the cost of more time. The experimental evaluation shows that the BS is the new state-of-the-art algorithm, especially for large problem instances.

The rest of the paper is organized as follows. Section 2 describes a greedy heuristic for the m–CLCS problem. In Sect. 3 the general search framework for the m–CLCS problem is presented. Section 4 describes the A* search, and in Sect. 5 the beam search is proposed. In Sect. 6, our computational experiments are presented. Section 7 concludes this work and outlines directions for future research.

2   A Fast Heuristic for the m–CLCS Problem

Henceforth we denote the length of a string s over a finite alphabet Σ by $|s|$, and the length of the longest string from the set of input strings $(s_1, \ldots, s_m)$ by n, i.e., $n := \max\{|s_1|, \ldots, |s_m|\}$. The j-th letter of a string s is denoted by $s[j]$, $j = 1, \ldots, |s|$, and for $j > |s|$ we define $s[j] = \varepsilon$, where ε denotes the empty string. Moreover, we denote the contiguous subsequence—that is, the substring—of s starting at position j and ending at position j' by $s[j, j']$, $j = 1, \ldots, |s|$, $j' = j, \ldots, |s|$. If $j > j'$, then $s[j, j'] = \varepsilon$. The concatenation of a string s and a letter $c \in \Sigma$ is written as $s \cdot c$. Finally, let $|s|_c$ be the number of occurrences of letter $c \in \Sigma$ in s. We make use of two data structures created during preprocessing to allow an efficient search:

- For each $i = 1, \ldots, m$, $j = 1, \ldots, |s_i|$, and $c \in \Sigma$, $Succ[i, j, c]$ stores the minimal position index x such that $x \geq j \wedge s_i[x] = c$ or -1 if c does not occur in s_i from position j onward. This structure is built in $O(m \cdot n \cdot |\Sigma|)$ time.

– For each $i = 1, \ldots, m$, $u = 1, \ldots, |P|$, $Embed[i, u]$ stores the right-most position x of s_i such that $P[u, |P|]$ is a subsequence of $s_i[x, |s_i|]$. If no such position exists, $Embed[i, u] := -1$. This table is built in $O(|P| \cdot m)$ time.

In the following we present GREEDY, a heuristic for the m–CLCS problem inspired by the well-known BEST–NEXT heuristic [11] for the LCS problem. GREEDY is pseudo-coded in Algorithm 1. The basic principle is straight-forward. The algorithm starts with an empty solution string $s := \epsilon$ and proceeds by appending, at each construction step, exactly one letter to s. The choice of the letter to append is done by means of a greedy function. The procedure stops once no more letters can be added. The basic data structure of the algorithm is a *position vector* $\mathbf{p}^s = (p_1^s, \ldots, p_m^s) \in \mathbb{N}^m$ which is initialized to $\mathbf{p}^s := (1, \ldots, 1)$ at the beginning. The superscript indicates that this position vector depends on the current (partial) solution s. Given $\mathbf{p}^s$, $s_i[p_i^s, |s_i|]$ for $i = 1, \ldots, m$ refer to the substrings from which letters can still be chosen for extending the current partial solution s. Moreover, the algorithm starts with a pattern position index $u := 1$. The meaning of u is that $P[u, |P|]$ is the substring of P that remains to be included as a subsequence in s. At each construction step, first, a subset $\Sigma_{\text{feas}} \subseteq \Sigma$ of letters is determined that can feasibly extend the current partial solution s, ensuring that the final outcome contains pattern P as a subsequence. More specifically, Σ_{feas} contains a letter $c \in \Sigma$ iff (i) c appears in all strings $s_i[p_i^s, |s_i|]$ and (ii) $s \cdot c$ can be extended towards a solution that includes pattern P. Condition (ii) is fulfilled if $u = |P| + 1$, $P[u] = c$, or $Succ[i, p_i^s, c] < Embed[i, u]$ for all $i = 1, \ldots, m$ (assuming that there is at least one feasible solution). These three cases are checked in the given order, and with the first case that evaluates to true, condition (ii) evaluates to true; otherwise, condition (ii) evaluates to false. Next, dominated letters are removed from Σ_{feas}. For two letters $c, c' \in \Sigma_{\text{feas}}$, we say that c *dominates* c' iff $Succ[i, p_i^s, c] \leq Succ[i, p_i^s, c']$ for all $i = 1, \ldots, m$. Afterwards, the remaining letters in Σ_{feas} are evaluated by the greedy function explained below, and a letter c^* that has the best greedy value is chosen and appended to s. Further, the position vector $\mathbf{p}^s$ is updated w.r.t. letter c^* by $p_i^s := Succ[i, p_i^s, c^*] + 1$, $i = 1, \ldots, m$. Moreover, u is increased by one if $c^* = P[u]$. These steps are repeated until $\Sigma_{\text{feas}} = \emptyset$, and the greedy solution s is returned.

The greedy function used to evaluate each letter $c \in \Sigma_{\text{feas}}$ is

$$g(\mathbf{p}^s, u, c) = \frac{1}{l_{\min}(\mathbf{p}^s, c) + \mathbb{1}_{P[u]=c}} + \sum_{i=1}^{m} \frac{Succ[i, p_i^s, c] - p_i^s + 1}{|s_i| - p_i^s + 1}, \qquad (1)$$

where $l_{\min}(\mathbf{p}^s, c)$ is the length of the shortest remaining part of any of the input strings when considering letter c appended to the solution string and thus consumed, i.e., $l_{\min} := \min\{|s_i| - Succ[i, p_i^s, c] \mid i = 1, \ldots, m\}$, and $\mathbb{1}_{P[u]=c}$ evaluates to one if $P[u] = c$ and to zero otherwise. GREEDY chooses at each construction step a letter that minimizes $g()$. The first term of $g()$ penalizes letters for which the $l_{\min}$ is decreased more and which are not the next letter from $P[u]$. The second term in Eq. (1) represents the sum of the ratios of characters that are skipped (in relation to the remaining part of each input string) when extending the current solution s with letter c.

Algorithm 1. GREEDY: a heuristic for the m–CLCS problem

1: **Input:** problem instance (S, P, Σ)
2: **Output:** heuristic solution s
3: $s \leftarrow \varepsilon$
4: $p_i^s \leftarrow 1$, $i = 1, \ldots, m$
5: $u \leftarrow 1$
6: $\Sigma_{\text{feas}} \leftarrow$ set of feasible and non-dominated letters for extending s
7: **while** $\Sigma_{\text{feas}} \neq \emptyset$ **do**
8:     $c^* \leftarrow \arg\min\{g(\mathbf{p}^s, u, c) \mid c \in \Sigma_{\text{feas}})\}$
9:     $s \leftarrow s \cdot c^*$
10:     **for** $i \leftarrow 1$ **to** m **do**
11:         $p_i^s \leftarrow Succ[i, p_i^s, c^*] + 1$
12:     **end for**
13:     **if** $P[u] = c^*$ **then**
14:         $u \leftarrow u + 1$ // consider next letter in P
15:     **end if**
16:     $\Sigma_{\text{feas}} \leftarrow$ set of feasible and non-dominated letters for extending s
17: **end while**
18: **return** s

---

3   State Graph for the m–CLCS Problem

This section describes the state graph for the m–CLCS problem, in which paths from a dedicated root node to inner nodes correspond to (meaningful) partial solutions, paths from the root to sink nodes correspond to complete solutions, and directed arcs represent (meaningful) extensions of partial solutions. Note that the state graph for the m–CLCS problem is an extension of the state graph of the 2–CLCS problem [6].

Given an m–CLCS problem instance $I = (S, P, \Sigma)$, let s be any string over Σ that is a common subsequence of all input strings S. Such a (partial) solution s induces a position vector $\mathbf{p}^s$ in a well-defined way by assigning a value to each p_i^s, $i = 1, \ldots, m$, such that $s_i[1, p_i^s - 1]$ is the smallest string among all strings in $\{s_i[1, k] \mid k = 1, \ldots, p_i^s - 1\}$ that contains s as a subsequence. Note that these position vectors are the same ones as already defined in the context of GREEDY. In other words, s induces a subproblem $I[\mathbf{p}^s] := \{s_1[p_1^s, |s_1|], \ldots, s_m[p_m^s, |s_m|]\}$ of the original problem instance. This is because s can only be extended by adding letters that appear in all strings of $s_i[p_i^s, |s_i|]$, $i = 1, \ldots, m$. In this context, let substring $P[1, k']$ of pattern string P be the maximal string among all strings of $P[1, k]$, $k = 1, \ldots, |P|$, such that $P[1, k']$ is a subsequence of s. We then say that s is a *valid (partial) solution* iff $P[k' + 1, |P|]$ is a subsequence of the strings in subproblem $I[\mathbf{p}^s]$, that is, a subsequence of $s_i[p_i^s, |s_i|]$ for all $i = 1, \ldots, m$.

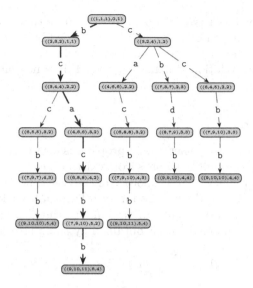

Fig. 1. State graph for the instance $(\{s_1 = \text{bcaacbdba}, s_2 = \text{cbccadcbbd}, s_3 = \text{bbccabcdbba}\}, P = \text{cbb}, \Sigma = \{\text{a}, \text{b}, \text{c}, \text{d}\})$. There are five non-extensible sink nodes (shown in gray). The longest path corresponds to the optimal solution $s = \text{bcacbb}$ with length six and leads to node $v = (\mathbf{p}^v = (9, 10, 11), l^v = 6, u^v = 4)$ (shown in blue). (Color figure online)

The state graph $G = (V, A)$ of our A* search is a *directed acyclic graph* where each node $v \in V$ stores a triple $(\mathbf{p}^v, l^v, u^v)$, with $\mathbf{p}^v$ being a position vector that induces subproblem $I[\mathbf{p}^v]$, l^v is the length of (any) valid partial solution (i.e., path from the root to node v) that induces $\mathbf{p}^v$, and u^v is the length of the longest prefix string of pattern P that is contained as a subsequence in any of the partial solutions that induce node v. Moreover, there is an arc $a = (v, v') \in A$ labeled with letter $c(a) \in \Sigma$ between two nodes $v = (\mathbf{p}^v, l^v, u^v)$ and $v' = (\mathbf{p}^{v'}, l^{v'}, u^{v'})$ iff (*i*) $l^{v'} = l^v + 1$ and (*ii*) subproblem $I[\mathbf{p}^{v'}]$ is induced by the partial solution that is obtained by appending letter $c(a)$ to the end of a partial solution that induces v. As mentioned above, we are only interested in meaningful partial solutions, and thus, for feasibly extending a node v, only the letters from Σ_{feas} can be chosen (see Sect. 2 for the definition of Σ_{feas}). An extension $v' = (\mathbf{p}^{v'}, l^{v'}, u^{v'})$ is therefore generated for each $c \in \Sigma_{\text{feas}}$ in the following way: $p_i^{v'} = Succ[i, p_i^v, c] + 1$ for $i = 1, \dots, m$, $l^{v'} = l^v + 1$, and $u^{v'} = u^v + 1$ in case $c = P[u^v]$, respectively $u^{v'} = u^v$ otherwise.

The *root* node of the state graph is defined by $r = (\mathbf{p}^r = (1, \dots, 1), l^r = 0, u^r = 1)$ and it thus represents the original problem instance. Sink nodes correspond to non-extensible states. A longest path from the root node to some sink node represents an optimal solution to the m–CLCS problem. Figure 1 shows as example the full state graph for the problem instance $(\{s_1 = \text{bcaacbdba}, s_2 = \text{cbccadcbbd}, s_3 = \text{bbccabcdbba}\}, P = \text{cbb}, \Sigma = \{\text{a}, \text{b}, \text{c}, \text{d}\})$. The root node, for example can only be extended by letters b and c, because letters a and d are dominated by the other two letters. Moreover, note that node $((6, 5, 5), 3, 2)$ (induced by partial solution bcc) can only be extended by letter b. Even though letter d is not dominated by letter b, adding letter d cannot lead to any feasible solution, because any solution starting with bccd does not have $P = $ cbb as a subsequence.

3.1   Upper Bounds

As any upper bound for the general LCS problem is also valid for the m–CLCS problem [8], we adopt the following ones from existing work on the LCS problem. Given a subproblem represented by a node v of the state graph, the upper bound proposed by Blum et al. [4] determines for each letter a limit for the number of its occurrences in any solution that can be built starting from a partial solution represented by v. The upper bound is obtained by summing these values for all letters from Σ:

$$\text{UB}_1(v) = \sum_{c \in \Sigma} \min_{i=1,\dots,m} \{|s_i[p_i^v, |s_i|]|_c\} \tag{2}$$

where $|s_i[p_i^v, |s_i|]|_c$ is the number of occurrences of letter c in $s_i[p_i^v, |s_i|]$. This bound is efficiently calculated in $O(m \cdot |\Sigma|)$ time by making use of appropriate data structures created during preprocessing; see [7] for more details.

A dynamic programming (DP) based upper bound was introduced by Wang et al. [15]. It makes use of the DP recursion for the classical LCS problem for all pairs of input strings $\{s_i, s_{i+1}\}, i = 1, \dots, m - 1$. In more detail, for each pair $S_i = \{s_i, s_{i+1}\}$, a *scoring matrix* M_i is recursively derived, where entry $M_i[x, y]$, $x = 1, \dots, |s_i| + 1$, $y = 1, \dots, |s_{i+1}| + 1$ stores the length of the longest common subsequence of $s_i[x, |s_i|]$ and $s_{i+1}[y, |s_{i+1}|]$. We then get the upper bound

$$\text{UB}_2(v) = \min_{i=1,\dots,m-1} M_i[p_i^v, p_{i+1}^v]. \tag{3}$$

Neglecting the preprocessing step, this bound can be calculated efficiently in $O(m)$ time. By combining the two bounds we obtain $\text{UB}(v) := \min\{\text{UB}_1(v), \text{UB}_2(v)\}$. This bound is *admissible* for the A* search, which means that its values never underestimate the optimal value of the subproblem that corresponds to a node v. Moreover, the bound is *monotonic*, that is, the estimated upper bound of any child node is never smaller than the upper bound of the parent node. Monotonicity is an important property in A* search, because it implies that no re-expansion of already expanded nodes [6] may occur.

4   A* Search for the m–CLCS Problem

A* search [10] is a well-known technique in the field of artificial intelligence. More specifically, it is a search algorithm based on the best-first principle, explicitly suited for path-finding in large possibly weighted graphs. Moreover, it is an informed search, that is, the nodes to be further pursued are prioritized according to a function that includes a heuristic guidance component. This function is expressed as $f(v) = g(v) + h(v)$ for all nodes $v \in V$ to which a path has already been found. In our context the graph to be searched is the acyclic state graph $G = (V, A)$ introduced in the previous section. The components of the priority function $f(v)$ are:

- the length $g(v)$ of a best-so-far path from the root r to node v, and
- an estimated length $h(v)$ of a best path from node v to a sink node (known as dual bound).

The performance of A* search usually depends on the tightness of the dual bound, that is, the size of the gap between the estimation and the real cost. In our A* search, $g(v) = l^v$ and $h(v) = \mathrm{UB}(v)$, and the search utilizes the following two data structures:

1. The set of all so far created (reached) nodes N: This set is realized as a nested data structure of sorted lists within a hash map. That is, $\mathbf{p}^v$ vectors act as keys of the hash-map, each one mapping to a list that stores all pairs (l^v, u^v) of nodes $(\mathbf{p}^v, l^v, u^v)$ that induce subproblem $I[\mathbf{p}^v]$. This structure was chosen to efficiently check if a specific node was already generated during the search and to keep the memory footprint comparably small.
2. The open list Q: This is a priority queue that stores references to all not-yet-expanded nodes sorted according to non-increasing values $f(v)$. The structure is used to efficiently retrieve the most promising non-expanded node at any moment.

The search starts by adding root node $r = ((1, \ldots, 1), 0, 1)$ to N and Q. Then, at each iteration, the node with highest priority, i.e., the top node of Q, is extended in all possible ways (see Sect. 3), and any newly created node v' is stored in N and Q. If some node v' is reached via the expanded node in a better way, its f-value is updated accordingly. Moreover, it is checked if v' is dominated by some other node from $N[v'] \subseteq N$, where $N[v']$ is the set of all nodes from N representing the same subproblem $I[\mathbf{p}^{v'}]$. If this is the case, v' is discarded. In this context, given $\hat{v}, \overline{v} \in N[v']$ we say that $\hat{v}$ *dominates* $\overline{v}$ iff $l^{\hat{v}} \geq l^{\overline{v}} \wedge u^{\hat{v}} \geq u^{\overline{v}}$. In the opposite case—that is, if any $v'' \in N[v']$ is dominated by v'—node v'' is removed from $N[v'] \subseteq N$ and Q. The node expansion iterations are repeated until the top node of Q is a sink node, in which case a path from the root node to this sink node corresponds to a proven optimal solution. Such a path is retrieved by reversing the path from following each node's predecessor from the sink node to the root node. Moreover, our A* search terminates without a meaningful solution when a specified time or memory limit is exceeded.

5 Beam Search for the m–CLCS Problem

It is well known from research on other LCS variants that *beam search* (BS) is often able to produce high-quality approximate solutions in this domain [8]. For those cases in which our A* approach is not able to deliver an optimal solution in a reasonable computation time, we therefore propose the following BS approach.

Before the start of the BS procedure, GREEDY is performed to obtain an initial solution s_{bsf}. This solution can be used in BS for pruning partial solutions (nodes) that provenly cannot be extended towards a solution better than s_{bsf}. Beam search maintains a set of nodes B, called the *beam*, which is initialized with the root node r at the start of the algorithm. Remember that this root node represents the empty partial solution. A single major iteration of BS consists of the following steps:

- Each node $v \in B$ is expanded in all possible ways (see the definition of the state graph) and the extensions are kept in a set V_{ext}. If any node $v \in V_{ext}$ is a complete node for which l^v is larger than $|s_{bsf}|$, the best-so-far solution s_{bsf} is updated accordingly.
- Application of function Prune(V_{ext}, UB_{prune}) (optional): All nodes from V_{ext} whose upper bound value is no more than $|s_{bsf}|$ are removed. UB_{prune} refers to the utilized upper bound function (see Sect. 3.1 for the options).
- Application of function Filter(V_{ext}, k_{best}) (optional): this function examines the nodes from V_{ext} and removes dominated ones. Given $v, v' \in V_{ext}$, we say in this context that v *dominates* v' iff $p_i^v \leq p_i^{v'}$, for all $i = 1, \ldots, m$ $\wedge\ u^v \geq u^{v'}$. Note that this is a generalization of the domination relation introduced in [4] for the LCS problem. Since it is time-demanding to examine the possible domination for each pair of nodes from V_{ext} if $|V_{ext}|$ is not small, the domination for each node $v \in V_{ext}$ is only checked against the best k_{best} nodes from V_{ext} w.r.t. a heuristic guidance function $h(v)$, where k_{best} is a strategy parameter. We will consider several options for $h(v)$ presented in the next section.
- Application of function Reduce(V_{ext}, β): The best at most β nodes are selected from V_{ext} to form the new beam B for the next major iteration; the *beam width* β is another strategy parameter.

These four steps are repeated until B becomes empty. Beam search is thus a kind of incomplete *breadth-first-search*.

5.1 Options for the Heuristic Guidance of BS

Different functions can be used as heuristic guidance of the BS, that is, for the function h that evaluates the heuristic goodness of any node $v = (p^{L,v}, l^v, u^v) \in V$. An obvious choice is, of course, the upper bound UB from Sect. 3.1. Additionally, we consider the following three options.

5.1.1 Probability Based Heuristic

For a probability based heuristic guidance, we make use of a DP recursion from [12] for calculating the probability $\Pr(p, q)$ that any string of length p is a subsequence of a *random string* of length q. These probabilities are computed in a preprocessing step for $p, q = 0, \ldots, n$. Remember, in this context that n is the length of the longest input string. Assuming independence among the input strings, the probability $\Pr(s \prec S)$ that a random string s of length p is a common subsequence of all input strings from S is $\Pr(s \prec S) = \prod_{i=1}^{m} \Pr(p, |s_i|)$. Given V_{ext} in some construction step of BS, the question is now how to choose the value p common for all nodes $v \in V_{\text{ext}}$ in order to take profit from the above formula in a sensible heuristic manner. For this purpose, we first calculate

$$p^{\min} = \min_{v \in V_{\text{ext}}} (|P| - u^v + 1), \qquad (4)$$

where P is the pattern string of the tackled m–CLCS instance. Note that the string $P[p^{\min}, |P|]$ must appear as a subsequence in all possible completions of all nodes from $v \in V_{\text{ext}}$, because pattern P must be a subsequence of any feasible solution. Based on $p^{\min}$, the value of p for all $v \in V_{\text{ext}}$ is then heuristically chosen as

$$p = p^{\min} + \min_{v \in V_{\text{ext}}} \left\lfloor \frac{\min_{i=1,\ldots,m} \{|s_i| - p_i^v + 1\} - p^{\min}}{|\Sigma|} \right\rfloor. \qquad (5)$$

The intention here is, first, to let the characters from $P[p^{\min}, |P|]$ fully count, because they will—as mentioned above—appear for sure in any possible extension. This explains the first term ($p^{\min}$) in Eq. (5). The second term is justified by the fact that an optimal m–CLCS solution becomes shorter if the alphabet size becomes larger. Moreover, the solution tends to be longer for nodes v whose length of the shortest remaining string from $I[\mathbf{p}^v]$ is longer than the one of other nodes. We emphasize that this is a heuristic choice which might be improvable. If p would be zero, we set it to one in order to break ties. The final probability-based heuristic for evaluating a node $v \in V_{\text{ext}}$ is then

$$H(v) = \prod_{i=1}^{m} \Pr(p, |s_i| - p_i^v + 1), \qquad (6)$$

and those nodes with a larger H–value are preferred.

5.1.2   Expected Length Based Heuristic

In [8] we derived an approximate formula for the expected length of a longest common subsequence of a set of uniform random strings. Before we extend this result to the m–CLCS problem, we state those aspects of the results from [8] that are needed for this purpose. For more information we refer the interested reader to the original article. In particular, from [8] we know that

$$\mathbb{E}[Y] = \sum_{k=1}^{l_{\min}} \mathbb{E}[Y_k], \qquad (7)$$

where $l_{\min} := \min\{|s_i| \mid i = 1, \ldots, m\}$, Y is a random variable for the length of an LCS, and Y_k is, for any $k = 1, \ldots, l_{\min}$, a binary random variable indicating whether or not there is an LCS with a length of at least k. $\mathbb{E}[\cdot]$ denotes the expected value of some random variable.

In the context of the m–CLCS problem, a similar formula with the following re-definition of the binary variables is used. Y is now a random variable for the length of an LCS that has pattern string P as a subsequence, and the Y_k are binary random variables indicating whether or not there is an LCS with a length of at least k having P as a subsequence. If we assume the existence of at least one feasible solution, we get $\mathbb{E}[Y] = |P| + \sum_{k=|P|+1}^{l_{\min}} \mathbb{E}[Y_k]$.

For $k = |P|, \ldots, l_{\min}$, let T_k be the set of all possible strings of length k over alphabet Σ. Clearly, there are $|\Sigma|^k$ such strings. For each $s \in T_k$ we define the event Ev_s that s is a subsequence of all input strings from S having P as a subsequence. For simplicity, we assume the independence among events Ev_s and $\mathrm{Ev}_{s'}$, for any $s, s' \in T_k$, $s \neq s'$. With this assumption, the probability that string $s \in T_k$ is a subsequence of all input strings from S is equal to $\prod_{i=1}^{m} \Pr(|s|, |s_i|)$. Further, under the assumption that (i) s is a uniform random string and (ii) the probabilities that s is a subsequence of s_i (denoted by $\Pr(s \prec s_i)$) for $i = 1, \ldots, m$, and the probability that P is a subsequence of s (denoted by $(\Pr(P \prec s))$ are independent, it follows that the probability $P^{\mathrm{CLCS}}(s, S, P)$ that s is a common subsequence of all strings from S having pattern P as a subsequence is equal to $\Pr(|P|, k) \cdot \prod_{i=1}^{m} \Pr(k, |s_i|)$. Moreover, note that, under our assumptions, it holds that $\Pr(P \prec s') = \Pr(P \prec s'') = \Pr(|P|, k)$, for any pair of sampled strings $s', s'' \in T_k$. Therefore, it follows that

$$\mathbb{E}[Y_k] = 1 - \prod_{s \in T_k} \left(1 - P^{\mathrm{CLCS}}(s, S, P)\right)$$

$$= 1 - \left(1 - \left(\prod_{i=1}^{m} \Pr(k, |s_i|)\right) \cdot \Pr(|P|, k)\right)^{|\Sigma|^k}. \tag{8}$$

Using this result, the expected length of a final m–CLCS solution that includes a string inducing node $v \in V$ as a prefix can be approximated by the following (heuristic) expression:

$$\mathrm{EX}^{\mathrm{CLCS}}(v) \overset{7,8}{=} |P| - u^v + (l_{\min} - (|P| - u^v + 1) + 1) -$$

$$\sum_{k=|P|-u^v+1}^{l_{\min}} \left(1 - \left(\prod_{i=1}^{m} \Pr(k, |s_i| - p_i^{\mathrm{L},v} + 1)\right) \cdot \Pr(|P| - u^v, k)\right)^{|\Sigma|^k}$$

$$= l_{\min}^v - \sum_{k=|P|-u^v+1}^{l_{\min}^v} \left(1 - \left(\prod_{i=1}^{m} \Pr(k, |s_i| - p_i^{\mathrm{L},v} + 1)\right) \cdot \Pr(|P| - u^v, k)\right)^{|\Sigma|^k},$$

$$\tag{9}$$

where $l_{\min}^v = \min\{|s_i| - p_i^{\mathrm{L},v} + 1 \mid i = 1, \ldots, m\}$. To calculate this value in practice, one has to take care of numerical issues, in particular the large power value $|\Sigma|^k$. We resolve it in the same way as in [8] by applying a Taylor series.

5.1.3 Pattern Ratio Heuristic

So far we have introduced three options for the heuristic function in beam search: the upper bound (Sect. 3.1), the probability based heuristic (Sect. 5.1.1) and the expected length based heuristic (Sect. 5.1.2). With the intention to test, in comparison, a much simpler measure we introduce in the following the pattern ratio heuristic that only depends on the length of the shortest string in $S[\mathbf{p}^v]$ and the length of the remaining part of the pattern string to be covered ($|P| - u^v + 1$). In fact, we might directly use the following function for estimating the goodness of any $v \in V$:

$$R(v) := \frac{\min_{i=1,\dots,m}(|s_i| - p_i^v + 1)}{|P| - u^v + 1}. \tag{10}$$

In general, the larger $R(v)$, the more preferable should be v. However, note that the direct use of (10) generates numerous ties. In order to avoid a large number of ties, instead of $R(v)$ we use the well-known k-norm $||v||_k^k = \sum_{i=1}^m \left(\frac{|s_i| - p_i^v + 1}{|P| - u^v + 1}\right)^k$, with some $k > 0$. Again, nodes $v \in V$ with a larger $||\cdot||_k$-values are preferable. In our experiments, we set $k = 2$ (Euclidean norm).

6 Experimental Evaluation

All algorithms were implemented in C++ using GCC 7.4, and the experiments were conducted in single-threaded mode on a machine with an Intel Xeon E5–2640 processor with 2.40 GHz and a memory limit of 32 GB. The maximal CPU time allowed for each run was set to 15 min, i.e., 900 s. We generated the following set of problem instances for the experimental evaluation. For each combination of the number of input strings $m \in \{10, 50, 100\}$, the length of input strings $n \in \{100, 500, 1000\}$, the alphabet size $|\Sigma| \in \{4, 20\}$ and the ratio $p' = \frac{|P|}{n} \in \{\frac{1}{50}, \frac{1}{20}, \frac{1}{10}, \frac{1}{4}, \frac{1}{2}\}$, ten instances were created, each one as follows. First, P is generated uniformly at random. Then, each string $s_i \in S$ is generated as follows. First, P is copied, that is, $s_i := P$. Then, s_i is augmented in $n - |P|$ steps by single random characters. The position for the new character is selected randomly between any two consecutive characters of s_i, at the beginning, or at the end of s_i. This procedure ensures that at least one feasible solution exists for each instance. The benchmarks are available at https://www.ac.tuwien.ac.at/files/resources/instances/m-clcs.zip. Overall, we thus created and use 900 benchmark instances.

We include the following six algorithms (resp. algorithm variants) in our comparison: (*i*) the approximation algorithm from [9] (APPROX), (*ii*) GREEDY from Sect. 2, and (*iii*) the four beam search configurations differing only in the heuristic guidance function. These BS versions are denoted as follows. BS-UB refers to BS using the upper bound, BS-PROB refers to the use of the probability based heuristic, BS-EX to the use of expected length based heuristic, and BS-PAT to the use of the pattern ratio heuristic. Moreover, we include the information of how many instances of each type were solved to optimality by the exact A^* search. Concerning the beam search, parameters β (the maximum number of nodes kept for the next iteration) and k_{best} (the extent of filtering) are crucial for obtaining good results. After tuning we selected $\beta = 2000$ and $k_{\text{best}} = 100$. Moreover, the tuning procedure indicated that function upper bound based pruning is indeed beneficial.

Table 1. Results for instances with $p' = \frac{|P|}{n} = \frac{1}{20}$.

$|\Sigma|$	m	n	Approx		Greedy		Bs-Ub		Bs-Prob		Bs-Ex		Bs-Pat		A*	
			$|s|$	$\bar{t}[s]$	$|s|$	$\bar{t}[s]$	$|s|$	$\bar{t}[s]$	$|s|$	$\bar{t}[s]$	$|s|$	$\bar{t}[s]$	$|s|$	$\bar{t}[s]$	$\#$	$\bar{t}[s]$
4	10	100	21.4	<0.1	30.8	<0.1	**34.5**	19.2	**34.5**	16.8	**34.5**	21.7	33.4	25.6	3	332.8
4	10	500	119.7	<0.1	162.3	<0.1	181.7	130.1	184.2	163.7	**185.1**	179.8	173.3	192.1	0	-
4	10	1000	244.4	0.1	330.9	0.1	365.7	288.5	372.7	346.7	**374.1**	339.2	343.8	391	0	-
4	50	100	18.7	<0.1	21.3	<0.1	24.3	11.5	24.7	13.3	**24.9**	15.1	24	19.8	0	-
4	50	500	111.1	0.1	127.1	0.1	137.9	98.5	141.2	109.4	**142.2**	115.4	134.2	162.8	0	-
4	50	1000	232.7	0.5	265	0.3	281	226.4	290.1	267.6	**291.3**	289.4	273	366.4	0	-
4	100	100	17.6	<0.1	18.5	<0.1	22.3	11.6	22.4	9.6	**22.5**	13.60	21.9	19.7	0	-
4	100	500	109.4	0.2	119.5	0.2	128.9	101.2	131.9	86.2	**132.4**	119.3	126.6	156	0	-
4	100	1000	227.5	0.8	248	0.9	263.7	244.2	272.0	218.1	**273.0**	232.2	259.2	301.8	0	-
20	10	100	6	<0.1	7.1	<0.1	*7.3	<0.1	*7.3	<0.1	*7.3	<0.1	*7.3	<0.1	10	<0.1
20	10	500	30.2	<0.1	40	<0.1	46.6	16.9	**47.0**	17.5	46.3	60.0	44.7	57	10	332.1
20	10	1000	56.6	0.1	81.2	0.1	95.7	37.9	**97.8**	45.5	95.4	185.4	87.9	146.3	0	-
20	50	100	*5.0	<0.1	*5.0	<0.1	*5.0	<0.1	*5.0	<0.1	*5.0	<0.1	*5.0	<0.1	10	<0.1
20	50	500	26.9	0.1	28.2	0.1	*29.9	1.8	*29.9	1.7	*29.9	1.3	*29.9	1.5	10	1.2
20	50	1000	53.1	0.5	58.2	0.5	62.4	17.6	**62.7**	17	62.5	8.6	60.4	34.4	0	-
20	100	100	*5.0	<0.1	*5.0	<0.1	*5.0	<0.1	*5.0	<0.1	*5.0	<0.1	*5.0	<0.1	10	<0.1
20	100	500	26.1	0.2	26.4	0.2	*27.3	0.3	*27.3	0.2	*27.3	0.3	*27.3	0.3	10	0.3
20	100	1000	52	1	54.7	0.8	57.2	14	*57.3	13.6	*57.3	9.4	56.4	17.7	10	86.0

Table 1 reports results for the instances with $p' = \frac{1}{20}$ and Table 2 those for the instances with $p' = \frac{1}{4}$. The remaining numerical results as well as the tuning process are, due to space limitations, reported in a supplementary document that can be downloaded from https://www.ac.tuwien.ac.at/files/resources/supplementary/clcs-suppl.pdf. The first three columns of each table indicate the instance characteristics. Then, for the six competitors we provide in each table row the obtained solution quality and computation time averaged over the 10 instances with the respective characteristics. The best result of each table row is shown in bold. Finally, for A^* search we provide in each table row the number of instances solved to optimality (out of 10) and the average runtime required to do so. A preceding asterisk indicates that the respective result is provenly optimal. The results allow to make the following observations.

– The m–CLCS problem tends to be most difficult to solve for short pattern strings—that is, low values of $|P|$—and for small alphabet sizes. With growing $|P|$ and $|\Sigma|$, the problem becomes easier. On the one side, this is indicated by the results of the A^* search. When $\frac{|P|}{n} = 1/20$ and $|\Sigma| = 4$, A^* can only solve three problem instances to optimality. When moving to instances

Table 2. Results for instances with $p' = \frac{|P|}{n} = \frac{1}{4}$.

$|\Sigma|$	m	n	Approx		Greedy		Bs-Ub		Bs-Prob		Bs-Ex		Bs-Pat		A*	
			$\bar{s}$	$\bar{t}[s][s]$	$\bar{s}$	$\bar{t}[s][s]$	$\bar{s}$	$\bar{t}[s][s]$	$\bar{s}$	$\bar{t}[s][s]$	$\bar{s}$	$\bar{t}[s][s]$	$\bar{s}$	$\bar{t}[s][s]$	#	$\bar{t}[s][s]$
4	10	100	28.6	<0.1	32.2	<0.1	*34.5	1.1	*34.5	0.9	*34.5	1.0	*34.5	1.5	10	0.2
4	10	500	134.3	<0.1	160.4	<0.1	179.3	45.6	**182.4**	48.8	181.1	98.0	168.6	97	1	660.8
4	10	1000	264.7	0.1	317.4	0.1	350.3	76.8	**361.7**	108	361.4	249.4	330.8	220.2	0	-
4	50	100	26.4	<0.1	26.9	<0.1	*27.5	<0.1	*27.5	<0.1	*27.5	<0.1	*27.5	<0.1	10	<0.1
4	50	500	130.1	0.1	139.5	0.1	146.2	33.6	**148.3**	28	146.3	19.9	142.7	55.9	0	-
4	50	1000	257.4	0.5	277.3	0.3	291.9	73.6	**296.4**	63.6	289.5	41.1	284.2	107.6	0	-
4	100	100	25.9	<0.1	26.2	<0.1	*26.5	<0.1	*26.5	<0.1	*26.5	<0.1	*26.5	<0.1	10	<0.1
4	100	500	128.9	0.2	135.8	0.2	140.4	24.6	**140.8**	34.8	140.3	17.4	137.3	45.9	0	-
4	100	1000	256.4	0.8	270.7	0.9	279.7	56.4	**282.5**	73.4	279.0	40.4	273.3	122	0	-
20	10	100	*25.0	<0.1	*25.0	<0.1	*25.0	<0.1	*25.0	<0.1	*25.0	<0.1	*25.0	<0.1	10	<0.1
20	10	500	*125.0	<0.1	*125.0	<0.1	*125.0	<0.1	*125.0	<0.1	*125.0	<0.1	*125.0	<0.1	10	<0.1
20	10	1000	*250.0	0.1	*250.0	0.1 0.1	*250.0	0.1	*250.0	0.1	*250.0	0.1	*250.0	0.1	10	0.1
20	50	100	*25.0	<0.1	*25.0	<0.1	*25.0	<0.1	*25.0	<0.1	*25.0	<0.1	*25.0	<0.1	10	<0.1
20	50	500	*125.0	0.1	*125.0	0.1	*125.0	0.2	*125.0	0.2	*125.0	0.1	*125.0	0.1	10	0.1
20	50	1000	*250.0	0.5	*250.0	0.5	*250.0	0.4	*250.0	0.5	*250.0	0.5	*250.0	0.5	10	0.5
20	100	100	*25.0	<0.1	*25.0	<0.1	*25.0	<0.1	*25.0	<0.1	*25.0	<0.1	*25.0	<0.1	10	<0.1
20	100	500	*125.0	0.3	*125.0	0.2	*125.0	0.3	*125.0	0.3	*125.0	0.2	*125.0	0.2	10	0.3
20	100	1000	*250.0	1	*250.0	0.8	*250.0	1.1	*250.0	1.1	*250.0	0.8	*250.0	1.1	10	1.0

with $|\Sigma| = 20$, A* search can already solve 70 instances to optimality. The corresponding numbers for instances with $\frac{|P|}{n} = 1/4$ are 31 (for $|\Sigma| = 4$) and 90 (for $|\Sigma| = 20$). In fact, in this last case 90 corresponds to all problem instances of this type. On the other side, the decreasing problem difficulty for growing $|P|$ and $|\Sigma|$ is also indicated by the differences between the results of the heuristic algorithms. In fact, for $\frac{|P|}{n} = 1/20$ and $|\Sigma| = 20$ all algorithms are able to solve all 90 problem instances to optimality.

- The reason for the problem difficulty to decrease with growing $|P|$ can be explained as follows. With growing $|P|$, the similarity between the input strings also grows. This results in a decrease of the search space size. Moreover, from [8] we know that EX-type guidance for BS becomes worse with a growing similarity of the input strings. And, in fact, this observation holds also in the case of the m–CLCS problem. For $\frac{|P|}{n} = 1/20$, Bs-Ex delivers in most cases better results than the other BS configurations. However, Bs-Ex seems to lose efficiency for $\frac{|P|}{n} = 1/4$ where Bs-Prob is generally the better choice.
- All our heuristic algorithms significantly improve over Approx, which is the only existing technique from the literature. This also holds for Greedy, which requires approximately the same computation time as Approx. Only when instances are easy to solve—that is, when n and m are small, and $|\Sigma|$ and $|P|$ are rather large—Approx is competitive with our algorithms.
- All versions of BS improve over Greedy. However, this comes at the price of significantly elevated computation times.
- Bs-Pat, which uses the most simplistic guidance heuristic, is clearly inferior to the other three BS variants in terms of solution quality for almost all instance types.

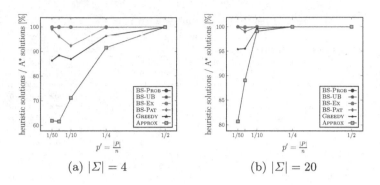

Fig. 2. Average fraction (in percent) of the length of heuristic solutions with respect to the length of the A* solutions.

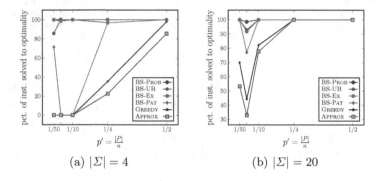

Fig. 3. Percentage of instances solved to optimality.

– The results of Bs-Ub are comparable with those of Bs-Ex only when $|P|$ and n are both small. This is because the upper bound is especially tight for smaller instances.

Finally, we want to shed some more light on the comparison of the heuristic techniques with A* search. For this purpose, from now on we only consider those instances that can be solved to optimality by A* search. The two plots in Fig. 2 show for each heuristic algorithm and for each value of $\frac{|P|}{n}$ (x-axis) the average fraction (in percent) of the length of heuristic solutions in respect to the length of the A* search solutions. A dot at 100% means that the length of the heuristic solution matches the length of the optimal solution from A* search. The plots show, in particular, that Bs-Prob, Bs-Ub and Bs-Ex always reach at least a value of 98%. Complementary, the two plots in Fig. 3 show for each heuristic algorithm and for each value of $\frac{|P|}{n}$ (x-axis) the percentage of instances that were solved to optimality. Bs-Prob fails to deliver an optimal solution for only one instance of type $m = 10$, $n = 500$, $|\Sigma| = 20$, and $\frac{|P|}{n} = 1/20$.

7   Conclusions and Future Work

We tackled the generalized constrained longest common subsequence problem. First, we presented an exact A* search algorithm. Moreover, apart from a simple greedy heuristic, we also introduced four different variants of beam search that differ in the heuristic function they use for selecting the partial solutions to be further expanded in the subsequent iteration. More specifically, we considered an upper bound, a probability based heuristic, an expected length based heuristic, and a simple greedy criterion. Our approaches are compared to an approximation algorithm from the literature, the only one so far available for the problem. In general, the BS variant using the expected length calculation heuristic is best when the pattern string is rather short, while the BS variant with the probability based heuristic is leading when the pattern string is rather long. Moreover, instances become more easily solvable the longer the pattern is.

Concerning future work, the general search framework derived for the CLCS problem can be further extended towards an arbitrary number of pattern strings. We also intend to search for possibly existing real-world benchmark sets for further testing the performances of the algorithms.

Acknowledgments. This work was partially funded by the Doctoral Program Vienna Graduate School on Computational Optimization, Austrian Science Foundation Project No. W1260-N35. This work was also supported by project CI-SUSTAIN funded by the Spanish Ministry of Science and Innovation (PID2019-104156GB-I00).

References

1. Abboud, A., Backurs, A., Williams, V.V.: Tight hardness results for LCS and other sequence similarity measures. In: Proceedings of FOCS 2015 - the 56th Annual Symposium on Foundations of Computer Science, pp. 59–78. IEEE (2015)
2. Arslan, A.N., Eğecioğlu, Ö.: Algorithms for the constrained longest common subsequence problems. Int. J. Found. Comput. Sci. **16**(06), 1099–1109 (2005)
3. Bezerra, F.N.: A longest common subsequence approach to detect cut and wipe video transitions. In: Proceedings of 17th Brazilian Symposium on Computer Graphics and Image Processing, pp. 154–160. IEEE Explore (2004). https://doi.org/10.1109/SIBGRA.2004.1352956
4. Blum, C., Blesa, M.J., López-Ibáñez, M.: Beam search for the longest common subsequence problem. Comput. Oper. Res. **36**(12), 3178–3186 (2009)
5. Chin, F.Y., De Santis, A., Ferrara, A.L., Ho, N., Kim, S.: A simple algorithm for the constrained sequence problems. Inf. Process. Lett. **90**(4), 175–179 (2004)
6. Djukanovic, M., Berger, C., Raidl, G.R., Blum, C.: An A* search algorithm for the constrained longest common subsequence problem. Technical report AC-TR-20-004, Algorithms and Complexity Group, TU Wien (2020). http://www.ac.tuwien.ac.at/files/tr/ac-tr-20-004.pdf
7. Djukanovic, M., Raidl, G.R., Blum, C.: A heuristic approach for solving the longest common square subsequence problem. In: Moreno-Díaz, R., Pichler, F., Quesada-Arencibia, A. (eds.) EUROCAST 2019. LNCS, vol. 12013, pp. 429–437. Springer, Cham (2020). https://doi.org/10.1007/978-3-030-45093-9_52

8. Djukanovic, M., Raidl, G.R., Blum, C.: A beam search for the longest common subsequence problem guided by a novel approximate expected length calculation. In: Nicosia, G., Pardalos, P., Umeton, R., Giuffrida, G., Sciacca, V. (eds.) LOD 2019. LNCS, vol. 11943, pp. 154–167. Springer, Cham (2019). https://doi.org/10.1007/978-3-030-37599-7_14

9. Gotthilf, Z., Hermelin, D., Lewenstein, M.: Constrained LCS: hardness and approximation. In: Ferragina, P., Landau, G.M. (eds.) CPM 2008. LNCS, vol. 5029, pp. 255–262. Springer, Heidelberg (2008). https://doi.org/10.1007/978-3-540-69068-9_24

10. Hart, P., Nilsson, N., Raphael, B.: A formal basis for the heuristic determination of minimum cost paths. IEEE Trans. Syst. Sci. Cybern. $\mathbf{4}$(2), 100–107 (1968)

11. Huang, K., Yang, C., Tseng, K.: Fast algorithms for finding the common subsequences of multiple sequences. In: Proceedings of ICS 2004 - the 3rd IEEE International Computer Symposium, pp. 1006–1011 (2004)

12. Mousavi, S.R., Tabataba, F.: An improved algorithm for the longest common subsequence problem. Comput. Oper. Res. $\mathbf{39}$(3), 512–520 (2012)

13. Storer, J.: Data Compression: Methods and Theory. Computer Science Press, Rockville (1988)

14. Tsai, Y.T.: The constrained longest common subsequence problem. Inf. Process. Lett. $\mathbf{88}$(4), 173–176 (2003)

15. Wang, Q., Pan, M., Shang, Y., Korkin, D.: A fast heuristic search algorithm for finding the longest common subsequence of multiple strings. In: Proceedings of the 24th AAAI Conference on Artificial Intelligence. AAAI Press (2010)

Optimization Problems in Tracking Control Design for an Underactuated Ship with Feedback Delay, State and Control Constraints

Olga Druzhinina[1,2] and Natalya Sedova[3(✉)]

[1] Federal Research Center "Computer Science and Control" of RAS, Moscow, Russia
ovdruzh@mail.ru
[2] V.A. Trapeznikov Institute of Control Sciences of RAS, Moscow, Russia
[3] Ulyanovsk State University, Ulyanovsk, Russia
sedovano@ulsu.ru

Abstract. The nonlinear trajectory tracking control problem for underactuated mechanical systems is considered on the example of a ship motion model. The control that implements the goal should be designed taking into account feedback delay as well as state and control constraints. The purpose is to propose a control law such that, on the one hand, it has the form of explicit feedback, on the other hand, its parameters can be found using standard software tools, provided that the numerical characteristics of the system are specified. In this case, uniform asymptotic stability of the desired motion of the system should be ensured.

The problem is solved by reducing to the stabilization problem for a triangular LPV system with the subsequent transformation into several optimization problems with standard numerical procedures for solving. The domain of attraction that fit into the prescribed region is estimated via sublevel sets of quadratic Lyapunov functions using the Razumikhin conditions and stability properties of cascaded systems. The developed algorithms are implemented based on standard procedures of computational software. The results obtained can be applied to other control problems.

Keywords: Optimization problem · Tracking control · Feedback delay · State and control constraints

1 Introduction

The classic approach to solving tracking problems for underactuated vehicles (in particular, underactuated ships) involves local linearization and decoupling

Supported by Basic Research Program I.7 "New Developments in Perspective Areas of Energetics, Mechanics and Robotics" of the Presidium of RAS.

N. Olenev et al. (Eds.): OPTIMA 2020, LNCS 12422, pp. 71–85, 2020.
https://doi.org/10.1007/978-3-030-62867-3_6

of the multi-variable model to steer the same number of degrees of freedom as the number of available control inputs. Alternative approach is related to linearization of the vehicle error dynamics around trajectories that lead to a time-invariant linear system combined with gain scheduling and (or) Linear Parameter Varying (LPV) design methodologies [11,13]. In addition, the approach based on Lyapunov's functions was significantly developed, in particular, in the problems of nonlinear trajectory-tracking controllers design for tracking the trajectory of underactuated ships [6,12,21]. In [1], the authors proposed a solution to the position tracking problem for a fairly general class of underactuated autonomous vehicles that is applicable to motion in either two or three dimensional spaces. An important class of control problems for underactuated vehicles is path-following problems that are usually considered when the vehicle is required to converge to and follow a path that is specified without a temporal law. The design of path-following controllers for aircraft and marine vehicles is considered, e.g., in [2,13].

The basic model of the ship motion in the horizontal plane is proposed in [9]. The model does not take into account the dynamics associated with the motion in heave, roll, and pitch. In addition, the limitations of the model are that it does not include the environmental forces due to wind, currents, and waves, and it is assumed that the inertia, added mass and damping matrices are diagonal. It is known that such a model is neither static feedback linearizable, nor can it be transformed into chained form; besides, no continuous or discontinuous static state-feedback law exists which makes the origin asymptotically stable [15]. In this paper, we use just the model from [15]. Generalizations and modifications of the model are studied, e.g., in [4,5,8,10,20].

Output-tracking control problems for the nonlinear model of the ship were considered in [5], were using feedback linearization and Lyapunov theory, tracking controllers were developed that stabilized the desired trajectories. The trajectories were, however, position trajectories, and the yaw angle was not controlled. In the case where only the position variables are controlled, the ship may turn around such that the desired position trajectory is followed backward. For such cases, it is useful to consider state-tracking instead of output-tracking. The state-tracking controller based on a nonlinear model was developed in [20] and yields global stability. Problems of semi-global asymptotic stability are discussed in [21].

In this paper, we consider the nonlinear trajectory tracking control problem for underactuated ship taking into account feedback delay as well as state and control constraints. We consider the ship control model as *a cascaded system*, which is "wide enough to cover a large number of applications while simple enough to allow criteria for stability which are easier to verify than finding a Lyapunov function for the closed-loop system" [17]. Numerous examples show that "simple (from a mathematical viewpoint) controllers can be obtained by aiming at giving the closed-loop system a cascaded structure" [17]. Based on a result for time-varying cascaded systems, in [18] the tracking error dynamics is decided into a cascade of two linear subsystems; a simple state-feedback control law is developed and proved to render the tracking error dynamics globally exponentially stable. In [15], a global solution to the tracking problem for an underactuated ship is presented, based on a result for cascaded systems. Along with tracking control problems, dynamic positioning is of theoretical and applied interest, see, e.g., [23,25]. For this problem, the use of a cascade structure is also effective. The paper [16] presents a simple PD-like globally asymptotically stabilizing controller for regulation and dynamic positioning of nonlinear ships with only noisy position measurements; the stability proof is based on a separation principle which is theoretically supported by results on cascaded nonlinear systems and standard Lyapunov theory.

In addition to the cascade structure, we use a special representation of the nonlinear system by a convex combination (with possibly time-varying weights) of linear "vertex" systems. Note that any nonlinear system can be exactly rewritten in that form on any compact set of the state space, by capturing the model nonlinearities in weights which held the convex-sum property [24]. The advantage of the representation is that linear methodologies can be developed, so that we can reduce the controller design to some convex optimization problems which are easy to solve numerically via available software.

The organization of this paper is as follows. In Sect. 2 we consider a stabilization problem for cascaded systems with state and control constraints and give a solution in the form of matrix inequalities. In Sect. 3 we describe the tracking error dynamics considered in [15], and reduce the constrained state-feedback tracking control problem to the problem considered in Sect. 2. Using the result of Sect. 2, we propose some convex optimization problems, as results of which the parameters of the stabilizing control, as well as an estimate of the domain of initial conditions, are obtained, such that the state and control constraints for the system are never violated. Some simulation results are presented in Sect. 4. Concluding remarks are given in Sect. 5.

2   Preliminaries and the Basic Result

In this section, our goal is to construct a delay feedback that stabilizes a differential system with state and control constraints.

This problem has been actively studied in the recent decades, especially useful results have been achieved here for linear time-invariant ODEs: LMI, SDP and other "standard" optimization problems are proposed for such systems, the solutions of which lead to finding controls with the required properties, e.g. [3,14,19]. The attractiveness of such an approach is that for these problems there are computational tools which are available and convenient to use.

So, what should be our regulator? It should: a) have the form of an explicit formula (a delay feedback of a given structure; herewith the delay can be time-variant, the only restriction is imposed on its maximum value); b) allow standard procedures of computational software to find the parameters; c) be suitable for systems that may be time-variant and nonlinear.

Assume that the problem of stabilization of the system is reduced to stabilization of the zero equilibrium: we call a control stabilizing if the zero solution of closed-loop system is uniformly asymptotically stable. Note that for time-varying nonlinear systems, from a robustness viewpoint, the most useful are uniform (global) asymptotic stability and uniform (local) exponential stability (see, e.g., [17]).

Also assume that the system can be decomposed into "cascaded" subsystems (the structure of a cascade system with two selected subsystems is shown in Fig. 1). Further, every "individual" subsystem is assumed to allow a representation by a convex combination (with possibly time-varying weights) of linear delay "vertex" systems. Systems of this kind include, for example, LPV systems, tensor-product type polytopic models, and so-called Takagi–Sugeno models.

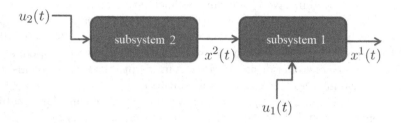

Fig. 1. The structure of a cascade system with two subsystems

2.1   Definitions

Let us write down the equations describing our system.

We use a fairly standard notation: R^n denotes the n-dimensional space of vectors $x = (x_1, \ldots, x_n)^\top$ with the norm $|x| = \sqrt{\sum_{i=1}^n x_i^2}$, $R^+ = [0, +\infty)$.

The system under consideration can be written in the following form:

$$
\begin{aligned}
\dot{x}^1(t) &= \sum_{i=1}^{p_1} \mu_1^i(\xi(t))(A_1^i x^1(t) + B_1 u_1(t)) + g(t, x^1(t), x^2(t)), \\
\dot{x}^2(t) &= \sum_{i=1}^{p_2} \mu_2^i(\xi(t))(A_2^i x^2(t) + B_2 u_2(t)),
\end{aligned}
\tag{1}
$$

where:

- $t \in R^+$, states $x^j(t)$ take values in $D_j \subset R^{n_j}$ with $n_1 + n_2 = n \geq 2$, $D_j = \{y \in R^{n_j} : T_j^{m_j} y \leq 1, (T_j^{m_j})^\top \in R^{n_j}, m_j = 1, \ldots, q_j\}$;
- inputs (controls) are bounded maps $u(\cdot) : R^+ \to U$ with $U = \{u = (u_1^\top, u_2^\top)^\top \in R^m : |u_j|^2 \leq h_j\}$;
- on the set $R^+ \times D_1 \times D_2$ the function g is assumed to be continuous and locally Lipschitz in uniformly in t, $g(t, x^1, 0) = 0$, $|g(t, x^1, x^2)|^2 \leq \Delta$;
- A_j^i and B_j^i are constant matrices ($j = 1, 2$, $i = 1, \ldots, p_j$);
- $\xi(t)$ denotes a piecewise continuous vector function whose values at the current time t depend on t, $x^1(t)$, and $x^2(t)$, and belong to a set $D_\xi \subset R^h$ ($h \geq 1$) whenever $t \in R^+$, $x^j(t) \in D_j$;

– the functions μ_j^i $(j = 1, 2,\ i = 1, \ldots, p_j)$ are continuous on the set D_ξ, and $\mu_j^i(\xi) \in [0, 1]$, $\sum_{i=1}^{p_j} \mu_j^i(\xi) = 1$ for all $\xi \in D_\xi$.

The properties of systems that are represented by a convex combination (with possibly time-varying weights) of linear systems, as well as the possibility of their use as models of nonlinear systems have been discussed by many authors, see, e.g. [24].

The advantage of using a model (1) to design control laws for the system is that linear controller design methodologies can be developed to achieve stability. In addition, the use of the stability properties of cascade systems (see, e.g. [7,17]) allows us to neglect the "interconnecting" term g when solving the stabilization problem. Therefore, the possibilities of linear techniques for such systems are expanding. The disadvantage is that such models usually fail to describe nonlinear systems globally. But practical stabilization problems often involve both state and control constraints. Thus, it is sufficient that the description of the form (1) is valid in the domain satisfying the constraints.

Here, we use quadratic Lyapunov functions and Razumikhin techniques [22] to obtain a state feedback that guarantees asymptotic stability as well as an estimate of the set of initial points from which the system can be driven to origin and meets the given constraints. In fact, we find an invariant subset of the domain of attraction that fit into the prescribed region. That subset is defined via sublevel sets of some Lyapunov functions.

Note that the results presented below are sufficient for the framework of the problem considered here, however, they can be extended (with some modifications) to the framework of systems with delay of a more general form than (1). We will look for control in the form of a delay state feedback:

$$u(t) = (u_1^\top(t), u_2^\top(t))^\top,$$
$$u_j(t) = \sum_{i=1}^{p_j} \mu_j^i(\xi(t)) K_j^i x^j(t - \tau_j(t)), \tag{2}$$

that stabilizes system (1) subject to the state and control constraints (K_j^i are assumed to be constant matrices, $\tau_j(t) : R^+ \to [0, T_j]$ are piecewise continuous delay functions with the upper bounds $T_j > 0$, $j = 1, 2$, $i = 1, \ldots, p_j$). The structure of the system closed by such a control is shown in Fig. 2.

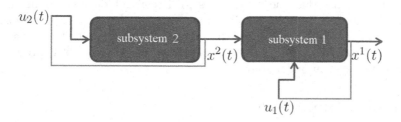

Fig. 2. The structure of the cascade system closed by feedback

2.2   The Basic Result

For M a real symmetric matrix M the inequality $M < 0$ ($M \leq 0$) means that the matrix is negative definite (semidefinite). Repeated blocks within symmetric matrices are replaced by $*$ for brevity.

For positive numbers a_j, c_j, T_j, γ, F, m_1, and m_2, matrices Q_1, Q_2, and M_{ij} of the appropriate dimensions, we introduce the notation: $\Phi_{ij} = \Omega_{ij} + T_j(a_j + c_j)Q_j$, $\Omega_{ij} = Q_j(A_j^i)^\top + A_j^i Q_j + B_j(M_{ij}^1 + M_{ij}^2) + (B_j(M_{ij}^1 + M_{ij}^2))^\top$, where E is the identity matrix. Now let us define the following matrix inequalities:

$$\begin{pmatrix} -a_j Q_j & A_j^i Q_j \\ * & -Q_j \end{pmatrix} \leq 0, \quad \begin{pmatrix} -c_j Q_j & B_j M_{ij} \\ * & -Q_j \end{pmatrix} \leq 0, \tag{3}$$

$$\begin{pmatrix} \frac{1}{T_j}\Phi_{ij} & B_j M_{ij} \\ * & -\frac{1}{2}Q_j \end{pmatrix} < 0, \tag{4}$$

$$\begin{pmatrix} \frac{h_j}{m_j}Q_j & (M_{ij})^\top \\ * & E \end{pmatrix} \geq 0, \quad \begin{pmatrix} \frac{1}{m_j}Q_j & Q_j(T_j^{m_j})^\top \\ * & 1 \end{pmatrix} \geq 0, \tag{5}$$

$$\Phi_{i1} + 2T_1 B_1 K_i^2 Q_1 (B_1 K_i^2)^\top + \gamma Q_1 + \frac{F}{\gamma m_1}E \leq 0. \tag{6}$$

We are now ready to state sufficient conditions to guarantee uniform asymptotic stability of the system (1) with the given feedback structure (2). The following theorem is proved by extending results from [7].

Theorem 1. *Suppose that for some positive numbers a_j, c_j, T_j, γ, m_1, m_2, and for matrices $Q_1 > 0$, $Q_2 > 0$, M_{ij} the following conditions hold:*

1. *there exist (nondecreasing) functions $f(\cdot) : R^+ \to R^+$ and $h(\cdot) : R^+ \to R^+$ such that $|g(t, x^1, x^2)|^2 \leq f(s_1)h(s_2)$ for $x \in \mathcal{M}(s_1, s_2) = \{x \in D_1 \times D_2 : (x^1)^\top Q_1^{-1} x^1 < s_1, (x^2)^\top Q_2^{-1} x^2 < s_2\}$;*
2. *matrix inequalities (3), (4), (5), and (6) hold for $j = 1, 2$, $i = 1, \ldots, p_j$, $m_j = 1, \ldots, q_j$, and $F = f(m_1)h(m_2)$.*

Then the control of the form (2) with $K_j^i = M_{ij}Q_j^{-1}$ is stabilizing for (1) and solutions starting in the set $\mathcal{M}(m_1, m_2)$ meet state and control constraints.

Despite the fact that the inequalities in Theorem 1 depend nonlinearly on the certain parameters, it is possible to construct algorithms in the form of sequences of optimization problems, the solution of which can be found e.g. using the standard tools of the MatLab package. Below we describe these problems as applied to the system under study.

3   Problem Statement

Now we will study the tracking problem for an underactuated ship. Following [15], we consider the motion in surge (forward), sway (sideways) and yaw (heading), see Fig. 3. We assume that:

- only surge control force and yaw control moment available, so need to control three degrees of freedom and have only two inputs available (an underactuated problem);
- the inertia, added mass and damping matrices are diagonal.

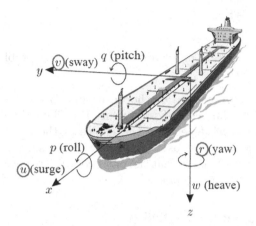

Fig. 3. Definition of state variables in surge, sway, heave, roll, pitch and yaw for a marine vessel [15]

Consider the ship dynamics which is described by [15]:

$$
\begin{aligned}
\dot{u} &= \frac{m_{22}}{m_{11}}vr - \frac{d_{11}}{m_{11}}u + \frac{1}{m_{11}}U_1, \\
\dot{v} &= -\frac{m_{11}}{m_{22}}ur - \frac{d_{22}}{m_{22}}v, \\
\dot{r} &= \frac{m_{11} - m_{22}}{m_{33}}uv - \frac{d_{33}}{m_{33}}r + \frac{1}{m_{33}}U_2, \\
\dot{z}_1 &= u + z_2 r, \\
\dot{z}_2 &= v - z_1 r, \\
\dot{z}_3 &= r
\end{aligned}
\tag{7}
$$

(for brevity, hereinafter, we omit the argument of unknown functions in differential equations), where z_1, z_2, and $z_3 = \psi$ denote the position and orientation of the ship in a frame with an earth-fixed origin having the x- and y-axis always oriented along the ship surge- and sway-axis; $m_{ii} > 0$ are given by the ship inertia and added mass effects; $d_{ii} > 0$ are given by the hydrodynamic damping; the available controls are the surge force U_1, and the yaw moment U_2.

Now, consider a feasible reference trajectory $(u^0, v^0, r^0, z_1^0, z_2^0, z_3^0)$, i.e., a trajectory satisfying under controls U_{10} and U_{20}, with constant u^0, v^0 and time-varying (bounded) $r^0(t)$, $r_{min} \le r^0(t) \le r_{max}$, and define the tracking errors:
$u_e = u - u^0$, $v_e = v - v^0$, $r_e = r - r^0$, $z_{1e} = z_1 - z_1^0$, $z_{2e} = z_2 - z_2^0$, $z_{3e} = z - z_3^0$,
$U_{1e} = U_1 - U_{10}$, $U_{2e} = U_2 - U_{20}$.

Then, we obtain the tracking error dynamics:

$$
\begin{aligned}
\dot{u}_e &= \tfrac{m_{22}}{m_{11}}(v_e r_e + v_e r^0(t) + v^0 r_e) - \tfrac{d_{11}}{m_{11}}u_e + \tfrac{1}{m_{11}}U_{1e}, \\
\dot{v}_e &= -\tfrac{m_{11}}{m_{22}}(u_e r_e + u_e r^0(t) + u^0 r_e) - \tfrac{d_{22}}{m_{22}}v_e, \\
\dot{r}_e &= \tfrac{m_{11}-m_{22}}{m_{33}}(v_e u_e + u_e v^0 + u^0 v_e) - \tfrac{d_{33}}{m_{33}}r_e + \tfrac{1}{m_{33}}U_{2e}, \\
\dot{z}_{1e} &= u_e + z_{2e}r_e + z_{2e}r^0(t) + z_2^0 r_e, \\
\dot{z}_{2e} &= v_e - z_{1e}r_e - z_{1e}r^0(t) - z_1^0 r_e, \\
\dot{z}_{3e} &= r_e.
\end{aligned}
\tag{8}
$$

Then the trajectory tracking problem is reduced to the problem of stabilizing the zero solution of (8).

Remark 1. It seems more natural to choose (x, y)-coordinates for the position of the ship in the (inertial) earth-fixed frame. But then the error coordinates $x_e = x - x^0$, $y_e = y - y^0$ depend on the choice of the frame. The change of coordinates from x, y to $z_1 = x\cos\psi + y\sin\psi$, $z_2 = -x\sin\psi + y\cos\psi$ (for both the ship and the reference) makes that the error coordinates become independent from the choice of the inertial frame.

4   Stabilizing Feedback Design

In this section we employ the approach of Sect. 2 for solving the state-feedback tracking problem proposed in the previous section.

First, we use a substitution similar to that proposed in [15]:

$$
U'_{2e} = (m_{11} - m_{22})(v_e u_e + u_e v^0 + u^0 v_e) + U_{2e}
$$

with a new input U'_{2e}. Then the equations for r_e and z_{3e} from (8) forms a linear system

$$
\begin{aligned}
\dot{r}_e &= -\tfrac{d_{33}}{m_{33}}r_e + \tfrac{1}{m_{33}}U'_{2e}, \\
\dot{z}_{3e} &= r_e.
\end{aligned}
\tag{9}
$$

After that, we substitute $r_e = 0$ and $z_{3e} = 0$ into the remaining equations of (8) to obtain the linear time-varying system

$$
\begin{aligned}
\dot{u}_e &= \tfrac{m_{22}}{m_{11}}r^0(t)v_e - \tfrac{d_{11}}{m_{11}}u_e + \tfrac{1}{m_{11}}U_{1e}, \\
\dot{v}_e &= -\tfrac{m_{11}}{m_{22}}r^0(t)u_e - \tfrac{d_{22}}{m_{22}}v_e, \\
\dot{z}_{1e} &= u_e + r^0(t)z_{2e}, \\
\dot{z}_{2e} &= v_e - r^0(t)z_{1e}.
\end{aligned}
\tag{10}
$$

As the result, we can write system (9), (10) as (1) with $x^1 = (u_e \ v_e \ z_{1e} \ z_{2e})^\top$, $x^2 = (r_e \ z_{3e})^\top$, $p_1 = 2$, $p_2 = 1$, $\mu_1^1(\xi(t)) \equiv \mu_1^1(t) = 2 - r^0(t)$, $\mu_1^2(\xi(t)) \equiv \mu_1^2(t) = r^0(t) - 1$, $\mu_2^1(\xi(t)) \equiv \mu_2^1(t) \equiv 1$,

$$
A_1^1 = \begin{pmatrix}
-\tfrac{d_{11}}{m_{11}} & \tfrac{m_{22}}{m_{11}}r_{min} & 0 & 0 \\
-\tfrac{m_{11}}{m_{22}}r_{min} & -\tfrac{d_{22}}{m_{22}} & 0 & 0 \\
1 & 0 & 0 & r_{min} \\
0 & 1 & -r_{min} & 0
\end{pmatrix},
$$

$$A_1^2 = \begin{pmatrix} -\frac{d_{11}}{m_{11}} & \frac{m_{22}}{m_{11}}r_{max} & 0 & 0 \\ -\frac{m_{11}}{m_{22}}r_{max} & -\frac{d_{22}}{m_{22}} & 0 & 0 \\ 1 & 0 & 0 & r_{max} \\ 0 & 1 & -r_{max} & 0 \end{pmatrix},$$

$$A_2^1 = \begin{pmatrix} -\frac{d_{33}}{m_{33}} & 0 \\ 1 & 0 \end{pmatrix},$$

$$B_1 = (1/m_{11}\ 0\ 0\ 0)^\top,\quad B_2 = (1/m_{33}\ 0)^\top,$$

$$g(t, x^1, x^2) = \left(\frac{m_{22}}{m_{11}}(v_e + v^0)r_e \ -\ \frac{m_{11}}{m_{22}}(u_e + u^0)r_e\ z_{2e} + z_2^0\ -(z_{1e} + z_1^0) \right)^\top.$$

Notice that, in such a formulation, it becomes necessary to verify that control constraints are met taking into account the substitution.

Let the constraints have the form: $|u_e| \le 1$, $|v_e| \le 1$, $|z_{1e}| \le 1$, $|z_{2e}| \le 1$, $|U_{1e}| \le 1$, $|U_{2e}| \le 1$. We find the parameters of the feedback U_{1e}, U'_{2e} in the form (2), solving the following problems successively:

1. Solve the following GEVP:

$$\min_{Q_1} \alpha_1 \text{ subject to } \alpha_1 < 0, Q_1 > 0, \Omega_{i1} < \alpha_1 Q_1 \ (i = 1, 2), \text{ and (5) with } j = 1.$$

   If this problem is feasible, we obtain matrices Q_1, M_{11}, M_{21}, and go to step 2.

2. With the values Q_1, M_{11}, and M_{21} solve the following GEVP:

$$\min_{Q_2} \alpha_2 \text{ subject to } \alpha_2 < 0, \ Q_2 > 0, \ \Omega_2 < \alpha_2 Q_2, \text{ and}$$

$$\begin{pmatrix} \frac{1}{4m_1}Q_1 - \frac{1}{2}Q_1\Lambda \cdot \Lambda Q_1 & 0 & L^\top \\ * & Q_2 & M_2^\top \\ * & * & 1 \end{pmatrix} > 0$$

   with $L = ((m_{22} - m_{11})v^0 \ (m_{22} - m_{11})u^0 \ 0 \ 0)Q_1$, $\Lambda = diag(m_{22} - m_{11}, m_{22} - m_{11}, 0, 0)$. Here the last inequality is feasible only if the following condition for m_1 is satisfied: $\frac{1}{4m_1}Q_1 - \frac{1}{2}Q_1\Lambda \cdot \Lambda Q_1 > 0$; we can use the gridding procedure to select an appropriate value of m_1. So we obtain matrices Q_2 and M_2.

3. Compute the minimal positive a_j, c_j meeting (3) $(j = 1, 2)$.

4. Compute the maximal τ_1^*, τ_2^* such that LMIs (4) are feasible for $(i, j) \in \{(1, 1), (2, 1), (1, 2)\}$.

5. Put $T_1 = q_1\tau_1^*$, $T_2 = q_2\tau_2^*$ $(q_1, q_2 \in (0, 1))$, and find some appropriate functions f and h for g (see the conditions of Theorem 1). Then solve the problem (see (6)):

$$b^* = \min_{0 < \gamma < \gamma_0} b: \begin{pmatrix} \Phi_{i1} + 2T_1B_1K_1^iQ_1(B_1K_1^i)^\top + \gamma Q_1 & E \\ * & -bm_1\gamma/f(m_1)E \end{pmatrix} < 0$$

$(i = 1, 2)$, where $\gamma_0 = -2\max_{i=1,2}\max_{k=1,\ldots,4} \Re\ \lambda_k(A_1^i + B_i^1K_1^i)$ ($\lambda_k(\cdot)$ are eigenvalues of a matrix). At last, compute $m_2 = h^{-1}(b^*)$.

As the result, we have the matrices P_1 and P_2, the numbers m_1, m_2, T_1, and T_2, and also the matrices K_1^1, K_1^2, K_2 such that conditions of Theorem 1 holds. So we obtain the feedback for (8):

$$U_{1e}(t) = \mu_1^1(t)K_1^1 x^1(t - \tau_1(t))) + \mu_1^2(t)K_1^2 x^1(t - \tau_1(t))),$$
$$U_{2e}(t) = K_2 x^2(t - \tau_2(t)) - (m_{11} - m_{22})(v_e(t)u_e(t) + u_e(t)v^0 + u^0 v_e(t)), \quad (11)$$

which is stabilizing for any piecewise continuous delay functions $\tau_j(t) : R^+ \rightarrow [0, T_j]$ $(j = 1, 2)$, and solutions of (8) starting in the set $\mathcal{M}(m_1, m_2)$ meet constraints.

5   Simulations

In this section, the effectiveness of the proposed control laws is verified via a simulation example.

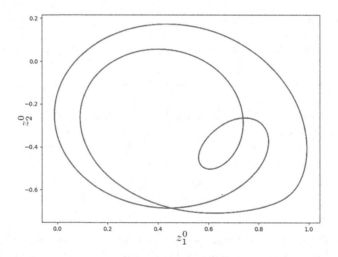

Fig. 4. Plot of the path for z_1^0, z_2^0

Consider an underactuated surface ship with the following model parameters:

$$m_{11} = 29, \ m_{22} = 31, \ m_{33} = 4, \ d_{11} = 0.9, \ d_{22} = 2.9, \ d_{33} = 0.5 \quad (12)$$

(the values are taken close to those used in [4]).

For trajectory tracking, the reference trajectory is taken with

$$u^0 = 0.5, \ v^0 = 0.8, \ 1 \le r^0(t) = \frac{1}{2}(3 + \sin(t)) \le 2, \quad (13)$$

the corresponding $z_1^0(t)$, $z_2^0(t)$ are calculated from the equations (7) with initial values $z_1^0(0) = 0.8$, $z_2^0(0) = -0.5$ (plot of the geometric path is shown in Fig. 4).

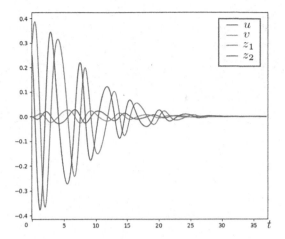

Fig. 5. Trajectories of system (8) with the parameters (12), the reference trajectory (13), and the control (11), for $T_1 = 0.03 < \tau_1^*$, $T_2 = 0.05 < \tau_2^*$, $\tau_j(t) = T_j(0.5 + 0.5cos(50t))$, $(u_e(0), v_e(0), r_e(0), z_{1e}(0), z_{2e}(0), z_{3e}(0)) = (0, 0, 0.002, 0.3, 0.25, -0.002) \in \mathcal{M}(m_1, m_2)$

Estimating g, we obtain $f(s) = f_1 + f_2s$, $f_1 = 1.25$, $f_2 = m_{22}/m_{11} + 1/\lambda_{\min}(P_1)$, $h(s) = s/\lambda_{\min}(P_2)$, and $m_2 = \min\{m_1, \lambda_{\min}(P_2)/b^*\}$ with $\lambda_{\min}(\cdot)$ be the minimal eigenvalue of a matrix.

Using suitable Matlab tools, we get:

$$P_1 = \begin{pmatrix} 7.6997 & -1.3499 & 1.0413 & -0.1867 \\ -1.3499 & 8.9695 & -0.1873 & 1.1474 \\ 1.0413 & -0.1873 & 0.1988 & -0.0260 \\ -0.1867 & 1.1474 & -0.0260 & 0.2049 \end{pmatrix},$$

$$P_2 = \begin{pmatrix} 5.9221 & 4.3568 \\ 4.3568 & 5.3248 \end{pmatrix},$$

$$K_1^1 = (-14.0446 \ \ 1.1155 \ -1.7878 \ \ 0.1606),$$

$$K_1^2 = (-16.3436 \ \ 1.3227 \ -2.1713 \ \ 0.2230),$$

$$K_2 = (-6.0470 \ -4.4501),$$

$m_1 = 1/35 = 0.0286$, $m_2 = 3.94 \cdot 10^{-4}$, $\tau_1^* = 0.0382$, $\tau_2^* = 0.1763$.

Simulation results show that the control obtained is indeed stabilizing. Typical trajectories of a closed system are presented in Fig. 5 (trajectories of $r_e(t)$ and $z_{3e}(t)$ in all the figures are omitted for the sake of greater clarity, their behavior is quite clear). Moreover, the obtained estimates of the maximum permissible values are, as expected, only sufficient: convergence to the trajectory and fulfillment of the constraints are preserved for larger values of delays (Fig. 6) and larger initial tracking errors (Fig. 7). However, an increase in feedback delay is

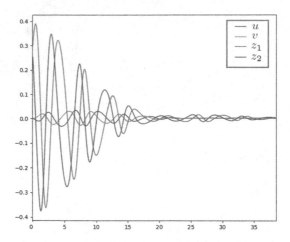

Fig. 6. Trajectories of system (8) with the same parameters as in Fig. 5, but with delay values T_1 and T_2 increased by 20 times

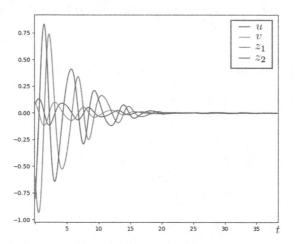

Fig. 7. Trajectories of system (8) with the same parameters as in Fig. 5, except for the initial point $(0.08, 0.1, 0.002, -0.6, -0.8, -0.002) \notin \mathcal{M}(m_1, m_2)$ $((x^1)^\top P_1 x^1 = 0.058,$ $(x^2)^\top P_2 x^2 = 2.53 \cdot 10^{-4})$

possible only up to a certain limit: for sufficiently large values of T_1, T_2 (with the other values of the parameters being the same), the system loses stability; an illustration see in Fig. 8.

It is interesting to compare the quality of the control (11) with the regulator proposed in [15]. That delay-free regulator has the following structure:

$$\tilde{U}_{1e}(t) = -k_1 u_e(t) + k_2 r^0(t) v_e(t) - k_3 z_{1e}(t) + k_4 r^0(t) z_{2e}(t),$$
$$\tilde{U'}_{2e}(t) = -k_5 r_e(t) - k_6 z_{3e}(t), \tag{14}$$

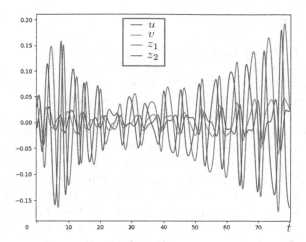

Fig. 8. Trajectories of system (8) with the same parameters as in Fig. 5, except for the initial point $(0.03, 0.05, 0.002, -0.01, -0.01, -0.002) \in \mathcal{M}(m_1, m_2)$ and the delay values $T_1 = 1$ and $T_2 = 2$

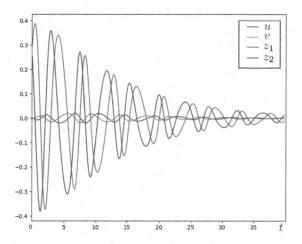

Fig. 9. Trajectories of system (8) with the parameters (12), the reference trajectory (13), and the control (14), for the same initial point as in Fig. 5

where constants k_1, $k_3 - k_6$ satisfy some inequalities (see [15, Proposition 6.3.2]), and k_2 is expressed through the others ones. Juxtaposing the two feedback, we put $k_1 = -((K_1^1)_1 + (K_1^2)_1)/2$, $k_4 = ((K_1^1)_4 + (K_1^2)_4)/3$, $k_5 = d_{33} - (K_2)_1$, $k_6 = -(K_2)_2$ (these values satisfy the conditions of [15, Proposition 6.3.2]), $k_2 = 0.04783$, $k_3 = 1.3$ (we select these values in accordance with the conditions of [15, Proposition 6.3.2]). Simulation results shows that the nature of the transition process under controls (11) and (14) is similar (see Fig. 9; compare with Fig. 5).

6   Conclusion

Representation of the system in the form of a cascade can greatly simplify the analysis and solution of control problems, including in problems of controlling movements of ships (see e.g. [7,16–18]). In addition, using a special representation permit to apply a linear controller design methodologies.

In this paper, the result that combines these two approaches is applied to the tracking control problem for an underactuated ship with feedback delay, state and control constraints. The convex optimization problems are proposed, as results of which the parameters of the control, as well as an estimate of the domain of initial conditions, are obtained, such that the state and control constraints are never violated. These optimization problems can be solved using known numerical procedures, e.g. standard tools of the MatLab package. The proposed approach can be extended to other problems of constructing controls for nonlinear systems by solving optimization problems.

References

1. Aguiar, A.P., Hespanha J.P.: Position tracking of underactuated vehicles. In: Proceedings of the 2003 American Control Conference, Denver, CO, USA, June 2003, vol. 3, pp. 1988–1993 (2003)
2. Al-Hiddabi, S., McClamroch, N.: Tracking and maneuver regulation control for nonlinear nonminimum phase systems: application to flight control. IEEE Trans. Contr. Syst. Tech. **10**(6), 780–792 (2002)
3. Balandin, D.V., Kogan, M.M.: Linear control design under phase constraints. Autom. Remote Control **70**, 958–966 (2009)
4. Bao-Li, M., Wen-Jing, X.: Global asymptotic trajectory tracking and point stabilization of asymmetric underactuated ships with non-diagonal inertia/damping matrices. Int. J. Adv. Robotic Sy. **10**(9), 336 (2013)
5. Berge, S., Ohtsu, K., Fossen, T.: Nonlinear control of ships minimizing the position tracking errors. Model. Identif. Control. **20**, 141–147 (1999)
6. Do, K.D., Jiang, Z.P., Pan, J.: Underactuated ship global tracking under relaxed conditions. IEEE Trans. Automat. Contr. **47**(9), 1529–1536 (2002)
7. Druzhinina, O.V., Sedova, N.O.: Analysis of stability and stabilization of cascade systems with time delay in terms of linear matrix inequalities. J. Comput. Syst. Sci. Int. **56**, 19–32 (2017)
8. Fredriksen, E., Pettersen, K.Y.: Global k-exponential way-point maneuvering of ships: theory and experiments. Automatica **42**, 677–687 (2006)
9. Fossen, T.I.: Guidance and Control of Ocean Vehicles. Wiley, Chichester (1994)
10. Fossen, T.I.: Handbook of Marine Craft Hydrodynamics and Motion Control. Wiley, Hoboken (2011)
11. Fossen, T.I., Pettersen, K.Y., Nijmeijer, H. (eds.): Sensing and Control for Autonomous Vehicles. LNCIS, vol. 474. Springer, Cham (2017). https://doi.org/10.1007/978-3-319-55372-6
12. Jiang, Z.-P.: Global tracking control of underactuated ships by Lyapunovs direct method. Automatica **38**, 301–309 (2002)
13. Kaminer, I., Pascoal, A., Hallberg, E., Silvestre, C.: Trajectory tracking controllers for autonomous vehicles: an integrated approach to guidance and control. J. Guid. Control Dyn. **21**(1), 29–38 (1998)

14. Khlebnikov, M.V., Shcherbakov, P.S.: Optimal feedback design under bounded control. Autom. Remote Control **75**(2), 320–332 (2014). https://doi.org/10.1134/S0005117914020118
15. Lefeber, A.A.J.: Tracking Control of Nonlinear Mechanical Systems. Universiteit Twente, Enschede (2000)
16. Loria, A., Fossen, T.I., Panteley, E.: A separation principle for dynamic positioning of ships: theoretical and experimental results. IEEE Trans. Control Syst. Technol. **8**(2), 332–343 (2000)
17. Loria, A., Panteley, E.: 2 Cascaded nonlinear time-varying systems: analysis and design. In: Lamnabhi-Lagarrigue, F., Loria, A., Panteley, E. (eds.) Advanced Topics in Control Systems Theory. LNCIS, vol. 311, pp. 23–64. Springer, London (2005). https://doi.org/10.1007/11334774_2
18. Panteley, E., Loria, A.: Growth rate conditions for uniform asymptotic stability of cascaded time-varying systems. Automatica **37**(3), 453–460 (2001)
19. Pesterev, A.V., Rapoport, L.B.: Construction of invariant ellipsoids in the stabilization problem for a wheeled robot following a curvilinear path. Autom. Remote Control **70**, 219–232 (2009)
20. Pettersen, K.Y., Nijmeijer, H.: Global practical stabilization and tracking for an underactuated ship - a combined averaging and backstepping approach. In: Proceedings of IFAC Conference System Structure Control, Nantes, France, July 1998, pp. 59–64 (1998)
21. Pettersen, K.Y., Nijmeijer, H.: Underactuated ship tracking control: theory and experiments. Int. J. Control **74**(14), 1435–1446 (2001)
22. Razumikhin, B.S.: On stability on systems with delay. Prikl. Mat. Mekh. **20**, 500–512 (1956). [in Russian]
23. Strand, J.P., Fossen, T.I.: Nonlinear passive observer design for ships with adaptive wave filtering. In: Nijmeijer, H., Fossen, T. (eds.) New Directions in nonlinear observer design. LNCIS, vol. 244, pp. 113–134. Springer, London (1999). https://doi.org/10.1007/BFb0109924
24. Tanaka, K., Wang, H.O.: Fuzzy Control Systems Design and Analysis: A Linear Matrix Inequality Approach. Wiley, Hoboken (2001)
25. Veremey, E.I. Dynamical correction of positioning control laws. In: Proceedings of 9th IFAC Conference on Control Applications in Marine Systems (CAMS-2103). Japan, Osaka, pp. 31–36 (2013)

Theorems of Alternative and Optimization

Yuri Evtushenko[1,2] and Alexander Golikov[1(✉)]

[1] Dorodnicyn Computing Centre, FRC "Computer Science and Control" of RAS,
Vavilov Street 40, 119333 Moscow, Russia
yuri-evtushenko@yandex.ru, gol-a@yandex.ru
[2] Moscow Institute of Physics and Technology (State Research University),
9 Institutskiy per., Dolgoprudny, Moscow Region 141701, Russia

Abstract. A new elementary proof of Farkas' theorem is proposed. The proof is based on the consideration of the always solvable problem of minimizing the residual of a system of linear equations/inequalities and the necessary and sufficient conditions for the minimum of this problem. Minimizing the residuals of an inconsistent system makes it easy to calculate the normal solution to a consistent system. The connection between Farkas' theorem and linear and quadratic programming is shown. A new version of the theorem on alternatives is proposed.

Keywords: Theorems of alternative · Systems of linear equations and inequalities · Linear programming · Quadratic programming · Normal solution · Duality

1 Introduction

It is well known that Farkas' theorem is of great importance not only to theory but also to computational methods. Currently, there are many different proofs of Farkas' theorem. We indicate only a few of them [1–8]. We especially note paper [9], which shows that it is possible not to use Farkas' theorem to prove the optimality conditions of linear programming (LP). Therefore, we show that the Farkas' theorem follows from the duality theory of LP.

The paper is structured as follows. In Sect. 2, a new elementary proof of Farkas' theorem is proposed. This proof is based on minimizing the residual of the original system of linear equations with non-negative variables. The proof uses the existence of a minimum for this problem and the optimality conditions. Earlier in [7,8], two problems of minimizing the residuals of alternative systems were used in the proof.

Section 3 shows the relationship between LP problems and Farkas' theorem, as well as the relationship between the quadratic penalty method and the quadratic regularization of the corresponding LP problems.

This work was supported in part by the Russian Foundation for Basic Research, project no. 17-07-00510.

N. Olenev et al. (Eds.): OPTIMA 2020, LNCS 12422, pp. 86–96, 2020.
https://doi.org/10.1007/978-3-030-62867-3_7

Section 4 provides a new version of Farkas' theorem when different matrices are present in the alternative systems.

Give some notations which will be used throughout the paper without explanations. Euclidean norm $\|\cdot\|$ is used everywhere. By a_+ denotes a vector a with all negative components replaced by zeros. By 0_i we will denote the zero vector of dimension i, and by 0_{nm}—the zero $n \times m$ matrix.

2   Farkas' Theorem and Quadratic Programming

Let a set X be defined by the system

$$Ax = b, \quad x \geq 0_n, \qquad (I)$$

where A is a matrix of dimension $m \times n$, a vector $b \in R^m$, $b \neq 0$. The alternative system defining the set U is written in the form

$$A^\top u \leq 0_n, \quad b^\top u = \rho > 0. \qquad (II)$$

Here and in what follows, ρ is an arbitrary fixed positive constant.

One, and only one of these systems, either (I) or (II), is always consistent, but never both. If the notation $b^\top u > 0$ is used in (II) the above statement is known as Farkas' theorem or Farkas' lemma (see, for example, [10]).

To solve the problem of solvability of systems (I) or (II) and to find a corresponding solution, it is enough to find solutions x^* or u^* to problems (1) and (2) below of minimization of quadratic function over the positive orthant and of unconstrained minimization of a convex piecewise quadratic function, respectively

$$F^1(x^*) = \min_{x \in R^n_+} \frac{1}{2}\|b - Ax\|^2, \qquad (1)$$

$$F^2(u^*) = \min_{u \in R^m} \frac{1}{2}\{(\rho - b^\top u)^2 + \|(A^\top u)_+\|^2\}. \qquad (2)$$

The minimized quadratic functions in the problems (1) and (2) are bounded from below by zero and by the Frank-Wolfe theorem are always solvable [11]. The problems (1) and (2) are the problems of finding the minimum residuals of systems (I) and (II).

The following statements always hold true

$$F^1(x^*)F^2(u^*) = 0, \quad F^1(x^*) + F^2(u^*) > 0.$$

This condition can be considered as another formulation of Farkas' theorem [7].

To identify the solvability of the system (I) or (II) and to find a solution to a consistent system, it is enough to find any solution x^* to the problem (1). If the optimal value of $F^1(x^*)$ is zero, then the system (I) is solvable at the point x^*. If $F^1(x^*) > 0$, then the system (I) is unsolvable. As will be shown below, in this case, the solution of the system (II) is easily calculated from x^*.

The simplest proof of Farkas' theorem is based on the always solvable problem (1) of minimizing the residuals of the system (I) and on the necessary and sufficient conditions for the minimum of this problem, which at the point x^* has the form

$$F_x^1(x^*) = -A^\top(b - Ax^*) \geq 0_n, \quad x^{*\top}(A^\top(b - Ax^*)) = 0, \quad x^* \geq 0_n. \tag{3}$$

We present the Farkas' theorem as follows:

Theorem 1. *One and only one of the systems is always consistent:*

$$Ax = b, \quad x \geq 0_n. \tag{I}$$

or

$$A^\top u \leq 0_n, \quad b^\top u > 0, \tag{II'}$$

Proof. Suppose that the system (I) is consistent and x^* is its solution. Then $b^\top u = (Ax^*)^\top u = x^{*\top}A^\top u \leq 0$ for all u such that $A^\top u \leq 0_n$. This implies the inconsistency of the system (II').

Now suppose that the system (I) is inconsistent. Let x^* be a solution to the minimization problem (1). We introduce the notation

$$u^* = b - Ax^* \tag{4}$$

for the vector of minimal residuals of an inconsistent system (I) and rewrite the conditions (3) for the problem (1) in the following form

$$A^\top u^* \leq 0_n, \quad x^{*\top}(A^\top u^*) = 0, \quad x^* \geq 0_n. \tag{5}$$

Taking into account that $\|u^*\| > 0$ we get from (4), (5):

$$b^\top u^* = (Ax^* + u^*)^\top u^* = x^{*\top}A^\top u^* + u^{*\top}u^* = 0 + \|u^*\|^2 > 0. \tag{6}$$

Thus, u^* is a solution to the system (II'). $\square$

We note that the above proof of Farkas's theorem is constructive. The solution x^* to minimization problem (1) is either some solution to system (I) or defines the unique solution u^* to system (II), which is equal to nonzero minimum residual vector inconsistent system (I): $u^* = b - Ax^*$.

We also note that the dual problem to (1) is not used in the theorem proof. Although in (6) the expression $\|u^*\|^2 = b^\top u^*$ is the equality of the optimal values of the objective function of the primal problem (1) and the dual quadratic programming problem taking into account the connection $u^* = b - Ax^*$ between their solutions.

Another constructive proof of Farkas's theorem was proposed in [7]. This proof focuses on calculating the normal solution of a solvable system by minimizing the residual of an inconsistent system. In this case, the problem of minimizing the residual of the system and the dual to it were considered simultaneously.

It is easy to prove that the systems (I) and (II) cannot be consistent at the same time. Only one of them is consistent. Each case was considered separately. To do this, we used the quadratic problems (1) and (2) of minimizing the residuals of systems (I) and (II) and, respectively, the dual problems of strictly concave quadratic programming

$$\max_{u \in \tilde{U}} \{b^\top u - \frac{1}{2}\|u\|^2\}, \ \tilde{U} = \{u \in R^m : A^\top u \leq 0_n\}, \tag{7}$$

$$\max_{x,t \in \tilde{X}} \{\rho t - \frac{1}{2}\|x\|^2 - \frac{1}{2}t^2\}, \ \tilde{X} = \{x \in R^n_+, t \in R^1 : Ax - bt = 0_m\}. \tag{8}$$

The dual quadratic problems (7) and (8) are also always solvable, since the primal quadratic problems (1) and (2) always have solutions [12].

We write down the necessary and sufficient optimality conditions for the primal problem (1)

$$A^\top(b - Ax) \leq 0_n, \ x^\top(A^\top(b - Ax)) = 0, \ x \geq 0_n,$$

and for the dual problem (7)

$$b - u - Ax = 0_m, \ A^\top u \leq 0_n, \ x^\top A^\top u = 0, \ x \geq 0_n.$$

From these optimality conditions it follows that the solution u^* to problem (7) can be expressed through solution x^* to problem (1) by equation $u^* = b - Ax^*$. Taking this equation into account, one can obtain $\|u^*\|^2 = b^\top u^*$ from the equality of the objective functions' optimal values. Therefore, if $\|u^*\| = \|b - Ax^*\| > 0$, i.e. system (I) is unsolvable ($X = \emptyset$), then $b^\top u^* > 0$ and, therefore, system (II) is solvable ($U \neq \emptyset$) since $A^\top u^* \leq 0$. Moreover, using u^* and x^*, the normal solution $\hat{u}^*$ (solution with minimal Euclidean norm) to the system (II) is easily calculated, i.e. the solution is found to be a strictly convex quadratic programming problem

$$\min_{u \in U} \frac{1}{2}\|u\|^2, \ U = \{u \in R^m : A^\top u \leq 0_n, \ b^\top u = \rho\}. \tag{9}$$

Then the dual problem to (9) is solvable and has the form

$$\max_{x \in R^n_+, t \in R^1} \{\rho t - \frac{1}{2}\|bt - Ax\|^2\}. \tag{10}$$

Here are the necessary and sufficient conditions for these problems. For primal problem (9) they have the form

$$u + Ax - bt = 0_m, \ A^\top u \leq 0_n, \ x^\top A^\top u = 0, \ x \geq 0_n, \ \rho - b^\top u = 0,$$

and for dual problem (10) they have the form

$$\rho - b^\top(bt - Ax) = 0, \ A^\top(bt - Ax) \leq 0_n, \ x^\top(A^\top(bt - Ax)) = 0, \ x \geq 0_n.$$

From these optimality conditions it follows that the normal solution $\hat{u}^*$ is expressed in terms of the solution $\tilde{x}$, $\tilde{t}$ to the dual problem (10) by the formula $\hat{u}^* = b\tilde{t} - A\tilde{x}$. Taking this formula into account, the equality of optimal values of the objective functions of the problems (9) and (10) takes the form $\rho\tilde{t} = \|\hat{u}^*\|^2$.

The necessary and sufficient optimality conditions for the problems (1), (7) and (9), (10) imply formulas that link their solutions to each other $\hat{u}^* = \tilde{t}u^*$, $\tilde{x} = \tilde{t}x^*$, $\tilde{t} = \rho/\|u^*\|^2$.

So, if $X = \emptyset$, then $\|u^*\| = \|b - Ax^*\| > 0$ and the normal solution $\hat{u}^*$ to the system (II) will be

$$\hat{u}^* = \rho(b - Ax^*)/\|b - Ax^*\|^2 = \rho u^*/\|u^*\|^2.$$

Let us now proceed to the consideration of mutually dual problems (2) and (8). Necessary and sufficient optimality conditions for primal problem (2) have the following form

$$-b(\rho - b^\top u) + A(A^\top u)_+ = 0_m,$$

and for dual (8) have the form

$$-\rho + t + b^\top u = 0, \quad -x + A^\top u \le 0_n, \quad x^\top(-x + A^\top u) = 0, \quad x \ge 0_n, \quad (11)$$
$$bt - Ax = 0_m. \quad (12)$$

From these optimality conditions it follows that the minimum residuals of the system (II), i.e. the solution x', t' to problem (8) are expressed in terms of the solution u' to problem (2) as follows $x' = (A^\top u')_+$, $t' = \rho - b^\top u'$. From the equality of the optimal values of the objective functions of problems (2) and (8), we have $\|x'\|^2 + t'^2 = \rho t'$. It follows that if the system (II) is incompatible, i.e. $\|x'\|^2 + t'^2 > 0$, then $t' > 0$. Then it follows from (11) and (12) that x'/t' is a solution to the system (I).

Using minimal residuals of inconsistent system (II), the normal solution $\hat{x}^*$ to consistent system (I) is calculated from following problem

$$\min_{x \in X} \frac{1}{2}\|x\|^2, \quad X = \{x \in R^n : Ax = b, x \ge 0_n\}. \quad (13)$$

The dual problem for (13) has the form

$$\max_{u \in R^m} \{b^\top u - \frac{1}{2}\|(A^\top u)_+\|^2\}. \quad (14)$$

Consider the necessary and sufficient optimization conditions for the problem (13)

$$x - A^\top u \ge 0_n, \quad x^\top(x - A^\top u) = 0, \quad x \ge 0_n, \quad b - Ax = 0_m \quad (15)$$

and for the problem (14)

$$b - A(A^\top u)_+ = 0_m. \quad (16)$$

So, if $U = \emptyset$, then $t' > 0$ and the normal solution $\hat{x}^*$ (the solution with the minimal Euclidean norm) to the system (I) is expressed in terms of the solution u' to problem (2) as follows

$$\hat{x}^* = (A^\top u')_+ / (\rho - b^\top u') = x'/t'.$$

As follows from the optimality conditions (15) and (16), the same solution $\hat{x}^*$ is found through the solution u'' to the dual problem (14) by the formula

$$\hat{x}^* = (A^\top u'')_+.$$

Thus, the problem of solvability of the system (I) or (II) is reduced to minimizing the residual of either system. If the norm of the minimum residual is nonzero, then the system is inconsistent and the residual can be used in simple formulas to find the normal solution to the consistent system.

3   Farkas' Theorem and Linear Programming

Let us provide the linear programming interpretation of Farkas' theorem. Here system (I) can be presented as a primal linear programming problem with its objective function coefficient vector identically equal to zero.

$$\min_{x \in R_+^n} \{0_n^\top x \ : \ Ax = b, \ x \geq 0_n\}. \qquad (P)$$

The dual to (P) problem is as follows:

$$\max_{u \in R^m} \{b^\top u \ : \ A^\top u \leq 0_n\}. \qquad (D)$$

It is common knowledge that for any pair of primal and dual LP problems, one of the following four cases always holds:

(1) both primal and dual problems have solutions;
(2) a primal problem is inconsistent and the dual one is unbounded;
(3) a primal problem is unbounded and the dual one is inconsistent;
(4) both primal and dual problems are inconsistent.

For problems (P) and (D) the latter two conditions cannot be fulfilled because the constraints in (D) are always consistent, vector $u = 0_m$ is feasible.

So, the first two cases are only possible.

In case 1) the optimal values of goal functions for problems (P) and (D) are equal to zero and inequality $b^\top u \leq 0$ holds for all feasible vectors u owing to the weak duality theorem. Hence it follows solvability of the system (I) and unsolvability of system (II)

$$A^\top u \leq 0_n, \quad b^\top u = \rho > 0. \qquad (II)$$

In case 2) system (I) is inconsistent and system (II) is consistent for any $\rho > 0$ due to unboundedness of the dual problem (D).

So one can obtain the simplest proof that (I) and (II) are alternative systems employing a specific type of linear programming problems (P) and (D) and linear programming duality theory.

Two pairs of mutually dual quadratic programming problems (1), (7) and (13), (14) are associated with the pair of mutually dual linear programming problems (P) and (D). These quadratic programming problems are obtained from linear programming problems using the quadratic penalty function (pen transformation) and quadratic regularization (reg transformation):

$$P_p: \quad \min_{x \in R_+^n} \; (0_n^\top x + \frac{1}{2}\|b - Ax\|^2).$$

$$D_r: \quad \max_{u \in \tilde{U}} \; (b^\top u - \frac{1}{2}\|u\|^2), \quad \tilde{U} = \{u \in R^m : A^\top u \leq 0_n\}.$$

The second pair of mutually dual quadratic problems has the form:

$$P_r: \quad \min_{x \in X} \; (0_n^\top x + \frac{1}{2}\|x\|^2), \quad X = \{x \in R^n : Ax = b, \quad x \geq 0_n\},$$

$$D_p: \quad \max_{u \in R^m} \; (b^\top u - \frac{1}{2}\|(A^\top u)_+\|^2).$$

The following diagram shows the relationship between the penalized and regularized problems (P) and (D)

$$
\begin{array}{ccccc}
P_p & \xleftarrow{\;pen\;} & P & \xrightarrow{\;reg\;} & P_r \\
\updownarrow & & \updownarrow & & \updownarrow \\
D_r & \xleftarrow{\;reg\;} & D & \xrightarrow{\;pen\;} & D_p
\end{array}
$$

The vertical arrows indicate the mutual duality of the corresponding problems.

Problem D_p (14) can be considered as an auxiliary problem of the penalty function method as applied to the problem (D). Note that in [4] the proof of Farkas' theorem is based on the use of a quadratic penalty, that is, the problem (14) is considered.

Problem (2) can be considered as an auxiliary problem of the Morrison method [13] (or parameterization method of the objective function) with its parameter being equal to $\rho > 0$ when applied to the problem (D).

4   New Version of Theorem of Alternative

One and the same matrix A and vector b have always been used in alternative linear systems. We show a different way of alternative systems involving the application of different matrices with various dimensions The consideration below represents a special case of the system (I) where matrix A has rank m with $m \leq n$. For the case concerned it will be shown that the system alternative to (I) can take a form different from (II), i.e. the alternative system can incorporate a matrix differing from A and a vector other than b.

If $m \leq n$ then the system

$$Ax = b \tag{17}$$

is always solvable but its solutions may fail to include any nonnegative ones. Let $\bar{X}$ denote the set of system (17) solutions. Note that set $\bar{X}$ is always nonempty in contrast to set X. The general solution of the system of linear equations (17) can be written in the form

$$x = \bar{x} - K^\top y, \tag{18}$$

where $\bar{x}$ is a particular solution of the system, and $K^\top y$ is the general solution of the homogeneous system $Ax = 0_m$, and $y \in R^\nu$. The matrix K can be chosen to be any $(\nu \times n)$ matrix such that its ν rows form a basis of the null space of A where $\nu = n - m$ is the defect of matrix A. Therefore, $AK^\top = 0_{m\nu}$. Here 0_{ij} denotes $(i \times j)$ matrix with zero entries.

Matrix K is not uniquely defined. It can be constructed in various ways. If we partition the matrix A as $A = [B \,|\, N]$, where B is nondegenerate, then we can represent K as $K = [-N^\top (B^{-1})^\top \,|\, I_\nu]$. If we reduce A by means of Gauss–Jordan transformations to the form $A = [I_m \,|\, N]$, then we can represent K as $K = [-N^\top \,|\, I_\nu]$ [14].

Let us determine the set Y as

$$Y = \{y \in R^\nu \ : \ \bar{x} - K^\top y \geq 0_n\}.$$

Equation (18) can be considered as an affine mapping from R^ν to R^n. Here the image of set Y is set X specified by the system (I). There exists a one-to-one correspondence between X and Y.

Indeed, for any $y \in Y$ Eq. (18) uniquely determines $x \in X$, i.e.

$$X = \bar{x} - K^\top Y \tag{19}$$

In case of a full-range overdetermined system (18) containing n linear equations and ν variables a pseudosolution y

$$y = (KK^\top)^{-1} K(\bar{x} - x) = (K^\top)^+ (\bar{x} - x) \tag{20}$$

always exists. Here $(K^\top)^+ = (KK^\top)^{-1}K$ is a pseudoinverse matrix. It solves (18) and is unique if and only if $\bar{x} - x \in \operatorname{im} K^\top$. This inclusion holds if and only if $x \in \bar{X}$. Thus, for any $x \in \bar{X}$, formula (20) determines an affine transformation that is the inverse of (18). Therefore, one can write

$$Y = (K^{\top})^{+}(\bar{x} - X). \tag{21}$$

So the following two systems

$$Ax = b, \quad x \geq 0_n, \tag{I}$$

$$K^{\top}y \leq \bar{x} \tag{I_y}$$

are either simultaneously solvable and interconnected by expressions (19) and (21) or simultaneously unsolvable if there exist no nonnegative general solution $x = \bar{x} - K^{\top}y$ to system (I).

By the Gale theorem [10] the following system determining set V will be alternative to the system (I_y),

$$Kv = 0_{\nu}, \quad -\bar{x}^{\top}v = \rho > 0, \quad v \geq 0_n. \tag{II_v}$$

System (I_y) is equivalent to system (I), and, simultaneously, system (II_v) is an alternative to the system (I).

The general solution to homogeneous system $Kv = 0_{\nu}$ can be expressed by matrix A as $v = -A^{\top}u$. By changing the variables $v = -A^{\top}u$, one can present system (II_v) as follows:

$$A^{\top}u \leq 0_n, \quad b^{\top}u = \rho > 0. \tag{II}$$

System (II) is an alternative to (I), hence to (I_y).

If set V is nonempty then set U determined by system (II) is nonempty too, the two sets having a one-one mapping expressed by:

$$V = -A^{\top}U, \quad U = -(A^{\top})^{+}V,$$

where pseudoinverse matrix $(A^{\top})^{+}$ is as follows: $(A^{\top})^{+} = (AA^{\top})^{-1}A$.

The following theorem of the alternative is an analog of the Farkas' and Gale theorems.

Theorem 2. *Suppose that the matrix $A \in R^{m \times n}$ has rank m, the matrix $K \in R^{\nu \times n}$ has rank $\nu = n - m$, $AK^{\top} = 0_{m \times \nu}$, $\|b\| \neq 0$, ρ is any positive number, and the vector $\bar{x}$ is an arbitrary solution to the system of linear equations $Ax = b$. Then:*

(1) either the system

$$Ax = b, \quad x \geq 0_n \tag{I}$$

or the system

$$Kv = 0_{\nu}, \quad -\bar{x}^{\top}v = \rho > 0, \quad v \geq 0_n \tag{II_v}$$

is consistent;

(2) either the system

$$K^\top y \leq \bar{x} \qquad\qquad (I_y)$$

   or the system

$$A^\top u \leq 0_n, \quad b^\top u = \rho > 0. \qquad (II)$$

   is consistent.

Proof. (1) Since systems (I) and (I_y) are equivalent in the sense of simultaneous solvability or unsolvability and system (I_y) is alternative to the system (II_v) by the Gale theorem, it follows that systems (I) and (II_v) are alternative.
(2) Since systems (II) and (II_v) are equivalent in the sense of simultaneous solvability or unsolvability and systems (II) and (I) are alternative by the Farkas theorem, it follows that system (II) is also alternative to the system (I_y). $\square$

The alternative systems interrelation can be represented as follows:

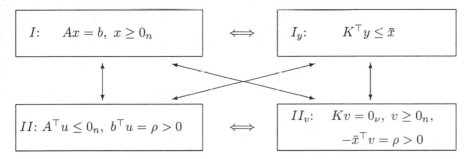

The double arrows correspond to simultaneously solvable/unsolvable systems and the ordinary ones stand for alternative systems.

References

1. Bartl, D.: A note on the short algebraic proof of Farkas' lemma. Linear and Multilinear Algebra **60**(8), 897–901 (2012)
2. Broyden, C.G.: On theorems of the alternative. Optim. Methods Softw. **16**(1–4), 101–111 (2001)
3. Dax, A.: An elementary proof of Farkas' lemma. SIAM Rev. **39**, 503–507 (1997)
4. Evtushenko, Y.G., Tret'yakov, A.A., Tyrtyshnikov, E.E.: New approach to Farkas' theorem of the alternative. Dokl. Math. **99**(2), 208–210 (2019)
5. Giannessi, F.: Theorems of the alternative and optimization. In: Floudas, C.A., Pardalos, P.M. (eds.) Encyclopedia of Optimization. Springer, Boston (2001). https://doi.org/10.1007/0-306-48332-7_521
6. Jaćmović, M.: Farkas' lemma of alternative. Teach. Math. **25**(27), 77–86 (2011)
7. Golikov, A.I., Evtushenko, Y.G.: Theorems of the alternative and their applications in numerical methods. Comput. Maths. Math. Phys. **43**(3), 354–375 (2003)

8. Evtushenko, Y.G., Golikov, A.I.: New perspective on the theorems of alternative. In: Di Pillo, G., Murli, A. (eds.) High Performance Algorithms and Software for Nonlinear Optimization. APOP, vol. 82, pp. 227–241. Springer, Boston, MA (2003). https://doi.org/10.1007/978-1-4613-0241-4_10
9. Forsgren, A., Wright, M.: An elementary proof of linear programming optimality conditions without using Farkas' lemma (2014). arXiv:1407.1240
10. Mangasarian, O.L.: Nonlinear Programming. SIAM, Philadelphia (1994)
11. Frank, M., Wolfe, P.: An algorithm for quadratic programming. Naval. Res. Log. Quart. **3**, 95–110 (1956)
12. Eremin, I.I.: On quadratic problems and fully quadratic problems in convex programming. Izv. Vyssh. Uchebn. Zaved., Mat. **42**(12), 22–28 (1998)
13. Morrison, D.: Optimization by least squares. SIAM J. Numer. Anal. **5**(l), 83–88 (1968)
14. Golikov, A.I., Evtushenko, Y.G.: Two parametric families of LP problems and their applications. Proc. Steklov Inst. Math. Supp. **1**, S52–S66 (2002)

P-Regularity Theory: Applications to Optimization

Yuri Evtushenko[1,2], Vlasta Malkova[1(✉)], and Alexey Tret'yakov[1,3,4]

[1] Dorodnicyn Computing Centre, FRC CSC RAS, Vavilov Street 40,
119333 Moscow, Russia
yuri-evtushenko@yandex.ru, vmalkova@yandex.ru, tret@ap.siedlce.pl
[2] Moscow Institute of Physics and Technology, Moscow, Russia
[3] Systems Research Institute, Polish Academy of Sciences,
Newelska 6, 01-447 Warsaw, Poland
[4] Faculty of Sciences, Siedlce University, 08-110 Siedlce, Poland

Abstract. We present recent advances in the analysis of nonlinear structures and their applications to nonlinear optimization problems with constraints given by nonregular mappings or other singularities obtained within the framework of the p-regularity theory developed over the last twenty years. In particular, we address the problem of description of the tangent cone to the solution set of the operator equation, optimality conditions, and solution methods for optimization problems.

Keywords: P-regularity · Nonlinear optimization · Nonregular mappings · Tangent cone

Several basic results of optimization theory are considered, such as the Lagrange theorem on optimality conditions in a problem with equality constraints, the Kuhn-Tucker theorem on optimality conditions in a problem with inequality constraints [8], Lyusternik theorem, sufficient optimality conditions for a mathematical programming problem, Farkas theorem and theorem on closed cone – from an unconventional point of view.

It is shown that in many cases the proof of these central results for optimization can be significantly simplified without using the ideas of separability, closedness of a cone, Farkas's theorem, convexity, limit technics, or the penalty function method, the principle of contraction mappings or the implicit function theorem, etc.

1 Elementary Proof of the Lagrange Theorem for the Problem with Equalities

We consider the problem

$$\min \varphi(x) \tag{1}$$

This work was supported in part by the Russian Foundation for Basic Research, project No. 17-07-00510.

N. Olenev et al. (Eds.): OPTIMA 2020, LNCS 12422, pp. 97–109, 2020.
https://doi.org/10.1007/978-3-030-62867-3_8

under conditions

$$F(x) = 0_m, \qquad (2)$$

where $F = (f_1, \ldots, f_m)^\top$, φ, $F \in C^2(\mathbb{R}^n)$, $f_i \colon \mathbb{R}^n \to R$, $i = 1, \ldots, m$ and 0_m is a column vector.

Let x^* denote the solution to the problem (1)–(2). Then the classical Lagrange theorem on the necessary optimality conditions [3, 7] holds.

Theorem 1. *Let φ, $F \in C^2(\mathbb{R}^n)$ and the Jacobi matrix*

$$F'(x^*) = \left(\frac{\partial f_i(x^*)}{\partial x_j} \right)_{\substack{i=\overline{1,m} \\ j=\overline{1,n}}}$$

be not degenerate. Then $\varphi'(x^*) \in \mathrm{Lin}\, \{f_i'(x^*),\ i = 1, \ldots, m\}$.

The well-known proofs of this theorem are very cumbersome and use facts such as the implicit function theorem [7], limit technics [1,3], Lustrenik's theorem [7], etc. We show how to avoid these difficulties in the proof and obtain the main result only on the basis of the Weierstrass theorem.

Proof. Suppose, on the contrary, that $\varphi'(x^*) \notin \mathrm{Lin}\, \{f_i'(x^*),\ i = 1, \ldots, m\}$. Then $\varphi'(x^*) = h + \bar{h}$, where $h \in \mathrm{Ker}\, F'(x^*)$, $\bar{h} \in \mathrm{Lin}\, \{f_i'(x^*),\ i = 1, \ldots, m\}$. Then for $\alpha \in (0, \varepsilon)$ and $\xi \in \mathbb{R}^m$, $\|\xi\| = 1$, $\| F\left(x^* - \alpha h + \alpha^{3/2} F'(x^*)^\top \xi\right) \|^2 \geq c\alpha^3$, and $\|F(x^* - \alpha h)\|^2 \leq c\alpha^4$, where $\varepsilon > 0$ is small enough and $c > 0$ is an independent constant.

This means, by Weierstrass's theorem, that there exists $\eta(\alpha) \colon R^1 \to \mathbb{R}^m$ such that

$$\min_{\|\eta\| \leq 1} \| F\left(x^* - \alpha h + \alpha^{3/2} F'(x^*)^\top \eta\right) \|^2$$
$$= \| F\left(x^* - \alpha h + \alpha^{3/2} F'(x^*)^\top \eta(\alpha)\right) \|^2$$

and $\|\eta(\alpha)\| \leq 1$. It means that

$$\left(\| F\left(x^* - \alpha h + \alpha^{3/2} F'(x^*)^\top \eta\right) \|^2 \right)'_{\eta=\eta(\alpha)} = 0_n$$

or

$$2F'\left(x^* - \alpha h + \alpha^{3/2} F'(x^*)^\top \eta(\alpha)\right) \cdot F'(x^*)^\top \cdot F\left(x^* - \alpha h + \alpha^{3/2} F'(x^*)^\top \eta(\alpha)\right) = 0_n,$$

which implies equality

$$F\left(x^* - \alpha h + \alpha^{3/2} F'(x^*)^\top \eta(\alpha)\right) = 0_n$$

and therefore

$$\varphi\left(x^* - \alpha h + \alpha^{3/2} F'(x^*)^\top \eta(\alpha)\right) =$$
$$\varphi(x^*) - \alpha\langle \varphi'(x^*), h \rangle + o(\alpha) < \varphi(x^*),$$

which contradicts the local minimality of the point x^* and proves the Theorem 1.

2 New Elementary Proof of the Kuhn-Tucker Theorem

We consider the problem

$$\min \varphi(x) \tag{3}$$

under conditions

$$g_i(x) \le 0, \qquad i = 1, \dots, m. \tag{4}$$

Here $x \in \mathbb{R}^n$, φ, $g_i \in C^1(\mathbb{R}^n)$, $i = 1, \dots, m$. The main result for characterizing a solution x^* of the problem (3), (4) is the Kuhn-Tucker theorem (KT), which establishes the necessary optimality conditions for this problem (3), (4).

Theorem 2 (in the form of Fritz John). *Let x^* be a solution of the problem (3), (4) and φ, $g_i \in C^1(\mathbb{R}^n)$, $i = 1, \dots, m$.*
Then there are non-negative factors $\lambda_0^*, \lambda_1^*, \dots, \lambda_m^*$ such that the conditions

$$\lambda_0^* \varphi'(x^*) + \sum_{i=1}^m \lambda_i^* g_i'(x^*) = 0_n, \tag{5}$$
$$\lambda_0^* + \lambda_1^* + \dots + \lambda_m^* = 1$$

are satisfied.

We denote the set of active in x^* constraints by $I(x^*) = \{i = 1, \dots, m \mid g_i(x^*) = 0\}$. The traditional proofs of this theorem are quite complex and use in their constructions results such as the implicit function theorem, the cone closure theorem, Farkas's lemma, separability theorems, limit theorems, etc. We give a proof that does not use the above fundamental results, using and relying only on the trivial lemma on the property of projection.

Lemma 1. *Let $\mathcal{K}$ be a closed convex set in $\mathbb{R}^n$.*
Then for the projection $p = \mathrm{Pr}_\mathcal{K} 0_n$ of the zero vector 0_n onto the set $\mathcal{K}$ the inequality holds

$$\langle z - p, p \rangle \ge 0 \tag{6}$$

for all $z \in \mathcal{K}$.

The proof of this lemma is trivial and we do not give it.
Let

$$\mathcal{K} = \Big\{ y \in \mathbb{R}^n \mid y = \lambda_0 \varphi'(x^*) + \sum_{i=1}^m \lambda_i g_i'(x^*),$$
$$\lambda_0 \ge 0, \quad \lambda_i \ge 0, \quad i = 1, \dots, m, \quad \lambda_0 + \lambda_1 + \dots + \lambda_m = 1 \Big\}.$$

Proof (of the KT theorem). We show that

$$0_n \in \mathcal{K}, \tag{7}$$

since relation (7) is equivalent to (5).
Indeed, if this is not so, i.e. $0_n \notin \mathcal{K}$, then since $\varphi'(x^*) \in \mathcal{K}$, Lemma 1 implies

$$\langle \varphi'(x^*), -p \rangle \le -\|p\|^2 < 0 \tag{8}$$

and, since $g_i'(x^*) \in \mathcal{K}$, $i = 1, \ldots, m$, we have

$$\langle g_i'(x^*), -p \rangle \le -\|p\|^2 < 0. \tag{9}$$

It follows that

$$\varphi(x^* - \alpha p) < \varphi(x^*)$$

and

$$g_i(x^* - \alpha p) \le g_i(x^*) \le 0, \qquad i = 1, \ldots, m,$$

for $\alpha > 0$ is sufficiently small. And this contradicts the minimality of the point x^*. The theorem is proved.

3   The Farkas Theorem and the Cone Closure Theorem

An important role in constructing optimality conditions and numerical methods is played by the so-called alternative theorems, one of which can be formulated in the following form.

Theorem 3 (Farkas). *Let the vectors η, $\xi_i \in \mathbb{R}^n$, $i = 1, \ldots, m$, and for each $x \in \mathbb{R}^n$ such that $\langle \xi_i, x \rangle \ge 0$, $i = 1, \ldots, m$, the inequality*

$$\langle \eta, x \rangle \ge 0, \tag{10}$$

hold. We call this the Farkas condition.
Then

$$\eta \in \text{Cone}\{\xi_i \,|\, i = 1, \ldots, m\} = \left\{ y \in \mathbb{R}^n \,|\, y = \sum_{i=1}^m \alpha_i \xi_i, \quad \alpha_i \ge 0, \quad i = 1, \ldots, m \right\}.$$

Proof. We introduce the function

$$\Phi(x) = \langle \eta, x \rangle + \sum_{i=1}^m \left(\langle \xi_i, x \rangle_- \right)^2,$$

here

$$a_- = \begin{cases} a, \, a \le 0 \\ 0, \, a \ge 0 \end{cases}, \qquad a \in \mathbb{R}^n.$$

Under Farkas conditions, there exists $y^* = \underset{x \in \mathbb{R}^n}{\text{Argmin}}\, \Phi(x)$. By Fermat's theorem, $\Phi'(y^*) = 0_n$ or $\eta + \sum_{i=1}^m 2\langle \xi_i, y^* \rangle_- \xi_i = 0_n$. Denoting by $-\lambda_i^* = 2\langle \xi_i, y^* \rangle_-$, $i = 1, \ldots, m$, we obtain $\eta = \sum_{i=1}^m \lambda_i^* x_i^*$, $\lambda_i^* \ge 0$, $i = 1, \ldots, m$, i.e. $\eta \in \text{Cone}\{\xi_i \,|\, i = 1, \ldots, m\}$.

Remark 1. The proof clearly implies the form of the Farkas factors λ_i^*.

The known proofs of this theorem rely mostly on separability theorems, limit technics, and, necessarily, the cone closure theorem. However, the existing proofs of the cone closure theorem are very cumbersome. Here is proof that is distinguished by its simplicity and grace. This proof is based on Farkas's lemma.

Theorem 4 (of cone closure). *The set*

$$\mathcal{K} = \{y \in \mathbb{R}^n \mid y = \sum_{i=1}^{m} \lambda_i a_i, \ a_i \in \mathbb{R}^n, \ \lambda_i \geq 0, \ i = 1, \ldots, m\}$$

is closed.

Proof. Let the sequence $\{y_i\}$, $i = 1, 2, \ldots$, be such that $y_i \in \mathcal{K}$, $i = 1, 2, \ldots$, $\|y_i\| \leq c$, $i = 1, 2, \ldots$, and $y_i \to y$ for $i \to \infty$. We need to show that $y \in \mathcal{K}$. Indeed $\forall x \in \mathbb{R}^n$ and such that $\langle a_j, x \rangle \geq 0$, $j = 1, \ldots, m$, it follows that $\langle y_i, x \rangle \geq 0$, $i = 1, 2, \ldots$. But then, going to the limit at $i \to \infty$, we get $\forall x \in \mathbb{R}^n$ and such that $\langle a_j, x \rangle \geq 0$, $j = 1, \ldots, m$, it follows

$$\lim_{i \to \infty} \langle y_i, x \rangle = \langle y, x \rangle \geq 0,$$

and by Farkas's theorem $y \in \text{Cone}\{a_j \mid j = 1, \ldots, m\}$, i.e. $\mathcal{K}$ is closed.

Thus, the Farkas theorem and the closed cone theorem are equivalent, i.e. one follows from the other. This is a new look at these fundamental results in optimization.

4 On the Reducibility of an Optimization Problem with Inequality Constraints to an Optimization Problem with Equalities

When solving optimization problems with inequality constraints of the type (3), (4), it is attractive to reduce the inequality constraints to the equality constraints by adding artificial variables. However, in order for the optimality condition for a problem with inequalities and problems with equalities to be equivalent, it is necessary to introduce the condition of strict complementarity [6]. Otherwise, the optimality conditions will not be equivalent, and therefore such a reduction, as stated in [3], is not of interest. Let us explain this. We introduce the Lagrange function for the problem (3), (4)

$$L(x, \lambda) = \varphi(x) + \sum_{i=1}^{m} \lambda_i g_i(x).$$

Then the necessary conditions for the Kuhn-Tucker minimum are as follows:

$$L'_x(x, \lambda) = \varphi'(x) + \sum_{i=1}^{m} \lambda_i g'_i(x) = 0 \tag{11}$$

plus the complementarity condition

$$\lambda_i g_i(x) = 0, \qquad i = 1, \ldots, m, \tag{12}$$

$$\lambda_i \geq 0, \qquad i = 1, \ldots, m. \tag{13}$$

The sufficient conditions for the problem (3), (4) will be as follows:
– if the necessary conditions (11)–(13) are satisfied at a feasible point x^* $(g_i(x^*) \leq 0, i = 1, \ldots, m)$ for some λ^* and

$$\langle L''_{xx}(x^*, \lambda^*)h, h \rangle > 0, \tag{14}$$

$\forall h \neq 0$ such that $\langle g'_i(x^*), h \rangle \leq 0, i \in I(x^*) = \{i \in \{1, \ldots, m\} \mid g_i(x^*) = 0\}$, then x^* is the local minimum point of the problem (3), (4) (see, for example, [9]).

A significant drawback of these conditions from the sufficient optimality conditions for the problem with equalities is that the point x^* must be admissible. The Eqs. (11)–(13) do not guarantee this, in contrast to the necessary conditions for the problem with equalities. Therefore, we modify the problem (3), (4) as a problem with equalities by introducing a new slack variable $s = (s_1, \ldots, s_m)^\top$.

$$\min \varphi(x) \atop g_i(x) + s_i^2 = 0, \qquad i = 1, \ldots, m. \tag{15}$$

The tasks (3), (4) and (15) are equivalent [5]. The necessary optimality conditions for problem (15) are as follows:

$$\varphi'(x) + \sum_{i=1}^{m} \lambda_i g'_i(x) = 0_n, \tag{16}$$

$$\lambda_1 \begin{pmatrix} 2s_1 \\ 0 \\ \vdots \\ 0 \end{pmatrix} + \ldots + \lambda_m \begin{pmatrix} 0 \\ 0 \\ \vdots \\ 2s_m \end{pmatrix} = 0_m = 2\lambda s, \tag{17}$$

$$g_i(x) + s_i^2 = 0, \qquad i = 1, \ldots, m, \tag{18}$$

where $\lambda s = (\lambda_1 s_1, \ldots, \lambda_m s_m)^\top$. Note that from the systems (16)–(18) it is impossible to determine the sign of the Lagrange multipliers $\lambda_i \geq 0, i = 1, \ldots, m$. But the admissibility of the solution x^* is no longer required – it follows from the system (18). We denote $\bar{g}_i(x, s) = g_i(x) + s_i^2, i = 1, \ldots, m$, and introduce the Lagrange function for the problem (15)

$$\bar{L}(x, s, \lambda) = \varphi(x) + \sum_{i=1}^{m} \lambda_i \left(g_i(x) + s_i^2\right). \tag{19}$$

Let $g(x) = (g_1(x), \ldots, g_m(x))^\top$, $s^2 = (s_1^2, \ldots, s_m^2)^\top$. We consider that in the solution x^* the constraints regularity condition is satisfied if the vectors $\{g'_i(x^*)\}$, $i \in I(x^*)$ are linearly independent. Then sufficient optimality conditions mean a positive definiteness of the matrix $\bar{L}''_{(x,s)}(x^*, s^*, \lambda^*)$ on the kernel Ker $\bar{g}'(x^*, s^*)$, i.e.

$$\langle \bar{L}''_{(x,s)}(x^*, \lambda^*, s^*)h, h \rangle \geq \alpha \|h\|^2, \qquad \alpha > 0 \tag{20}$$

$\forall h \in$ Ker $\bar{g}'(x^*, s^*)$. Hence, by the way, the non-negativity of $\lambda_i^* \geq 0$, $i = 1, \ldots, m$ follows. For clarity, we write that

$$\bar{L}''_{(x,s)}(x^*, \lambda^*, s^*) = \begin{pmatrix} L_{xx}(x^*, \lambda^*) & 0_{nm} \\ 0_{mn} & 2\lambda^* \end{pmatrix}.$$

However, the matrix $\bar{L}''(x^*, s^*, \lambda^*)$ will not be positive definite on the kernel of $\bar{g}'(x^*, s^*)$ if the strict complementarity condition fails, i.e. λ_i^* and s_i^* are equal to zero at the same time for $i = 1, \ldots, m$. That is, the sufficient conditions (14) and (20) are not equivalent. This is due to the fact that it is possible to weaken the necessary optimality conditions of the 2-th order for the problem with equalities, and, therefore, to construct weaker sufficient conditions. Let us explain the following. Let there exist $h_2 \in$ Ker $F'(x^*)$ for the problem (1)–(2) such that

$$\langle L''_{xx}(x^*, \lambda^*)h_2, h_2 \rangle = 0 \qquad \left(L(x, \lambda) = \varphi(x) + \sum_{i=1}^m \lambda_i f_i(x) \right) \tag{21}$$

and, therefore, second order sufficient conditions not fulfilled. Let Ker $F'(x^*) = \mathbb{R}_1^n$ and $\mathbb{R}_2^n = (\mathbb{R}_1^n)^\perp$. Then $\mathbb{R}^n = \mathbb{R}_1^n + \mathbb{R}_2^n$. The following theorem will be true.

Theorem 5 (necessary and sufficient optimality conditions of a special kind). *Let φ, $F \in C^3(\mathbb{R}^n)$ and for any $h_2 \in$ Ker $F'(x^*)$ ($\|h\| = 1$) such that $L''(x^*, \lambda^*)[h_2]^2 = 0$ or $(L''(x^*\lambda^*)h_2 = 0)$ the conditions are satisfied*

1.

$$\|F''(x^*)[h_2]^2\| \geq \alpha > 0 \tag{22}$$

2. and there are $h_1 \in \mathbb{R}_2^r$ and $c(h_2) \geq c > 0$, $\|h_1\| = 1$, satisfying the relation

$$F'(x^*)[h_1] + \frac{1}{2}F''(x^*)[c(h_2)h_2]^2 = 0. \tag{23}$$

Then, (necessity) if x^* is a solution of (1)–(2), then $\exists \lambda^*$ such that $L'(x^*, \lambda^*) = 0$ and

$$L''(x^*, \lambda^*)[h_1]^2 \geq 0. \tag{24}$$

Moreover, (sufficiency) if $\forall h_1$, h_2 satisfying (22), (23), there exists $\beta > 0$ such that $L'(x^*, \lambda^*) = 0$ and

$$L''_{xx}(x^*, \lambda^*)[h_1]^2 \geq \beta > 0, \tag{25}$$

then x^* is the local minimum of the problem (1), (2).

Remark 2. Moreover, the element h_1 may not belong to Ker $F'(x^*)$.

Proof. Necessity. Let us show that $\forall t \in (0, \varepsilon)$, for $\varepsilon > 0$ sufficiently small, there exists $\omega(t) \colon \mathbb{R} \to \mathbb{R}^n$ such that the arc

$$\gamma(t) = x^* + c_2 t h_2 + t^2 h_1 + \omega(t) \in M(x^*) = \{x \in \mathbb{R}^n \mid F(x) = F(x^*) = 0\}$$

and $\|\omega(t)\| = o(t^2)$. Really

$$\begin{aligned}
F(x^* + c_2 t h_2 + t^2 h_1) &= F(x^*) + F'(x^*)[c_2 t h_2 + t^2 h_1] \\
&\quad + \frac{1}{2} F''(x^*)[c_2 t h_2 + t^2 h_1]^2 + o(t^2) \\
&= t^2 F'(x^*) h_1 + \frac{1}{2} F''(x^*)[c_2 t h_2]^2 + o(t^2) = o(t^2).
\end{aligned} \tag{26}$$

Since $F'(x^*)$ is non-degenerate, we can apply the principle of contraction mappings [7] to the map

$$\Phi_t(x) = x - (F'(x^*))^{-1} F(x^* + c_2 t h_2 + t^2 h_1 + x), \tag{27}$$

which implies the existence of $\omega(t)$ such that

$$F\left(x^* + c_2 t h_2 + t^2 h_1 + \omega(t)\right) = 0 \tag{28}$$

and

$$\|\omega(t)\| = o(t^2). \tag{29}$$

Next, we apply the standard technique to the relation $L(x^* + c_2 t h_2 + t^2 h_1 + \omega(t), \lambda^*) \geq L(x^*, \lambda^*)$ we get $L''_{xx}(x^*, \lambda^*)[h_1]^2 \geq 0$. The necessity is proven.

Sufficiency. Assuming the contrary, i.e. that there exists a sequence of points $\{x_k\} \in M(x^*)$ and $x_k \to x^*$ for $k \to \infty$ such that $\varphi(x_k) < \varphi(x^*)$, using the proof of necessity and the standard technique of proving the sufficiency of conditions, we come to a contradiction with the relation (25).

Example 1. We consider the problem (15) in the form

$$\begin{aligned}
&\min \varphi(y) \\
&g_i(y) \leq 0, \qquad 1, \ldots, m,
\end{aligned} \tag{30}$$

for $F(x) = g(y) + s^2 = 0_m$. Here the role of x is played by (y, s) (to avoid confusion with the latest entries, we replaced x with y in the problem (3), (4)). If the point (y^*, s^*), $s^* = 0$ is a solution to the problem (3), (4), then for $h_2 = (0, \bar{s})^\top$ obviously there is $h_1 = (\bar{y}, 0)^\top$ such that

$$F'(x^*)[h_1] + \frac{1}{2} F''(x^*)[h_2]^2 = 0 \Leftrightarrow g'(y^*)\bar{y} + \frac{1}{2} 2 \bar{s}^2 = 0. \tag{31}$$

Since optimality conditions are local in nature, we consider only active constraints, i.e. $i \in I(x^*)$ and, without loss of generality, we consider $r = m$. Obviously, the condition (22) is satisfied if $\bar{s} \neq 0$. And the condition (23) is fulfilled due to the regularity of the constraints $\{g_i(x)\}$, $i = 1, \ldots, m$, at the point x^*, and $c_2 = 1$. Thus, the Theorem 5 is applicable to the problem (15), since $L''_{xx}(x^*, \lambda^*)[h_2]^2 = 0$, although $h_2 = (0, \bar{s})^\top \in \operatorname{Ker} F'(x^*) = \{(\bar{y}, \bar{s}) \mid g'(y^*)\bar{y} + 0 \cdot \bar{s} = 0\}$. Moreover, the following equivalence theorem holds.

Theorem 6 (on equivalence of optimality conditions). *Let* $\varphi, g \in C^3(\mathbb{R}^n)$. *Then the optimality conditions* (11), (12), (14) *and* (25) *are equivalent.*

Proof. Obviously, one does need to check only a sufficient part of the optimality conditions. Suppose that the conditions (11)–(14) are satisfied. Then, taking $\bar{s} \neq 0$, we get $\bar{y} = \{g'(y^*)\}^{-1}(-\bar{s}^2)$ from the condition (23). Moreover, $L''_{xx}(y^*, s^*\lambda^*)[y]^2 \geq \beta \|\bar{y}\|^2$, since the sufficient condition (14) is satisfied. Here $\bar{y}$ plays the role of h_1 in the Theorem 5 and therefore (25) holds, which means that the sufficient conditions of the Theorem 5 are satisfied. In one direction, the theorem is proved.

Now suppose that the sufficient conditions (25) for the problem (30) are satisfied. Then $\forall h$ such that $g'(y^*)h < 0_m$ will be $\bar{s} \neq 0$ and, therefore, (25) holds for $h_1 = h$, i.e. $L''_{xx}(y^*, 0, \lambda^*)[h]^2 \geq \beta \|h\|^2$, which means that $L''_{xx}(y^*, \lambda^*)[h]^2 \geq \beta \|h\|^2$, (where $L(y, \lambda)$ plays the role of the function $L(x, \lambda) = \varphi(x) + \sum_{i=1}^{m} \lambda_i g_i(x)$ in the problem (3), (4)) and finally, the sufficient conditions (14) for the problem (3), (4). The theorem is proved.

5   2-Factor Method for Solving Degenerate Systems of Nonlinear Equations

For the numerical solution of systems of equations, in particular those arising from the necessary optimality conditions, the basic construction for constructing various iterative processes is the Newton method [6]. However, if for the system of nonlinear equations

$$\Phi(x) = 0_n \tag{32}$$

in the solution x^* the matrix $\Phi'(x^*)$ is degenerate, then the application of the Newton method, the iterative scheme of which has the form

$$x_{k+1} = x_k - \{\Phi'(x_k)\}^{-1} \Phi(x_k), \qquad k = 0, 1, \ldots, \tag{33}$$

becomes ineffective or even impossible. In this case, the so-called 2-factor method is applied, which ensures convergence, and moreover, the quadratic rate of convergence. In the general case, the scheme of the 2-factor method for solving a degenerate system of nonlinear equations (32) for $x \in \mathbb{R}^n$ has the following form:

$$x_{k+1} = x_k - \{\Phi'(x_k) + P\Phi''(x_k)h\}^{-1} (\Phi(x_k) + P\Phi'(x_k)h), \qquad k = 0, 1, \ldots, \tag{34}$$

where P is the orthogonal projection matrix on $(\operatorname{Im} \Phi'(x^*))^{\perp}$, and the vector h, $\|h\| = 1$ is some fixed vector such that the matrix $\{\Phi'(x^*) + P\Phi''(x^*)h\}$ is non-degenerate, i.e. the map $\Phi(\cdot)$ is 2-regular on the vector h at the solution point x^* of the system (32). In this case, the following theorem holds.

Theorem 7. *Let $\Phi \in C^3(\mathbb{R}^n)$ and $\Phi(\cdot)$ is 2-regularly at the solution x^* on the vector h. Then for $x_0 \in U_\varepsilon(x^*)$ the sequence $\{x_k\}$ defined by the (34) scheme converges to the solution x^*, and the estimate for the convergence rate is true*

$$\|x_{k+1} - x^*\| = c\|x_k - x^*\|^2, \qquad k = 0, 1, \ldots, \qquad (35)$$

where $\varepsilon > 0$ is quite small.

For a proof, see, for example, [2]. Let us show how the 2-factor method can be used for the numerical solution of optimization problems with inequality constraints when using artificial variables.

6   Concept of 2-Regularity and 2-Factor Newton's Method for Solving a Degenerate Kuhn-Tucker System

When solving optimization problems with inequality constraints by introducing artificial variables, it is necessary to introduce conditions of strict complementarity so that the KT system is non-degenerate. Otherwise, the applicability of highly efficient methods such as Newton cannot be guaranteed. In this section, we consider the effectiveness of this approach from the point of view of applying the 2-factor method for solving a system of degenerate optimality conditions and prove that the KT system is 2-regular.

So, we have the problem (3), (4)

$$\min \varphi(x)$$

for

$$g_i(x) \leq 0, \ldots 1 = 1, \ldots, m,$$

which we replace with the problem (15)

$$\min \varphi(x)$$
$$g_i(x) + s_i^2 = 0, \ldots 1 = 1, \ldots, m.$$

We introduce the Lagrange function

$$L(x, s, \lambda) = \varphi(x) + \langle g(x) + s^2, \lambda \rangle, \qquad \lambda \in \mathbb{R}^m,$$

and denote $u = (x, s, \lambda)$. Then we have

$$L'(u) = \begin{pmatrix} L_x(u) \\ 2\lambda s \\ g(x) + s^2 \end{pmatrix}, \qquad (36)$$

$$L''(u) = \begin{bmatrix} L_{xx}(u) & 0_{nm} & (g_x(x))^\top \\ 0_{mn} & 2\lambda & 2s \\ g_x(x) & 2s & 0_{mm} \end{bmatrix}. \qquad (37)$$

We introduce a system of ℓ nonlinear equations, where $\ell = n + 2m$

$$L'(u) = 0_\ell. \tag{38}$$

Let $u^* = (x^*, s^*, \lambda^*)^\top \in \mathbb{R}^\ell$ be its solution. If $\lambda^* \geq 0_m$, $L'(u^*) = 0_\ell$, then at the point (x^*, λ^*) the necessary minimum conditions for the problem (3), (4) are satisfied. If we add the sufficient conditions (14), then the point x^* will be a local solution to the problem (3), (4). However, the presence of the inequality $\lambda^* \geq 0_m$ does not allow us to obtain a complete system of equalities. Therefore, we replace $\lambda^* \geq 0_m$ with the system of equalities

$$\lambda^\top - y^2 = 0_m, \tag{39}$$

where $y \in \mathbb{R}^m$, $y^2 = (y_1^2, \ldots, y_m^2)^\top$ and consider the extended system

$$\Psi(z) = \begin{pmatrix} L'(u) = 0_\ell \\ \lambda^\top - y^2 = 0_m \end{pmatrix} = 0_r, \tag{40}$$

where $z = (x, s, \lambda, y)^\top$, $r = n + 3m$. Thus, we reduced the optimization problem with inequalities to a system of equalities. Moreover, in the solution (x^*, λ^*) the necessary conditions of the KT are satisfied, and now $\lambda^* \geq 0$. Let be

$$\Psi'(z) = \begin{pmatrix} L_{xx}(u) & 0_{nm} & (g_x(x))^\top & 0_{nm} \\ 0_{mn} & 2\lambda & 2s & 0_{mm} \\ g_x(x) & 2s & 0_{mm} & 0_{mm} \\ 0_{mn} & I_m & 0_{mm} & 2y \end{pmatrix}. \tag{41}$$

Here I_m is the identity matrix of dimension $m \times m$. Newton's method for solving the system (41) has the form

$$z_{k+1} = z_k - (\Psi'(z_k))^{-1} \Psi(z_k). \tag{42}$$

If the solution of z^* does not satisfy the strict complementarity condition, then the matrix $\Psi'(z^*)$ is degenerate and the method (42) does not guarantee convergence. We construct the so-called 2-factor Newton method for solving the system (40), which ensures the quadratic convergence rate. Following [2,4], we introduce the mapping

$$\Phi(z) = \Psi(z) + P\Psi'(z)h \tag{43}$$

and 2-factor operator

$$\Phi'(z) = \Psi'(z) + P\Psi''(z)h, \tag{44}$$

where $r \times r$ the matrix P projects the vector $\Psi''(z^*)h$ onto $(\operatorname{Im}\Psi'(z^*))^\perp$, the r-dimensional vector h having the unit norm is chosen from the non-degeneracy condition of the $\Phi'(z^*)$. Note that the matrix P can be chosen, generally speaking, simply from the non-degeneracy condition of 2-factor operator $\Phi'(z^*)$.

Theorem 8. *Let $\varphi, g \in C^3(\mathbb{R}^n)$, there exists a vector z^* satisfying the system (40), at the point (x^*, λ^*) sufficient optimality conditions are satisfied (14) and constraints are regular. Then the 2-factor matrix $\Phi'(z^*)$ is non-degenerate.*

Proof. Let us show that the map $\Psi(z)$ is 2-regular at point z^* on the vector $h^\top = (0_n^\top, 0_m^\top, e_0^\top, e_1^\top)$, where $e_0 \in \mathbb{R}^m$ and each i-th component of the vector e_0 is equal to one, if $i \in I_0(x^*) = \{i \in I(x^*) \mid \lambda_i^* = 0, \, s_i^* = 0\}$, and it is equal to zero otherwise, and $e_1 \in \mathbb{R}^m$ and each i-th component of the vector e_1 is equal to one, if $\lambda_i^* = 0$, and is equal to zero otherwise. Define the matrix

$$\Phi'(z^*) = \Psi'(z^*) + P\Psi''(z^*)h =$$
$$\begin{pmatrix} L_{xx}(z^*) & 0_{nm} & (g_x(x^*))^\top & 0_{nm} \\ 0_{mn} & 2\lambda^* + 2J_0(x^*) & 2s^* & 0_{mm} \\ (g_x(x^*))^\top & 2s^* & 0_{mm} & 0_{mm} \\ 0_{mn} & I_m & 0_{mm} & 2y^* + 2J_1(x^*) \end{pmatrix}, \tag{45}$$

where the $r \times r$-diagonal matrix P has the i-th diagonal element equal to zero, if $1 \le i \le k$ or $k + m + 1 \le i \le \ell$, and for $k + 1 \le i \le n + m$, the i-th diagonal element is equal to one, if $i - n \in I_0(x^*)$, and is equal to zero otherwise. For $\ell + 1 \le i \le r$, the i-th diagonal element is equal to one, if $i - \ell \in I_0(x^*)$, and is equal to zero otherwise. Moreover, the $m \times m$ diagonal matrix $J_0(x^*)$ has the j-th diagonal element equal to either zero or one, if $j \in I_0(x^*)$. In turn, the $m \times m$ diagonal matrix $J_1(x^*)$ has the j-th diagonal element equal to unity, if $j \in I_0(x^*)$, and zero otherwise. It follows that for $h = (0_n^\top, 0_m^\top, e_0^\top, e_1^\top)^\top$ we get

$$P\Psi''(z^*)h = \begin{pmatrix} 0_{nn} & 0_{nm} & 0_{nm} & 0_{nm} \\ 0_{mn} & 2J_0(x^*) & 0_{nm} & 0_{mm} \\ 0_{mn} & 0_{mm} & 0_{mm} & 0_{mm} \\ 0_{mn} & 0_{mn} & 0_{mm} & 2J_1(x^*) \end{pmatrix}$$

and the validity of the formula (45), as well as the non-degeneracy of the matrix $\Phi'(z^*)$.

Obviously, the 2-factor method applied to the (40) system has the form

$$z_{k+1} = z_k - \left(\Psi'(z_k) + P\Psi''(z_k)h\right)^{-1} \left(\Psi(z_k) + P\Psi'(z_k)h\right), \qquad k = 0, 1, \dots. \tag{46}$$

Theorem 9. *Let $\Psi \in C^3(\mathbb{R}^r)$ and for $h \in \mathbb{R}^r$, $\|h\| \ne 0$, there exists $\left(\Psi'(z^*) + P\Psi''(z^*)h\right)^{-1}$. Then the 2-factor method (46) converges to the solution z^* of the system (40) with quadratic rate, that is,*

$$\|z_{k+1} - z^*\| \le c\|z_k - z^*\|^2, \qquad k = 0, 1, \dots,$$

where $z_0 \in U_\varepsilon(z^*)$, $\varepsilon > 0$ is quite small, and $c > 0$ is an independent constant.

Proof. It follows from the fact that the 2-factor method (46) is the usual Newton method applied to the system

$$\Psi(z) + P\Psi'(z)h = 0, \tag{47}$$

since the point z^* is the solution of the system (40), it will be simultaneously the solution of the system (47) and for it all the conditions for the convergence of the Newton method with quadratic rate are satisfied.

References

1. Ashmanov, S.A., Timokhov, A.V.: Optimization Theory in Tasks and Exercises (in Russian). Fizmatlit, Moscow (1991)
2. Belash, K.N., Tret'yakov, A.A.: Methods for solving degenerate problems. USSR Comput. Math. Math. Phys. **28**(4), 90–94 (1988)
3. Bertsekas, D.P.: Nonlinear Programming. Athena Scientific, Belmont (1999)
4. Brezhneva, O.A., Tret'yakov, A.A.: The p-factor-Lagrange methods for degenerate nonlinear programming. Numer. Funct. Anal. Optim. **28**(9–10), 1051–1086 (2007)
5. Evtushenko, Y.G., Tretyakov, A.A.: pth-order approximation of the solution set of nonlinear equations. Comput. Math. Math. Phys. **53**(12), 1763–1780 (2013)
6. Evtushenko, Y.G.: Methods for Solving Extreme Problems and Their Application in Optimization Systems (in Russian). Nauka, Moscow (1982)
7. Ioffe, A.D., Tihomirov, V.M.: Theory of Extremal Problems. North-Holland, New York (1979)
8. Kuhn, H.W., Tucker, A.W.: Nonlinear programming. In: Proceedings of 2nd Berkeley Symposium, pp. 481–492. University of California Press, Berkeley (1951)
9. Polyak, B.T.: Introduction to Optimization (in Russian). Nauka, Moscow (1983)

A Given Diameter MST on a Random Graph

Edward K. Gimadi[1,2] , Aleksandr S. Shevyakov[2],
and Alexandr A. Shtepa[2(✉)]

[1] Sobolev Institute of Mathematics, prosp. Akad. Koptyuga, 4, Novosibirsk, Russia
gimadi@math.nsc.ru
[2] Novosibirsk State University, Pirogova, 1, Novosibirsk, Russia
shevash.97@gmail.com, shoomath@gmail.com
http://www.nsu.ru/
http://www.math.nsc.ru/

Abstract. We give a new approximation polynomial time algorithm for one of the intractable problem of finding given-diameter Minimum Spanning Tree (MST) on n-vertex complete graph with randomly weighted edges. A significant advantage of this algorithm is that it turned out to be well suited for finding several edge-disjoint MST of a given diameter. A probabilistic analysis was performed under conditions that edge weights of given graph are identically independent uniformly distributed random variables on an segment $[a_n; b_n]$, $a_n > 0$. Sufficient conditions of asymptotic optimality are presented. It is also noteworthy that the new algorithmic approach to solve the problem of finding a given-diameter MST both on directed and undirected graphs.

Keywords: Given-diameter minimum spanning tree · Approximation algorithm · Probabilistic analysis

1 Introduction

The Minimum Spanning Tree (MST) problem is a one of the classic discrete optimization problems. Given weighted graph $G = (V, E)$, MST is to find a spanning tree of a minimal weight. The polynomial solvability of the problem was shown in the classic algorithms by Boruvka (1926), Kruskal (1956) and Prim (1957). These algorithms have time complexities $O(m \log n)$, $O(m \log m)$ and $O(n^2)$, respectively, where $m = |E|$ and $n = |V|$. It is interesting to note that the mathematical expectation of weight MST on a random graph can be unexpectedly small. For example, on a complete graph with weights of edges from class of random variables with uniform distribution on a segment $[0; 1]$, the

Supported by the program of fundamental scientific researches of the SB RAS No. I.5.1., project No. 0314-2019-0014, and by the Russian Foundation for Basic Research, project No. 20-31-90091.

N. Olenev et al. (Eds.): OPTIMA 2020, LNCS 12422, pp. 110–121, 2020.
https://doi.org/10.1007/978-3-030-62867-3_9

weight of a MST w.h.p. (with high probability) is close to the constant 2,02...
[3]. Similar results were obtained in [1], [2].

A generalization of the problem is a diameter-bounded MST problem. The
diameter of a tree is the maximum number of edges within the tree connecting
a pair of vertices. For this problem, given a graph and a number $d = d_n$, the
goal is to find a spanning tree of minimal total weight in the graph having its
diameter bounded above by given number d, or from below by given number d.
Both problems are NP-hard in general. The bounded above MST problem is
polynomially solvable for diameters two or three, and NP-hard for any diameter
between 4 and $(n-1)$, even for the edge weights equal to 1 or 2 [4, p. 206]. The
MST problem bounded from below is NP-hard, because its particular case for
$d = n-1$ is the problem "Hamiltonian Path" [4].

In the papers [6–8,10] a bounded MST problem with a graph diameter
bounded either from below or above was studied.

Recently, we began to study another modification of the problem, when the
diameter of the desired spanning tree is a given number. The work [11] gives
a probabilistic analysis of an effective algorithm for solving a given-diameter
MST problem in the case of directed complete graph. Unfortunately, the algorithm analysis, presented in this work becomes unacceptable for a problem on
undirected graph. The appearance of the difficulty of probabilistic analysis in
the case of the undirected graph arises from the need to take into account the
possible dependence between different objects (random variables) in the course
of the algorithm. In current paper we consider a given-diameter MST problem
(d-MST) on the complete graph. We introduce a polynomial time algorithm to
solve this problem and provide conditions for this algorithm to be asymptotically optimal. A probabilistic analysis is performed under conditions that edge
weights of given graph are identical independent uniformly distributed random
variables.

By $F_A(I)$ and $OPT(I)$ we denote respectively the approximate (obtained
by some approximation algorithm A) and the optimum value of the objective
function of the problem on the input I. An algorithm A is said to have *estimates
(performance guarantees)* $(\varepsilon_A(n), \delta_A(n))$ on the set of random inputs of the n-
sized problem, if

$$\mathbf{P}\Big\{F_A(I) > \big(1 + \varepsilon_A(n)\big)OPT(I)\Big\} \le \delta_A(n), \tag{1}$$

where $\varepsilon_A(n)$ is an estimation of *the relative error* of the solution obtained by
algorithm A, $\delta_A(n)$ is an estimation of *the failure probability* of the algorithm,
which is equal to the proportion of cases when the algorithm does not hold the
relative error $\varepsilon_A(n)$ or does not produce any answer at all.

Following [5], we say that an approximation algorithm A is called *asymptotically optimal* on the class of input data of the problem, if there exist such
performance guarantees that for all inputs I of size n

$$\varepsilon_A(n) \to 0 \text{ and } \delta_A(n) \to 0 \quad \text{as } n \to \infty.$$

Let's denote $UNI(a_n; b_n)$ a class of random variables with uniform distribution on a segment $[a_n; b_n]$, $a_n > 0$.

One of the first results of the asymptotically optimal approach to the MST problem with diameter bounded from below on complete graphs with n vertices and weights of edges from the $UNI(a_n; b_n)$ was presented in [8].

2　Algorithms for Finding MST with Given Diameter

2.1　Algorithm for Solving d-Directed MST Problem

In [11] for solving d-Directed MST problem the algorithm was proposed, which on Stage 1 builds a simple d-edges path P, and then on Stage 2 every unvisited vertex is connected by the shortest possible edge to the path P (see Fig. 1). Some sufficient conditions of asymptotical optimality of algorithm were obtained.

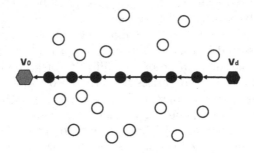

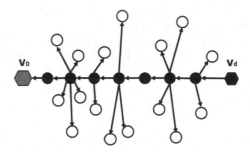

Fig. 1. Example of the work of the Algorithm in [11].

As mentioned above, the analysis of this algorithm becomes unacceptable for d-Undirected MST problem, because during the process of the algorithm in the case of undirected graph, we lose the property of independence of random variables, considered on Stage 2, since these variables are functions of the edge weights already considered in Stage 1.

2.2 Algorithm for Solving d-Undirected MST Problem

Algorithm [9] is based on dividing the vertices of the undirected graph into two subsets, in each of which the corresponding half of the path is constructed, and then the vertices, that do not fall into the path, are connected to the internal vertices of the path located in another subset (see Fig. 2).

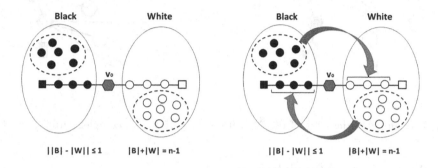

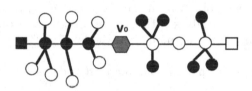

Fig. 2. Example of the work of the algorithm in [9].

2.3 A New Algorithm $\mathcal{A}$ for Finding d-MST on Arbitrary Complete Graph

Now we give a new Algorithm $\mathcal{A}$ for solving d-MST problem on an arbitrary complete graph.

Stage 1.$\mathcal{A}$. Choose in the graph G an arbitrary $(d+1)$-vertex subgraph G'.
Stage 2.$\mathcal{A}$. Starting at arbitrary vertex in G' on the set $V(G')$ of its vertices construct Hamiltonian path of a length $d = d_n$, using the approach "go to the nearest unvisited vertex".
Stage 3.$\mathcal{A}$. Connecting the remaining $(n-d-1)$ vertices to the nearest $(d-1)$ inner vertices of the path, we get the spanning tree $T_{\mathcal{A}}$.

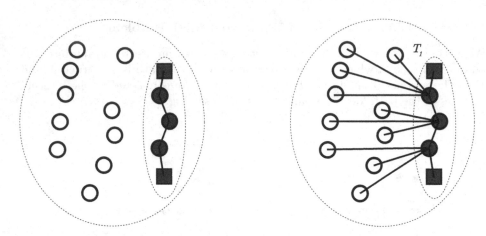

Fig. 3. Example of the work of the Algorithm $\mathcal{A}$ on 15-vertex complete graph, $d = 4$.

3 Analysis of Algorithm $\mathcal{A}$

The algorithm has polynomial time complexity $\mathcal{O}(n^2)$, since the construction of the path P in Stage 1$\mathcal{A}$ and Stage 2$\mathcal{A}$ is done by greedy procedure in time $\mathcal{O}\big((d)^2\big)$, and in the Stage 3$\mathcal{A}$ it takes $\mathcal{O}\big(d(n-d)\big)$ comparison operations.

We perform the probabilistic analysis under conditions that weights of graph edges are random variables η from the class $\mathrm{UNI}(a_n; b_n)$, namely, are identical independent random variables with uniform distribution on a set $[a_n; b_n]$, $0 < a_n < b_n < \infty$. Obviously, normalized variables $\xi = \frac{\eta - a_n}{b_n - a_n}$ belong to the class $\mathrm{UNI}(0; 1)$.

Further we suppose that the parameter d is defined on the set of values in two ranges

$$\text{Case 1: } \ln n \le d < \frac{n}{\ln n} \quad \text{and} \quad \text{Case 2: } \frac{n}{\ln n} \le d < n.$$

We denote random variable equal to minimum over k variables from the class $\mathrm{UNI}(a_n; b_n)$ (from $\mathrm{UNI}(0; 1)$) by η_k (ξ_k, correspondingly).

For simplicity, we further assume that n is odd, and the parameter d is even.

According to the description of Algorithm $\mathcal{A}$, the weight $W_{\mathcal{A}}$ of the constructed spanning tree $T_{\mathcal{A}}$ is a random value, such that

$$W_{\mathcal{A}} = \sum_{k=1}^{d} \eta_k + (n - d - 1)\eta_{d-1} = (n-1)a_n + (b_n - a_n)W'_{\mathcal{A}},$$

where

$$W'_{\mathcal{A}} = \sum_{k=1}^{d} \xi_k + (n - d - 1)\xi_{d-1}.$$

Lemma 1.

$$\mathbf{E}W'_{\mathcal{A}} \le \left(\ln d + \frac{n-d-1}{d} \right).$$

Proof. Since $\mathbf{E}\xi_k = 1/(k+1)$, and

$$\sum_{k=1}^{d} \mathbf{E}\xi_k = \sum_{k=1}^{d+1} \frac{1}{k} - 1 \le \ln d,$$

we have

$$\mathbf{E}W'_{\mathcal{A}} = \mathbf{E}\left(\sum_{k=1}^{d} \xi_k + (n-d-1)\xi_{d-1} \right) = \sum_{k=1}^{d+1} \frac{1}{k} - 1 + \frac{n-d-1}{d} \le \ln d + \frac{n-d-1}{d}.$$

Lemma 2. *Algorithm $\mathcal{A}$ for solving the d-MST problem on n-vertex complete graph with weights of edges from $UNI(a_n; b_n)$ has the following estimates of the relative error ε_n and the failure probability δ_n:*

$$\varepsilon_n = (1 + \lambda_n) \frac{(b_n - a_n)}{(n-1)a_n} \widetilde{\mathbf{E}W}'_{\mathcal{A}}, \qquad (2)$$

$$\delta_n = \lambda_n \widetilde{\mathbf{E}W}'_{\mathcal{A}}, \qquad (3)$$

where $\lambda_n > 0$, $\widetilde{\mathbf{E}W}'_{\mathcal{A}}$ is some upper bound for expectation $\mathbf{E}W'_{\mathcal{A}}$.

Proof.

$$\mathbf{P}\left\{ W_{\mathcal{A}} > (1 + \varepsilon_n)OPT \right\} \le \mathbf{P}\left\{ W_{\mathcal{A}} > (1 + \varepsilon_n)(n-1)a_n \right\}$$

$$= \mathbf{P}\left\{ (n-1)a_n + (b_n - a_n)W'_{\mathcal{A}} > (1 + \varepsilon_n)(n-1)a_n \right\}$$

$$= \mathbf{P}\left\{ W'_{\mathcal{A}} - \mathbf{E}W'_{\mathcal{A}} > \frac{\varepsilon_n(n-1)a_n}{(b_n - a_n)} - \mathbf{E}W'_{\mathcal{A}} \right\}$$

$$= \mathbf{P}\left\{ \widetilde{W}'_{\mathcal{A}} > \frac{\varepsilon_n(n-1)a_n}{(b_n - a_n)} - \mathbf{E}W'_{\mathcal{A}} \right\}$$

$$\le \mathbf{P}\left\{ \widetilde{W}'_{\mathcal{A}} > \frac{\varepsilon_n(n-1)a_n}{(b_n - a_n)} - \widetilde{\mathbf{E}W}'_{\mathcal{A}} \right\} = \lambda_n \widetilde{\mathbf{E}W}'_{\mathcal{A}} = \delta_n.$$

Further for the probabilistic analysis of Algorithm $\mathcal{A}$ we use the following probabilistic statement

Petrov's Theorem [12]. *Consider independent random variables $X_1, \ldots, X_n$. Let there be positive constants T and $h_1, \ldots, h_n$ such that for all $k = 1, \ldots, n$ and $0 \le t \le T$ the following inequalities hold:*

$$\mathbf{E}e^{tX_k} \le \exp\left\{ \frac{h_k t^2}{2} \right\}. \qquad (4)$$

Set $S = \sum_{k=1}^{n} X_k$ and $H = \sum_{k=1}^{n} h_k$. Then

$$\mathbf{P}\{S > x\} \le \begin{cases} \exp\left\{ -\frac{x^2}{2H} \right\}, & \text{if } 0 \le x \le HT, \\ \exp\left\{ -\frac{Tx}{2} \right\}, & \text{if } x \ge HT. \end{cases}$$

Theorem 1. *Let the parameter $d = d_n$ be defined so that*

$$\ln n \leq d < n, \tag{5}$$

Then Algorithm $\mathcal{A}$ solves the d-MST problem on n-vertex arbitrary complete undirected graph with weights of edges from $UNI(a_n; b_n)$ asymptotically optimal w.h.p., if

$$\frac{b_n}{a_n} = \begin{cases} o(d), & \text{if } \ln n \leq d < \frac{n}{\ln n}, \\ o(\frac{n}{\ln n}), & \text{if } \frac{n}{\ln n} \leq d < n. \end{cases} \tag{6}$$

Proof. We will carry a proof for two cases of possible values of the parameter d: $\ln n \leq d < \frac{n}{\ln n}$ and $\frac{n}{\ln n} \leq d < n$.

$$\textbf{Case 1:} \quad \ln n \leq d < \frac{n}{\ln n}.$$

Lemma 3. *In the case $d < \frac{n}{\ln n}$ the following upper bound for $\mathbf{EW}'_{\mathcal{A}}$ holds:*

$$\widetilde{EW}'_{\mathcal{A}} = \frac{2(n-1)}{d}.$$

Proof. Given the fact, that $\ln d \leq \ln n$ and $d < \frac{n}{\ln n}$ we have:

$$\mathbf{EW}'_{\mathcal{A}} \leq \ln d + \frac{n-d-1}{d} \leq \ln n + \frac{2(n-1)}{d} - \left(\frac{n-1}{d} + 1\right)$$
$$\leq \frac{2(n-1)}{d} - \left(\frac{n-1}{d} + 1 - \ln n\right) \leq \frac{2(n-1)}{d} - \left(\frac{n-1}{n}\ln n + 1 - \ln n\right)$$
$$= \frac{2(n-1)}{d} - \left(1 - \frac{\ln n}{n}\right) \leq \frac{2(n-1)}{d} = \widetilde{EW}'_{\mathcal{A}}.$$

According to this Lemma and the formula (2) for the relative error we have

$$\varepsilon_n = (1 + \lambda_n)\frac{(b_n - a_n)}{(n-1)a_n}\widetilde{EW}'_{\mathcal{A}} = (1 + \lambda_n)\frac{(b_n - a_n)}{(n-1)a_n}\frac{2(n-1)}{d} \leq 2(1 + \lambda_n)\frac{b_n/a_n}{d}.$$

Let, within the framework of Case 1, the parameter λ_n from Lemma 2 takes the value $\lambda_n = \sqrt{\frac{\ln n}{n-d-1}}$. Since $\ln n \leq d < \frac{n}{\ln n}$, it is true: $\lambda_n < 1$, and we see, that $\varepsilon_n \to 0$ under the condition

$$\frac{b_n}{a_n} = o(d_n).$$

Now using Petrov's Theorem and Lemma 3, we can estimate the failure probability:

$$\delta_n = \mathbf{P}\{W'_{\mathcal{A}} > \lambda_n\widetilde{EW}'_{\mathcal{A}}\} = \mathbf{P}\left\{W'_{\mathcal{A}} > \lambda_n\frac{2(n-1)}{d}\right\}.$$

Define constants $h_e = 1/d^2$ for all edges, which are included to the spanning tree $T_{\mathcal{A}}$ at Stage 3.$\mathcal{A}$. Let remaining edges have constants equal 0. In this case we have $H = \frac{n-d-1}{d^2}$.

Set $T = d$; $x = \lambda_n\frac{2(n-1)}{d}$.

Taking into account the values λ_n, T, H and x, the following inequality is satisfied:

$$TH = \frac{n-d-1}{d} \geq \lambda_n \frac{2(n-1)}{d} = x.$$

According to Petrov's Theorem, we have an estimate for the failure probability of Algorithm $\mathcal{A}$:

$$\delta_n = \mathbf{P}\{\widetilde{W}'_{\mathcal{A}} > x\} \leq \exp\left\{-\frac{x^2}{2H}\right\}.$$

Now show that

$$\frac{x^2}{2H} \geq 2\ln n.$$

Indeed, since $n - d \geq n(1 - \frac{1}{\ln n})$, and taking into account the inequality (5), we get

$$\frac{x^2}{2H} = \frac{\left(\lambda_n \frac{2(n-1)}{d}\right)^2}{\frac{2(n-d-1)}{d^2}} \geq \frac{\left(\lambda_n \frac{2(n-d-1)}{d}\right)^2}{\frac{2(n-d-1)}{d^2}}$$

$$= 2(n-d-1)\lambda_n^2 = 2(n-d-1)\frac{\ln n}{(n-d-1)} = 2\ln n.$$

From this it follows that

$$\delta_n = \mathbf{P}\{W'_{\mathcal{A}} > x\} \leq \exp\left\{-\frac{x^2}{2H}\right\} \leq \exp(-2\ln n) = \frac{1}{n^2} \to 0,$$

as $n \to \infty$.

So in the Case 1 Algorithm $\mathcal{A}$ solves the d-MST problem on n-vertex arbitrary undirected complete graph with weights of edges from the class $\mathrm{UNI}(a_n; b_n)$ asymptotically optimal.

$$\textbf{Case 2}: \quad \frac{n}{\ln n} \leq d < n.$$

Lemma 4. *In the Case 2 ($n/\ln n \leq d < n$) the following inequality is correct:*

$$EW'_{\mathcal{A}} \leq 2\ln n = \widetilde{EW}'_{\mathcal{A}}.$$

Proof.
For all d, $n/\ln n \leq d < n$ the following inequality holds:

$$\frac{n-d-1}{d} \leq \ln n. \tag{7}$$

According to the Lemma 1, and taking into account the inequality (7), we have

$$EW'_{\mathcal{A}} \leq \left(\ln d + \frac{n-d-1}{d}\right) \leq 2\ln n = \widetilde{EW}'_{\mathcal{A}}.$$

Lemma 4 proved.

According to the Lemma 4 and formula (2), for the relative error we have

$$\varepsilon_n = (1+\lambda_n)\frac{(b_n - a_n)}{(n-1)a_n}\widetilde{\mathbf{EW}}'_{\mathcal{A}} = (1+\lambda_n)\frac{(b_n - a_n)}{(n-1)a_n}\cdot 2\ln n \le \frac{2n(1+\lambda_n)}{n-1}\frac{b_n/a_n}{n/\ln n}.$$

Within considered Case 2, we set $\lambda_n = 1$. From this we see, that $\varepsilon_n \to 0$ under condition

$$\frac{b_n}{a_n} = o\Big(\frac{n}{\ln n}\Big).$$

Now using Petrov's Theorem and Lemma 4 , estimate the failure probability

$$\delta_n = \mathbf{P}\{W'_{\mathcal{A}} > \lambda_n\widetilde{\mathbf{EW}}'_{\mathcal{A}}\} = \mathbf{P}\{W'_{\mathcal{A}} > 2\lambda_n\ln n\}.$$

Set constants $h_e = 1/d^2$ for all edges, which are included to the spanning tree $T_{\mathcal{A}}$ at Stage 3.$\mathcal{A}$. Let remaining edges have constants equal 0. So $H = \frac{n-d-1}{d^2}$. Define $T = 2$ and $x = 2\lambda_n\ln n$.

Taking into account the values λ_n, T, H and x, the following inequality is true:

$$TH = \frac{2(n - d - 1)}{d^2} \le 2\lambda_n\ln n = x.$$

According to Petrov's Theorem, we have an estimate for the failure probability of Algorithm $\mathcal{A}$:

$$\delta_n = \mathbf{P}\{\widetilde{W}'_{\mathcal{A}} > x\} \le \exp\Big\{-\frac{Tx}{2}\Big\}.$$

Since $T = 2$, $\lambda_n = 1$, $x = 2\lambda_n\ln n$, we have $\frac{Tx}{2} = 2\ln n$. From this it follows that

$$\delta_n = \mathbf{P}\{W'_{\mathcal{A}} > x\} \le \exp\Big\{-\frac{Tx}{2}\Big\} = \exp(-2\ln n) = \frac{1}{n^2} \to 0,$$

as $n \to \infty$. So in the Case 2 Algorithm $\mathcal{A}$ also solves the d-MST problem on n-vertex complete graph with weights of edges from UNI$(a_n; b_n)$ asymptotically optimal.

We conclude, that within the values of the parameter d for both cases, under conditions (6) we have estimates of the relative error $\varepsilon_n \to 0$ and the failure probability $\delta_n \to 0$ as $n \to \infty$.

Theorem 1 is completely proved.

4 An Algorithm for Finding Several Edge-Disjoined Spanning Trees with a Given Diameter (m-d-MST Problem)

Given a complete weighted n-vertex graph $G = (V, E)$, the m-d-MST problem is to find m edge-disjoint spanning trees $T_1, \ldots, T_m \subset E$ with a given diameter $d = d_n \le n/m$, such that minimize their total weight.

To construct an algorithm for finding several edge-disjoint trees with given diameter, it turned out to be convenient to use a new algorithm $\mathcal{A}$ for constructing a single such tree.

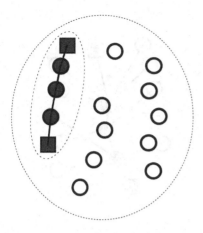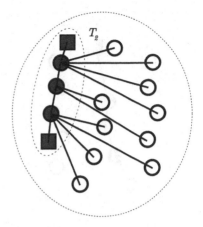

Fig. 4. Example of the work of the Algorithm $\widetilde{\mathcal{A}}$ on 15-vertex complete graph, $d = 4$. Constructing of the spanning tree T_2.

4.1 Algorithm $\widetilde{\mathcal{A}}$

The approach is to build trees sequentially. At the beginning of the algorithm, we set $V' = V$ and $E' = E$.

Repeat m times the following Common Stage with four steps:

Common Stage $i = 1, ..., m$

Step 1. Choose in the vertex set V' an arbitrary $(d+1)$-vertex subset V_i.

Step 2. Starting at arbitrary vertex in the subgraph $G(V_i)$, construct a Hamiltonian path P_i of a length $d = d_n$, using the approach "go to the nearest unvisited vertex".

Step 3. The remaining $(n - d - 1)$ vertices of the set V' connect to the nearest $(d - 1)$ inner vertices of the path P_i. This path and the edges connected to it form a spanning tree T_i.

Step 4. Remove the vertices of the path P_i from the vertex set V' and edges of the spanning tree T_i from the edge set E'.
If $i \leq m$ then return to the Step 1. Otherwise, the construction of all m edge-disjoint spanning trees $T_1, ..., T_m$ is completed.

Figures 3, 4 and 5 illustrate the construction by the Algorithm $\widetilde{\mathcal{A}}$ of three edge-disjoint given-diameter $d = 4$ spanning trees on a 15-vertex complete graph.

Figure 3 is the example of constructing first spanning tree T_1.

4.2 Analysis of Algorithm $\widetilde{\mathcal{A}}$

We assume that the weights of the edges are independent and identically distributed random reals with uniform distribution function defined on $[a_n; b_n]$.

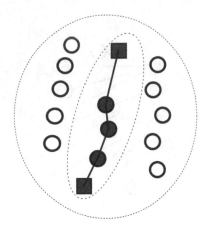 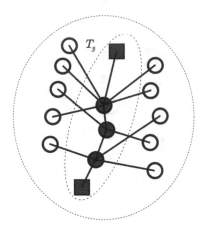

Fig. 5. Example of the work of the Algorithm $\widetilde{\mathcal{A}}$ on 15-vertex complete graph, $d = 4$. Constructing of the spanning tree T_3.

Theorem 2. *Algorithm $\widetilde{\mathcal{A}}$ solves the m-d-MST problem within the running time $\mathcal{O}(n^2)$. On complete directed graphs with n vertices and weights of edges from the $UNI(a_n; b_n)$ it has the failure probability $\delta_{\widetilde{\mathcal{A}}}(n) = \mathcal{O}\left(\frac{m}{n}\right)$ and works asymptotically optimal under conditions $m = o(n)$ and*

$$\frac{b_n}{a_n} = \begin{cases} o(d), & if \ \ln n \le d < \min\left(\frac{n}{m}, \frac{n}{\ln n}\right), \\ o(\frac{n}{\ln n}), & if \ \frac{n}{\ln n} \le d < \frac{n}{m}. \end{cases} \qquad (8)$$

Proof. The time complexity of Algorithm $\widetilde{\mathcal{A}}$ consists of the m-fold repeating the Steps 1–2 (takes $\mathcal{O}(d^2)$ time) and the Step 3 (takes $\mathcal{O}((n-d)d)$ running time). So, the total time complexity is $\mathcal{O}(n^2)$, since given $m \le n/d$ it is true $m\left(d^2 + (n-d)d\right) \le n^2$.

Theorem 2 is completely proved.

So we construct the algorithm $\widetilde{\mathcal{A}}$ for arbitrary complete graph (directed or undirected graph), but we prove the property of asymptotic optimality only for directed graph's case. Thus to illustrate the work of algorithm $\widetilde{\mathcal{A}}$ on directed graph we should replace each edge by directed arc in Figs. 3, 4 and 5 according to the order of construction of the tree, from the previous vertex to the next one.

As for the probability of the algorithm failure, the following should be noted. Actually, Algorithm $\widetilde{\mathcal{A}}$ repeats m times the steps of Algorithm $\mathcal{A}$ for finding single spanning tree. But for the case of finding m spanning trees, it is necessary for the algorithm to work correctly every time. This worsens (increased) the estimate of the failure probability of by the factor of m.

Some difference between condition (8) of asymptotic optimality in the case under consideration from similar condition (6) in the case of searching for single spanning tree is caused by taking into account inequality $m \le n/d$.

5   Conclusion

It would be interesting to investigate

(a) The Random d-MST problem on input data with infinite support like exponential or truncated-normal distribution;
(b) The Random Bounded MST problem on discrete distributions;
(c) Asymptotic optimality of the problem of finding several edge-disjoined spanning trees with a given or bounded diameter on undirected graph.

References

1. Angel, O., Flaxman, A.D., Wilson, D.B.: A sharp threshold for minimum bounded-depth and bounded-diameter spanning trees and Steiner trees in random networks. arXiv:0810.4908v2 [math.PR] (2011)
2. Cooper, C., Frieze, A., Ince, N., Janson, S., Spencer, J.: On the length of a random minimum spanning tree. Comb. Probab. Comput. **25**(1), 89–107 (2016)
3. Frieze, A.: On the value of a random MST problem. Discrete Appl. Math. **10**, 47–56 (1985)
4. Garey, M.R., Johnson, D.S.: Computers and Intractability. Freeman, San Francisco (1979)
5. Gimadi, E.K., Glebov, N.I., Perepelitsa, V.A.: Algorithms with estimates for discrete optimization problems. Problemy Kibernetiki **31**, 35–42 (1975) (in Russian)
6. Gimadi, E.K., Istomin, A., Shin, E.: On algorithm for the minimum spanning tree problem bounded below. In: Proceedings of DOOR 2016, Vladivostok, Russia, 19–23 September 2016, vol. 1623, pp. 11–17. CEUR-WS (2016)
7. Gimadi, E.K., Istomin, A.M., Shin, E.Y.: On given diameter MST problem on random instances. In: CEUR Workshop Proceedings 2018, pp. 159–168 (2019)
8. Gimadi, E.K., Serdyukov, A.I.: A probabilistic analysis of an approximation algorithm for the minimum weight spanning tree problem with bounded from below diameter. In: Inderfurth, K., Schwödiauer, G., Domschke, W., Juhnke, F., Kleinschmidt, P., Wäascher, G. (eds.) Operations Research Proceedings 1999. ORP, vol. 1999, pp. 63–68. Springer, Heidelberg (2000). https://doi.org/10.1007/978-3-642-58300-1_12
9. Gimadi, E., Shevyakov, A., Shin, E.: Asymptotically optimal approach to a given diameter undirected MST problem on random instances. In: Proceedings of 15-th International Asian School-Seminar "Optimization Problems of Complex System" (OPCS-2019), Novosibirsk, Akademgorodok, Russia, 26–30 August 2019, pp. 48–52. IEEE Xplore (2019). https://doi.org/10.1109/OPCS.2019.8880223
10. Gimadi, E.K., Shin, E.Y.: Probabilistic analysis of an algorithm for the minimum spanning tree problem with diameter bounded below. J. Appl. Ind. Math. **9**(4), 480–488 (2015). https://doi.org/10.1134/S1990478915040043
11. Gimadi, E.K., Shin, E.Y.: On given diameter MST problem on random input data. In: Bykadorov, I., Strusevich, V., Tchemisova, T. (eds.) MOTOR 2019. CCIS, vol. 1090, pp. 30–38. Springer, Cham (2019). https://doi.org/10.1007/978-3-030-33394-2_3
12. Petrov, V.V.: Limit Theorems of Probability Theory. Sequences of Independent Random Variables. Clarendon Press, Oxford, 304 p. (1995)

Method of Parametric Correction in Data Transformation and Approximation Problems

Victor Gorelik[1,2]([envelope]) [iD] and Tatiana Zolotova[3] [iD]

[1] CC FRC CSC RAS, Vavilova Street 40, 119333 Moscow, Russia
vgor16@mail.ru
[2] Moscow State Pedagogical University,
M. Pirogovskaya Street 1/1, 119991 Moscow, Russia
[3] Financial University under the Government of RF,
Leningradsky Prospekt 49, 125993 Moscow, Russia
tgold11@mail.ru

Abstract. The linear regression analysis problems are considered under the assumption of the presence of noise in the output and input variables. To estimate the measure of approximation of the initial data, the l_1 metric is used, which has a probabilistic rationale as a maximum likelihood method for two-sided exponential noise distribution. This approximation problem can also be interpreted as an improper (has no solution) interpolation problem, for which it is required to change (correct) optimally the positions of the points so that they all lie on the same hyperplane. In addition, the case of preliminary linear data transformation is considered, while the original information matrix is subject to correction. Therefore, the arising problems belong to the new class of parametric correction. It is shown that these approximation (corrections) problems can be reduced to a set of a finite number of linear programming problems.

Keywords: Data processing · Regression analysis · Parametric correction · Maximum likelihood method · Two-sided exponential distribution

1 Introduction

The task of regression analysis is to select a function that, in a certain sense, accurately describes the given experimental table data. Since an exact description is, as a rule, impossible, it can be interpreted as an improper (without solution) interpolation problem. The latter consists in constructing a function $f : X \to Y$ from some class Φ such that the initial data $(x^1, y^1), \ldots, (x^m, y^m)$ satisfy the system of equations

$$y^i = f(x^i), \ i = 1, ..., m, \ f \in \Phi. \tag{1}$$

Geometrically this means that the surface described by this function passes through the given points. Since the data are usually obtained experimentally

N. Olenev et al. (Eds.): OPTIMA 2020, LNCS 12422, pp. 122–133, 2020.
https://doi.org/10.1007/978-3-030-62867-3_10

and include measurement inaccuracies or noise, this problem, as a rule, has no solution for a fixed class of functions. In this case, the optimal correction (approximation) problem is considered: it is required to find a function that satisfies the condition (1) for some other data $[X_H, y_h]$, and this data is in some sense closest to the original $[X, y]$. This problem is formalized by introducing the proximity measure as some matrix norm and finding the correction matrix $A = [X_H, y_h] - [X, y]$, which is minimal in the sense of this norm.

The methods of matrix correction (based on the total set of the initial data) are now widely used in solving incompatible and unstable systems of linear algebraic equations and inequalities and improper linear programming problems (for example, [1–5]). The use of correction methods in problems of processing experimental data in the presence of noise when measuring not only output but also input variables is the content of the total least squares method (see [6–8]). This method has a probabilistic justification as the maximum likelihood method in the case of the normal distribution of errors both in measuring the vector argument x and the values of the function y.

Often when modeling stochastic processes in complex systems, particularly in economics and finance, the distribution of random variables does not correspond to the normal law. This is due to the phenomenon, that the kurtosis value is greater for statistical distributions based on real data than for the normal law. Such distributions of random variables have "heavy tails", i.e. the corresponding distribution density decreases more slowly as compared to the normal one under $|x| \to \infty$. This is confirmed both by the type of empirical densities and by standard statistical criteria. It is generally accepted that a distribution has a heavy right tail if the probability that a random variable exceeds a sufficiently large x has a value of the order of x^{-a} (for example, the distribution of Student or Pareto [9]).

The least squares method (including the total one) loses its probabilistic justification if the noise distribution in the statistical data differs from the normal one. In [10], examples of real financial indicators are given that are well described by the two-sided Laplace distribution (note that classically under the name of Laplace a one-sided exponential distribution is conceived). This distribution does not have a heavy tail in the sense indicated above, but it nevertheless decreases more slowly compared to the normal law. It was shown in [11] that the maximum likelihood method, using the hypothesis of an exponential noise distribution, leads as an approximation measure to the polyhedral norm l_1 of the matrix (by which we mean the sum of the moduli of its entries, that is, the matrix is considered as "expanded" vector).

Particularly for the class of affine functions a geometric interpretation means that in the space R^{n+1} it is required to find such a hyperplane

$$L : y = \langle a, x \rangle + b,$$

which is located closest among all hyperplanes to given points in the sense of the sum of distances in the metric l_1.

It is known, that the solution of the matrix correction problem in accordance with the norm l_1 can be reduced to solving a number of linear programming problems (various versions of the corresponding methods are contained, for example, in [11–13]).

In this paper, this result is generalized to the problem of constructing a linear regression for data that must be linearly transformed at first. This leads to new more complex tasks of correction (approximation) and the need to develop methods of parametric correction. It is shown that such problems can also be reduced to linear programming (LP). The developed methods are applied to a practical task with real data. A widespread procedure preceding the construction of regression is data transformation, which consists in optimizing their representations and formats (normalization, standardization, etc.).

Normalization is a procedure for preprocessing the input data, during which the values of the attributes (explanatory variables) in the training or test samples are reduced to a certain specified range (for example, [0,1]). Normalization is necessary when the input vector contains values that differ from each other by several orders. Normalization can be done by linear transformation with multiplication of the information matrix X on the right by a diagonal matrix, consisting of various scaling factors, and the addition of a matrix that changes the origin.

Standardization is a data transformation with a Gaussian distribution that subtracts the average value and divides the result by the standard deviation of the data sample. This results in a data transformation with an average value of zero and a standard deviation of 1. The resulting distribution is called the standard Gaussian distribution or the standard normal distribution. However, a similar transformation can be applied to other laws of data distribution, for example, to the Laplace distribution.

Using suitable linear transformations, you can also change the properties of multi-dimensional distributions. So, if the data presented by the matrix X are correlated, and Σ is the covariance matrix, then to eliminate the correlation of the data, we can use the decomposition $\Sigma = AA^T$ and the transformation $A^{-1}X$, which leads to the identity covariance matrix.

Considerable attention in the modern literature on regression analysis is given to the problem of outliers – atypical data in the sample (see, for example, [14]). Such data can also be smoothed out using linear transformations.

Thus, many (but certainly not all) data transformation procedures can be represented as a linear transformation of the information matrix, i.e. multiplying it on the left or right by a specially selected matrix of the corresponding dimension, as well as adding the corresponding matrix. It turns out that the choice of the option of multiplying the information matrix affects the way to solve the resulting optimization problem, therefore, they are further considered separately.

2   The Generalized Problem of Constructing a Linear Regression in the Metric l_1 with Left-Sided Multiplication of the Information Matrix

So, the mathematical formulation of the regression problem is as follows. The dependence of the numerical value of the variable y, reflecting a certain property of a process or phenomenon on some factors or features x is given by a set of points $y^i \in R$, $x^i \in R^n$, $i = 1, \ldots, m$. We introduce the information matrix

$$Z = \begin{pmatrix} -y^1 & x_1^1 & \cdots & x_n^1 \\ -y^2 & x_1^2 & \cdots & x_n^2 \\ \cdots & \cdots & \cdots & \cdots \\ -y^m & x_1^m & \cdots & x_n^m \end{pmatrix}.$$

It is assumed that observation errors (noise) occur both in the input data (explanatory variables) and in the output data (explained variable). Therefore, we will subject all entries of this matrix to a linear transformation and an additive correction. As special cases, it is not difficult to consider correction (approximation) options only according to input or output data.

In this section, we will consider the option of the initial data linear transformation of the form

$$D(Z) = D_0 + D_1 Z, \tag{2}$$

where Z and D_0 are the matrices of dimension $m \times (n + 1)$, D_1 - the non-degenerate matrix of dimension $m \times m$.

The use of the linear dependence hypothesis (explained by explanatory variables) supposes that after transformation (2), all points must lie on one hyperplane, i.e. satisfy the matrix equation

$$D(Z)\bar{a} = -be, \tag{3}$$

where $e = (1, 1, \ldots, 1)^T \in R^m$, $\bar{a} = (a_0, a_1, \ldots, a_n)^T$, $a_0 = 1$, $b \in R$.

Due to the presence of noise in the initial data, the overdetermined system of Eq. (3) with respect to unknown $\bar{a}, b$ is, as a rule, incompatible. The approximation problem is equivalent to the minimal correction of the initial data, at which the given system becomes compatible.

If it is necessary to preprocess the initial data, the procedure for constructing a regression usually consists of two stages. First, the initial information matrix is transformed, and then the approximation problem is solved for the obtained information matrix. However, the hypothesis of the presence of noise with a certain distribution law applies to the initial data obtained experimentally. Therefore, an idea is proposed consisting in the joint use of transformation and approximation procedures. Note that the presence of the free term b in the class of affine functions essentially means the introduction of an additional dummy variable that takes the value 1. Therefore, this value cannot be corrected.

We formulate the statement of the data correction problem, for which direct correction of the entries of the matrix Z is possible, and the entries of the matrix D undergo a change according to (2).

Here, as a minimized criterion for the correction value, we will use the matrix norm l_1. Denote the correction matrix of the entries of the matrix Z by H. This matrix must satisfy the condition $D_0\bar{a} + D_1(Z - H)\bar{a} = -eb$ or

$$H\bar{a} = D_1^{-1}D_0\bar{a} + Z\bar{a} + D_1^{-1}eb. \qquad (4)$$

Consider the problem of finding the minimum matrix H by norm l_1 for which the system of Eq. (4) is compatible:

$$\min_{\bar{a},b,H}\{\|H\|_1 \mid H\bar{a} = D_1^{-1}D_0\bar{a} + Z\bar{a} + D_1^{-1}eb\}, \qquad (5)$$

where $\|H\|_1$ is the norm l_1 of the matrix H.

In what follows, we need the notation: $(\cdots)_i$ - the i-th component of the vector in brackets.

Theorem 1. *The problem (5) has a solution if and only if there is such a solution (u^*, v^*, w^*, j^*) to the problem*

$$\min_{u,v,w,j}\{\langle e, u\rangle | u \geq \pm(D_1^{-1}D_0 v + Zv + D_1^{-1}ew),$$
$$u \geq 0, \quad |v_i| \leq 1, \quad i = 0, ..., n, \quad |v_j| = 1\}, \qquad (6)$$

that $v_0^* \neq 0$. Then

$$\bar{a}^* = \frac{v^*}{v_0^*}, \qquad b^* = \frac{w^*}{v_0^*}$$

is a solution to the problem (5), the entries of the correction matrix H^* are determined by the formulas

$$h_{ij^*}^* = \frac{(D_1^{-1}D_0\bar{a}^* + Z\bar{a}^* + D_1^{-1}eb^*)_i}{\max_{0\leq j\leq n}|\bar{a}_j^*|}, \quad h_{ij}^* = 0, j \neq j^*, \quad i = 1, ..., m, \qquad (7)$$

and $\|H^*\|_1 = \langle e, u^*\rangle$.

Proof. The minimum correction matrix H according to the norm l_1 for fixed $\bar{a}, b$ is determined, obviously, by the formulas

$$|h_{ij_0}| = \frac{|(D_1^{-1}D_0\bar{a} + Z\bar{a} + D_1^{-1}eb)_i|}{\max_{0\leq j\leq n}|\bar{a}_j|}, \quad h_{ij} = 0, \ j \neq j_0, \ i = 1, ..., m, \qquad (8)$$

where index $j_0 = \arg\max_{0\leq j\leq n}|\bar{a}_j|$.

Indeed, the i-th component of the vector $H\bar{a}$ of the left side of the formula (4) is a linear function of the entries of the i-th row of the matrix H, and the coefficients are the components of the vector $\bar{a}$. This linear function is equal to some constant, i.e. satisfies one equality type constraint. The minimum sum of the moduli of the entries of the i-th row of the matrix H is achieved when all of them are equal to zero except for the entry with the largest modulus coefficient.

Moreover, this element is equal to the ratio of this constant to the modulus of the corresponding coefficient, and its sign is determined by the sign of the constant.

Thus, for fixed $\bar{a}, b$ the minimum according to the norm l_1 correction matrix H satisfying (4) has one nonzero column, whose moduli of entries are calculated by formulas (8). The task of minimal correction is reduced to minimizing the sum of the modules of these entries of the matrix H with respect to $\bar{a}, b$.

We introduce scalar variables r, w and the vector v

$$r = (\max_{0 \le j \le n} |\bar{a}_j^*|)^{-1}, \quad w = rb, \quad v = r\bar{a}.$$

The components of the vector v must satisfy the conditions $-1 \le v_j \le 1$, and there exists an index j_0 such that $v_{j_0} = 1$ or -1. We also introduce the variables u_i, which are no less than the expressions on the right-hand side of formulas (8) and the m-dimensional vector $u = (u_1, ..., u_m)$. In the new variables we get the problem of mathematical programming

$$\min_{u,v,w} \{\langle e, u \rangle | u \ge \pm (D_1^{-1} D_0 v + Zv + D_1^{-1} ew), \quad u \ge 0,$$
$$-1 \le v_i \le 1, \; i = 0, ..., n, \; v_j = 1 \lor v_j = -1\}. \tag{9}$$

This is a linear, partially integer programming problem. It can be reduced to solving standard LP problems. To do this, we sequentially set $v_j = \pm 1$ for $j = 0, \ldots, n$ and choose from the resulting LP problems that which gives the smallest value of the criterion $\langle e, u \rangle = \sum_{i=1}^m u_i$. Thus, we arrive at the problem (6).

By the found solution to this problem v^* and w^* taking into account the requirement $a_0 = 1$ we find the vector $\bar{a}^* = \frac{v^*}{v_0^*}$ and the scalar $b^* = \frac{w^*}{v_0^*}$ (in the case $v_0^* \ne 0$) and according to formulas (7) the entries of the correction matrix (the sign of the entry is determined by the sign of the expression inside the module). In the case $v_0^* = 0$ the problem (5) has no solution. Indeed, suppose the contrary, i.e. that problem (5) has a solution. Then by inverse transformations of this solution we construct a solution to the problem (6). Moreover, by the conditions of the problem (6), there exists $v_j^* \ne 0$, therefore, $v_0^* \ne 0$. The theorem is proved.

Note that, due to the uniformity of constraints in (9), for values $v_j = \pm 1$ the solutions are identical. Therefore, it suffices to solve only $n + 1$ LP problems.

3 The Generalized Problem of Constructing a Linear Regression in the Metric l_1 with Right-Sided Multiplication of the Information Matrix

In this section, we will consider the option of linear transformation of the initial data of the form

$$D(Z) = D_0 + ZD_2, \tag{10}$$

where Z and D_0 are the matrices of dimension $m \times (n + 1)$, D_2 - the non-degenerate matrix of dimension $(n + 1) \times (n + 1)$.

The linear dependence hypothesis (explained by explanatory variables) supposes that after transformation (10) all points must lie on one hyperplane, i.e. satisfy the matrix Eq. (3).

We formulate the statement of the data correction problem, for which direct correction of the entries of the matrix Z is possible, and the entries of the matrix D undergo a change according to (10).

Here we will also use the matrix norm l_1 as a minimized criterion for the correction value. The correction matrix H of the entries of the matrix Z must satisfy the condition $D_0\bar{a} + (Z - H)D_2\bar{a} = -eb$ or

$$HD_2\bar{a} = D_0\bar{a} + ZD_2\bar{a} + eb. \tag{11}$$

Consider the problem of finding the minimum matrix H by norm l_1 for which the system of Eqs. (11) is compatible:

$$\min_{\bar{a},b,H}\{\|H\|_1 \mid HD_2\bar{a} = D_0\bar{a} + ZD_2\bar{a} + eb\}. \tag{12}$$

In this case the technique of multiplication by the inverse matrix used in the previous section does not work.

Theorem 2. *The problem (12) has a solution if and only if there is such a solution (u^*, v^*, w^*, j^*) to the problem*

$$\min_{u,v,w,j}\{\langle e, u\rangle | u \geq \pm(D_0D_2^{-1}v + Zv + ew),$$
$$u \geq 0,\ |v_i| \leq 1,\ i = 0, ..., n,\ |v_j| = 1\}, \tag{13}$$

that $v_0^* \neq 0$. Then

$$\bar{a}^* = \frac{D_2^{-1}v^*}{v_0^*}, \qquad b^* = \frac{w^*}{v_0^*}$$

is a solution to the problem (13), the entries of the correction matrix H^* are determined by the formulas

$$h_{ij^*}^* = \frac{(D_0\bar{a}^* + ZD_2\bar{a}^* + eb^*)_i}{\max_{0\leq j\leq n}|(D_2\bar{a}^*)_j|},\ h_{ij}^* = 0,\ j \neq j^*,\ i = 1, ..., m, \tag{14}$$

and $\|H^*\|_1 = \langle e, u^*\rangle$.

Proof. We will substitute the variables: $\tilde{a} = D_2\bar{a}$. In the new variables we obtain the equation

$$H\tilde{a} = D_0D_2^{-1}\tilde{a} + Z\tilde{a} + eb.$$

The minimum correction matrix H according to the norm l_1 for fixed $\tilde{a}, b$ is determined, obviously, by the formulas

$$|h_{ij_0}| = \frac{|(D_0D_2^{-1}\tilde{a} + Z\tilde{a} + eb)_i|}{\max_{0\leq j\leq n}|\tilde{a}_j|},\ h_{ij} = 0,\ j \neq j_0,\ i = 1, ..., m, \tag{15}$$

where index $j_0 = \arg\max\limits_{0 \le j \le n} |\tilde{a}_j|$.

We introduce scalar variables r, w and the vector v

$$r = (\max\limits_{0 \le j \le n} |\tilde{a}_j^*|)^{-1}, \quad w = rb, \quad v = r\tilde{a}.$$

The components of the vector v must satisfy the conditions $-1 \le v_j \le 1$, and there exists an index j_0 such that $v_{j_0} = 1$ or -1. We also introduce the variables u_i, which are no less than the expressions on the right-hand side of formulas (15) and the m-dimensional vector $u = (u_1, ..., u_m)$. In the new variables we get the problem of mathematical programming

$$\min_{u,v,w} \{\langle e, u \rangle \mid u \ge \pm(D_0 D_2^{-1} v + Zv + ew), \ u \ge 0,$$
$$-1 < v_i \le 1, \ i = 0, ..., n, \ v_j = 1 \vee v_j = -1\}.$$

This is a linear, partially integer programming problem. It can be reduced to solving standard LP problems. To do this, we sequentially set $v_j = \pm 1$ for $j = 0, \ldots, n$ and choose from the resulting LP problems that which gives the smallest value of the criterion $\langle e, u \rangle = \sum_{i=1}^{m} u_i$. Thus, we arrive at the problem (13).

By the found solution to this problem v^* and w^* taking into account the requirement $a_0 = 1$ we find the vector $\tilde{a}^* = \frac{v^*}{v_0^*}$ and the scalar $b^* = \frac{w^*}{v_0^*}$. Regression coefficients are determined by the formula $\bar{a} = D_2^{-1}\tilde{a}$, and by formulas (14) – entries of the correction matrix (the sign of the entry is determined by the sign of the expression inside the module). In the case $v_0^* = 0$ the problem (12) has no solution (proof by contradiction as in Theorem 1). The theorem is proved.

Due to the homogeneity of the constraints, it suffices to solve $n + 1$ LP problems.

4   An Example of Data Transformation and Approximations When Building a Demographic Trend

Consider a practical example based on real data taken from the Rosstat website [15]. We will analyze the birth rate in the Federal Regions, depending on two factors: urban population and income.

Table 1 shows data for 2018 for the Federal Regions of the Russian Federation: birth rate, proportion of urban population, population income.

The analysis of indicators shows that the North Caucasus Region stands out among other regions, it has one of the lowest per capita incomes, the lowest percentage of the urban population, and the birth rate is much higher than the values in other regions. Thus, we can assume that in this case we are dealing with the presence of an outlier. On the one hand, this region is an element of the overall picture and should be considered when building a demographic trend; on the other hand, it is very atypical for the Russian Federation. Apparently, here the birth rate is determined by other important factors (traditions, religion, large homesteads, etc.).

Table 1. Analyzed socio-economic indicators for the Russian Regions for 2018.

Federal regions	Total fertility rate (number born per 1000 people population)	Share of urban population (%)	Per capita income (per month)
Central region	10.4	82.2	40.843
Northwestern region	11.1	84.4	33.89
South region	11.1	62.6	26.928
North caucasian region	15	49.8	24.017
Volga region	11.1	71.9	25.87
Ural region	12.6	81.4	32.944
Siberian region	12.3	73.1	23.925
Far eastern region	12.1	75.8	37.07

Using the initial data, we construct a linear regression in the metric l_1 without data transformation, i.e. $D_0 = (0)_{8\times3}$ (zero matrix), and $D_1 = E_{8\times8}$ (identity matrix). The minimum value of the norm of the correction matrix is $\langle e, u \rangle = 5.997$. The regression equation has the form

$$y = -0.116x_1 + 0.006x_2 + 20.563.$$

We exclude the North Caucasus Federal Region and construct a linear regression in the metric l_1 without data transformation, i.e. $D_0 = (0)_{7\times3}$ and $D_1 = E_{7\times7}$. The minimum value of the norm of the correction matrix is $\langle e, u \rangle = 3.980$. The regression equation has the form

$$y = 0.071x_1 - 0.151x_2 + 10.698.$$

Thus, the exclusion of this region completely changes the result: the regression coefficients change signs. This confirms the preliminary consideration about the presence of an outlier.

Using the initial data, we construct a linear regression in the metric l_1 with left-sided multiplication of the information matrix. As matrices of linear data transformation, we take

$$D_0 = \begin{pmatrix} 0 & 0 & 0 \\ 0 & 0 & 0 \\ 0 & 0 & 0 \\ -10 & 49.8 & 20.017 \\ 0 & 0 & 0 \\ 0 & 0 & 0 \\ 0 & 0 & 0 \\ 0 & 0 & 0 \end{pmatrix}, \quad D_1 = \begin{pmatrix} 1 & 0 & 0 & 1/\rho & 0 & 0 & 0 & 0 \\ 0 & 1 & 0 & 1/\rho & 0 & 0 & 0 & 0 \\ 0 & 0 & 1 & 1/\rho & 0 & 0 & 0 & 0 \\ 0 & 0 & 0 & 1/\rho & 0 & 0 & 0 & 0 \\ 0 & 0 & 0 & 1/\rho & 1 & 0 & 0 & 0 \\ 0 & 0 & 0 & 1/\rho & 0 & 1 & 0 & 0 \\ 0 & 0 & 0 & 1/\rho & 0 & 0 & 1 & 0 \\ 0 & 0 & 0 & 1/\rho & 0 & 0 & 0 & 1 \end{pmatrix},$$

where $\rho = 84.4$. Then

$$D_0 + D_1 Z = \begin{pmatrix} -10.58 & 82.79 & 41.13 \\ -11.28 & 84.99 & 34.17 \\ -11.28 & 63.19 & 27.21 \\ -10.18 & 50.39 & 24.30 \\ -11.28 & 72.49 & 25.15 \\ -12.78 & 81.99 & 33.23 \\ -12.48 & 73.69 & 24.21 \\ -12.28 & 76.39 & 37.35 \end{pmatrix}.$$

The minimum value of the norm of the correction matrix is $\langle e, u \rangle = 4.429$. The regression equation has the form

$$y = 0.098x_1 - 0.165x_2 + 9.247.$$

Using the initial data, we construct a linear regression in the metric l_1 with right-sided multiplication of the information matrix. As matrices of linear data transformation, we take

$$D_2 = \begin{pmatrix} 10 & 0 & 0 \\ 0 & 1 & 0 \\ 0 & 0 & 1 \end{pmatrix}, \quad D_0 = \begin{pmatrix} 100 & 0 & 0 \\ 100 & 0 & 0 \\ 100 & 0 & 0 \\ 146 & 0 & 0 \\ 100 & 0 & 0 \\ 100 & 0 & 0 \\ 100 & 0 & 0 \\ 100 & 0 & 0 \end{pmatrix}.$$

Then

$$D_0 + Z D_2 = \begin{pmatrix} -4 & 82.2 & 40.84 \\ -11 & 84.4 & 33.89 \\ -11 & 62.6 & 26.93 \\ -4 & 49.8 & 24.02 \\ -11 & 71.9 & 25.87 \\ -26 & 81.4 & 35.94 \\ -23 & 73.1 & 23.93 \\ -21 & 75.8 & 37.07 \end{pmatrix}.$$

The minimum value of the norm of the correction matrix is $\langle e, u \rangle = 4.143$. The regression equation has the form

$$y = x_1 - 1.926x_2 + 0.113.$$

The results of calculation show that data smoothing allows you to save the general view of the trend and, at the same time, improve the value of the approximation criterion, while the complete exclusion of the outlier point radically changes the form of the trend. Of course, the choice of linear transformation is empirical, which makes it difficult to compare objectively with other approaches, which are also essentially heuristic (see, for example, [14]). It is necessary to find a compromise between the requirements of stability of the general form of the trend and a decrease in the approximation measure.

5   Conclusion

This paper considered the improper interpolation problem and the approach to solving it, based on the linear transformation of the information matrix and the matrix correction of the resulting system of linear equations.

As an approximation measure, we used the polyhedral norm of the matrix l_1, which has a probabilistic justification as the maximum likelihood method for a two-sided exponential distribution of noise in the input and output data. The proposed method leads to the solution of LP problems. Note that this approach is computationally simpler than the total least squares method, which leads to more complicated computational problems of finding eigenvalues and vectors or singular-value decomposition of large-dimensional matrices even without data transformation (how to combine it with data transformation is still unclear).

Note that the option of transformation data with right-hand multiplication by a matrix allows us to normalize each column of the initial information matrix separately. So, it is convenient when measuring data with very different ranges of values. The option of transformation data with left-sided multiplication by a matrix allows to normalize each row of the initial information matrix separately, i.e. solve the problem of outliers. On the other hand, with the appropriate transformation, you can also give more weight to points with a large deviation from the regression line, i.e. achieve the effect that the use of the l_2 norm gives.

References

1. Eremin I.I.: Theory of Linear Optimization. Inverse and Ill-Posed Problems Series. VSP, Utrecht, Boston, Koln, Tokyo (2002)
2. Gorelik, V.A.: Matrix correction of a linear programming problem with inconsistent constraints. Comput. Math. Math. Phys. **11**(41), 1615–1622 (2001)
3. Gorelik, V.A., Erohin, V.I.: Optimal Matrix Correction of Inconsistent Systems of Linear Algebraic Equations by Minimal Euclidean Norm. CC RAS, Moscow (2004)
4. Gorelik, V.A., Murav'eva, O.V.: Methods of Correction of Improper Problems and Their Application to Optimization and Classification. CC RAS, Moscow (2012)
5. Erokhin, V.I., Krasnikov, A.S., Khvostov, M.N.: Matrix corrections minimal with respect to the Euclidean norm for linear programming problems. Autom. Remote Control **2**(73), 219–231 (2012)
6. Back, A.: The matrix-restricted total least squares problem. Signal Process **87**(10), 2303–2312 (2007)
7. Markovsky, I.: Bibliography on total least squares and related methods. Stat. Inter. **2**, 1–6 (2010)
8. Markovsky, I., Van Huffel, S.: Overview on total least squares methods. Signal Process. **87**, 2283–2302 (2007)
9. Shiryaev, A.N.: Essentials of Stochastic Finance: Facts, Models, Theory. World Scientific Publishing Co., Pte. Ltd., Singapore (1999)

10. Gorelik, V.A., Zolotova, T.V.: Issues the formation of an optimal stock portfolio of Russian companies with a probabilistic risk function. Financ. J. **3**, 35–44 (2016). Research financial institute
11. Gorelik, V.A., Trembacheva (Barkalova) O.S.: Solution of the linear regression problem using matrix correction methods in the l_1 metric. Comput. Math. Math. Phys. **2**(56), 200–205 (2016)
12. Osborne, M., Watson, J.: An analysis of the total approximation problem in separable norms, and an algorithm for the total l_1 problem. SIAM J. Sci. Stat. Comput. **2**(6), 410–424 (1985)
13. Rosen, J.B., Park, H., Glick, J.: Total least norm formulation and solution for structured problems. SIAM J. Matrix Anal. Appl. **1**(17), 110–128 (1996)
14. Shibzukhov, Z.M.: On the principle of empirical risk minimization based on averaging aggregation functions. Doklady Math. **96**(2), 494–497 (2017). https://doi.org/10.1134/S106456241705026X
15. Regions of Russia. Socio-economic indicators (2018). https://www.gks.ru/storage/mediabank/Reg-pok18.pdf. Accessed 30 Oct 2019

On the Use of Decision Diagrams for Finding Repetition-Free Longest Common Subsequences

Matthias Horn[1]([✉]), Marko Djukanovic[1], Christian Blum[2], and Günther R. Raidl[1]

[1] Institute of Logic and Computation, TU Wien, Vienna, Austria
{horn,djukanovic,raidl}@ac.tuwien.ac.at
[2] Artificial Intelligence Research Institute (IIIA-CSIC),
Campus UAB, Bellaterra, Spain
christian.blum@iiia.csic.es

Abstract. We consider the repetition-free longest common subsequence (RFLCS) problem, where the goal is to find a longest sequence that appears as subsequence in two input strings and in which each character appears at most once. Our approach is to transform a RFLCS instance to an instance of the maximum independent set (MIS) problem which is subsequently solved by a mixed integer linear programming solver. To reduce the size of the underlying conflict graph of the MIS problem, a relaxed decision diagram is utilized. An experimental evaluation on two benchmark instance sets shows the advantages of the reduction of the conflict graphs in terms of shorter total computation times and the number of instances solved to proven optimality. A further advantage of the created relaxed decision diagrams is that heuristic solutions can be effectively derived. For some instances that could not be solved to proven optimality, new state-of-the-art results were obtained in this way.

Keywords: Repetition-free longest common subsequence · Decision diagram · Maximum independent set

1 Introduction

The *longest common subsequence* (LCS) problem asks for the longest string which is a subsequence of a set of input strings. A subsequence is a string that can be obtained from another string by possibly deleting characters. For instance a longest common subsequence of the two input strings ABCDBA and ACBDBA is ABDBA. The LCS problem has applications in bioinformatics, where strings often

This project is partially funded by the Doctoral Program "Vienna Graduate School on Computational Optimization", Austrian Science Foundation (FWF) Project No. W1260-N35. Furthermore, this work was also supported by project CI-SUSTAIN funded by the Spanish Ministry of Science and Innovation (PID2019-104156GB-I00).

N. Olenev et al. (Eds.): OPTIMA 2020, LNCS 12422, pp. 134–149, 2020.
https://doi.org/10.1007/978-3-030-62867-3_11

represent segments of RNA or DNA [13, 16, 18]. Other fields where the LCS problem appears are text editing [17], data compression, file comparison [2, 19], and the production of circuits in field programmable gate arrays [8]. If the number of input strings m is constant, the problem is solvable by *dynamic programming* (DP) in $O(n^m)$ time, where n is the length of the longest input string [13]. Otherwise, if the number of input strings is arbitrary, the problem is $\mathcal{NP}$-hard. An additional constraint which arises in the context of gene duplication in the domain of genome rearrangement and which we consider in this work is that each character may appear in a *common subsequence* (CS) at most once. This problem, first introduced by Adi et al. [1] and denoted as the *repetition-free LCS* (RFLCS) problem, is usually considered for two input strings and is even then APX-hard [1].

This work builds upon the work of Blum et al. [6], where instances of the RFLCS problem are transformed to instances of the *maximum independent set* (MIS) problem. Hereby, an independent set of the underlying conflict graph of the MIS problem corresponds to a *repetition-free common subsequence* (RFCS) of the RFLCS instance. To solve the MIS problem the *integer linear programming* (ILP) solver CPLEX is applied. The performance of the ILP solver depends to a large extent on the size of the conflict graph. Therefore, in [6] the size of the conflict graph is reduced by filtering redundant nodes based on lower and upper bounds. This boosts the range of instances that can be solved to optimality as well as the quality of heuristic solutions obtained for larger instances. In this way, numerous new state-of-the-art results were obtained.

Contributions. To reduce the size of the conflict graph even further we compile a relaxed *multivalued decision diagram* (MDD) for the RFLCS problem, yielding a performance improvement of the subsequently applied ILP solver. In the last decade, *decision diagrams* (DDs) have been recognized as a powerful tool for combinatorial optimization problems; see [3] for a comprehensive survey. In particular, *relaxed DDs* may provide compact representations of discrete relaxations. Besides allowing for new interference techniques in constraint programming and novel branching schemes, they may also provide tight dual bounds. In case of the RFLCS problem it is further possible to effectively derive heuristic solutions directly from the relaxed MDD. This has the advantage that if the ILP solver is not able to solve an instance to proven optimality within a given time limit then the compiled relaxed MDD may be able to provide a tighter upper bound as the ILP solver does and/or may be able to deliver a better heuristic solution.

After an overview of related work in Sect. 2 we give a formal problem definition in Sect. 3. The MIS problem and a corresponding ILP model are described in Sect. 4. Decision diagrams for the RFLCS are introduced in Sect. 5, and Sect. 6 describes the incremental refinement of relaxed MDDs. Section 7 provides experimental results, showing that the suggested approach yields to a performance improvement in terms of average computation times, number of instances solved to optimality, and average solution quality.

2   Related Work

As already mentioned, the current work builds upon the approach of Blum et al. [6]. Besides the RFLCS, Blum et al. also consider the *longest arc-preserving common subsequence* (LAPCS) problem [15], where additional dependencies among characters must be respected in a solution, as well as the *longest common palindromic subsequence* (LCPS) problem [11], where the resulting sequence must also be a palindrome. All these LCS variants where solved by transforming instances to instances of the MIS problem. Moreover, the equivalent maximum clique problem of the complement of the conflict graph is solved heuristically by the LSCC-BMS solver as well as exactly by the LMC solver. Both solvers are currently among the leading solvers for the maximum clique problem.

In the literature LCS related problems with additional constraints are well known for almost 40 years and research in that field is still active due the practical relevance and computational difficulties. Besides RFLCS and the already mentioned LAPCS and LCPS problems other considered variants are, for instance, the constrained longest common subsequence problem [20] or the generalized constrained longest common subsequence problem [10]. For further problem variants we refer to survey papers such as [7].

The RFLCS problem in particular was tackled by several heuristic approaches [1,4,9]. The so far best heuristic is a *construct, merge, solve and adapt* (CMSA) metaheuristic combined with beam search as proposed by Blum and Blesa [5]. The authors showed that this approach can outperform other heuristics as well as the CPLEX solver applied to an ILP model of the RFCLS problem.

3   Problem Definition

The RFLCS problem considers a set of two input strings $S = \{s_1, s_2\}$ over a finite alphabet Σ. The goal is to find the longest subsequence which is common for both input strings s_1 and s_2 such that there is no character which occurs more than once. The character at position i is denoted by $s[i]$. A matching $m = (m_1, m_2)$ is a pair of positions s.t. $s_1[m_1] = s_2[m_2]$ and the corresponding character is denoted by $c(m) = s_1[m_1]$. Hence, the character $c(m)$ of a matching m is a possible candidate to appear in a common sequence (CS). A matching m *dominates* a matching n, denoted as $m \succeq n$, if $m_1 \leq n_1 \wedge m_2 \leq n_2$, meaning that in a possible CS $c(m)$ may appear before $c(n)$. Therefore, a CS can be represented by a sequence of matchings $(m_1, m_2, \ldots)$ s.t. $c(m_1), c(m_2), \ldots$ maps to the CS and each matching of the sequence dominates each subsequent matching of the sequence. This observation is important since relaxed MDDs for the RFLCS problem will encode such sequences of matchings. If for two matchings m and n neither $m \succeq n$ nor $n \succeq m$ holds then $c(m)$ and $c(n)$ cannot appear together in a CS which will be henceforth referred to as m and n are *in conflict*, denoted as $n \curlyvee m$. Figure 1a shows an example of a RFLCS instance with input strings $s_1 = $ ABCDBA and $s_2 = $ ACBDBA and an optimal solution of ACDB. In this example, matching m_1 dominates matching m_2 and m_3 whereas matching m_2 is in conflict with matching m_3.

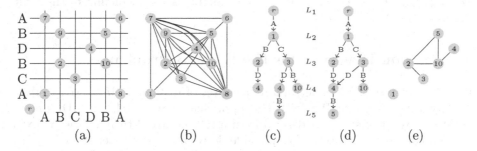

(a)	(b)	(c)	(d)	(e)

Fig. 1. (a) Example of a RFLCS instance with input strings $s_1 = $ ABCDBA and $s_2 = $ ACBDBA. Gray circles correspond to the matchings $M = \{m_1, \ldots, m_{10}\}$ of the instance. (b) Corresponding MIS instance. (c) Exact MDD $\mathcal{D}_M$ with mat$(\mathcal{D}_M) = \{m_1, \ldots, m_5, m_{10}\}$. The state of each node u is partially indicated by $m(u)$. (d) Relaxed MDD where nodes associated with matching m_4 in layer L_3 are merged. (e) MIS instance obtained from matchings mat$(\mathcal{D}_M)$.

4   Integer Linear Program and Independent Set Model

An instance of the RFLCS problem can be solved by transforming it into an instance of the MIS problem. Thereby, each matching corresponds to a node of the underlying conflict graph of the MIS problem. An edge is added between two nodes if the corresponding matchings are in conflict or they refer to the same character; see Fig. 1b for an example. A solution of the MIS instance corresponds to a solution of the RFLCS instance and vice versa, since only matchings are selected that are not in conflict with each other and can therefore appear in the same CS and for each character there is at most one matching selected. The resulting CS can be derived from the set of selected matchings by a topological sort considering the domination relationship.

We solve the MIS instance by a corresponding ILP model. Let M be the set of all matchings of the RFLCS instance and thus nodes of the MIS instance. We use a binary decision variable x_m for each matching $m \in M$ indicating whether the matching is selected (=1) for the solution or not (=0). The model is:

$$\text{ILP}(M) = \max \quad \sum_{x_m \in M} x_m \tag{1a}$$

$$\text{s.t.} \quad x_m + x_n \leq 1 \qquad\qquad m, n \in M : m \curlyvee n \tag{1b}$$

$$\sum_{x_m \in M_a} x_m \leq 1 \qquad\qquad a \in \Sigma \tag{1c}$$

$$x_m \in \{0, 1\} \qquad\qquad m \in M \tag{1d}$$

The number of selected matchings is maximized. Inequalities (1b) ensure that the CS constraints are satisfied, i.e., no conflicting matchings are selected together. The *repetition-free* (RF) constraint is realized by Inequalities (1c), where set $M_a = \{m \in M \mid c(m) = a\}$ contains all matchings corresponding to the same character $a \in \Sigma$.

5   Decision Diagrams for the RFLCS Problem

We use a relaxed MDD to derive a reduced set of matchings $M' \subseteq M$ to subsequently solve the model ILP(M'). Our approach compiles relaxed MDDs in an iterative way s.t. if set M' is derived from a relaxed MDD then another relaxed MDD is compiled w.r.t. set M' to possible derive an even smaller set $M'' \subseteq M'$. This procedure is repeated until some termination criterion is fulfilled.

A MDD w.r.t. a set of matchings M is a directed acyclic multi-graph $\mathcal{D}_M = (V, A)$ with one root node $\mathbf{r}$. All nodes are partitioned into at most $|\Sigma| + 1$ layers $L_1, \ldots, L_{|\Sigma|+1}$, where L_i, $i > 0$ contains only nodes that are reachable from $\mathbf{r}$ over exactly $i - 1$ arcs and L_1 is a singleton containing only $\mathbf{r}$. An arc $\alpha = (u, v) \in A(\mathcal{D}_M)$ is always directed from a source node u in some layer L_i to a target node v in a subsequent layer L_{i+1}. Each arc α is associated with a matching $\mathrm{mat}(\alpha) \in M$ that represents the assignment of character $c(\mathrm{mat}(\alpha)) \in \Sigma$ to the i-th position of a CS. For convenience we write $c(\alpha)$ for $c(\mathrm{mat}(\alpha))$. Any directed path $\varphi = (\alpha_1, \alpha_2, \ldots)$ originating from $\mathbf{r}$ identifies a sequence of characters $(c(\alpha_1), c(\alpha_2), \ldots)$ and thus a (partial) solution. A node without any further outgoing arcs is a *sink node*. An *exact* MDD encodes precisely the set of all feasible solutions. Due to the NP-hardness of the RFLCS problem such exact MDDs tend to have exponential size.

Therefore we consider more compact *relaxed* MDDs which encode supersets of all feasible solutions. In such a relaxed MDD nodes of an exact MDD are superimposed (*merged*) so that at each layer a maximum allowed number of nodes, called width, is not exceeded. We do this merging in such a way that any path from the root still represents a CS, but RF constraints may be violated. The length of the longest path in such a relaxed MDD then represents an upper bound to the length of a RFLCS.

To compile a MDD, a DP formulation of the considered problem is usually the starting point [14]. Each node $u \in V(\mathcal{D}_M)$ is associated to a state of the DP formulation. For the RFLCS problem, the DP formulation is defined as follows. Consider for a matching $m \in M$ the set $\mathrm{D}_M(m) = \{m' \in M \setminus \{m\} \mid m \succeq m'\}$ of possible successor matchings of m that may appear in the same CS after m. Note that set $\mathrm{D}_M(m)$ can be efficiently pre-computed for each $m \in M$. Then a state $(m(u), P(u), S(u))$ associated to node u consists of

- a matching $m(u)$ whose successor matchings $\mathrm{D}_M(m(u))$ represent the remaining matchings to consider further,
- set $P(u) \subseteq \Sigma$ containing all letters that may still be appended to the CS,
- set $S(u) \subseteq \Sigma$ containing all letters that appear on some paths from $\mathbf{r}$ to u.

The root state is $(\boldsymbol{m}_r, \Sigma, \emptyset)$ with the artificial matching $\boldsymbol{m}_r = (-1, -1)$, and all characters may be appended to it. Note that $D_M(\boldsymbol{m}_r) = M$. An arc $\alpha = (u, v)$ corresponds to a transition from state $(\boldsymbol{m}(u), P(u), S(u))$ to state $(\boldsymbol{m}(v), P(v), S(v))$ that is achieved by appending character $c(\alpha)$ to the CS w.r.t. to the remaining matchings $D_M(\boldsymbol{m}(u))$. Instead of considering all matchings from $D_M(\boldsymbol{m}(u))$ as possible outgoing transitions, we consider only matchings that can appear *directly* after $\boldsymbol{m}(u)$ in a longest CS, i.e., matchings from the subset $\mathrm{ND}(u) = \{\boldsymbol{m}' \in D_M(\boldsymbol{m}(u)) \mid \nexists \boldsymbol{m}'' \in D_M(\boldsymbol{m}(u)) \setminus \{\boldsymbol{m}'\} : c(\boldsymbol{m}'') \notin S(u) \wedge \boldsymbol{m}'' \succeq \boldsymbol{m}'\}$ which are not dominated by any other matching in $D_M(\boldsymbol{m}(u))$. The transition function to obtain the successor state $(\boldsymbol{m}(v), P(v), S(v))$ by considering matching $\mathrm{mat}(\alpha) \in D_M(u)$ is defined as

$$\tau((\boldsymbol{m}(u), P(u), S(u)), \mathrm{mat}(\alpha)) = \quad (2)$$
$$\begin{cases} (\mathrm{mat}(\alpha), P(u) \setminus \{c(\alpha)\}, S(u) \cup \{c(\alpha)\}) & \text{if } c(\alpha) \in P(u) \wedge \mathrm{mat}(\alpha) \in \mathrm{ND}(u) \\ \hat{0} & \text{otherwise} \end{cases}$$

where $\hat{0}$ represents the infeasible state. Note that no node $\hat{0}$ is created in $\mathcal{D}_M$ and the respective arcs are also skipped.

Moreover, a state $(\boldsymbol{m}(u), P(u), S(u))$ may be replaced by a *strengthened state* $(\boldsymbol{m}(u), P'(u), S(u))$, where $P'(u) = \{a \in P(u) \mid \exists \boldsymbol{m}' \in D_M(\boldsymbol{m}(u)) : a = c(\boldsymbol{m}')\} \subset P(u)$ without excluding any feasible solutions.

So far, we considered exact MDDs. For relaxed MDDs we have to define a state merger which computes the state of merged nodes. To still encode all feasible CSs in the relaxed MDD, only nodes of the same layer and with the same associated matching are merged. Let U be a subset of nodes s.t. all nodes are associated to matching $\boldsymbol{n}$, i.e. $\forall u \in U : \boldsymbol{m}(u) = \boldsymbol{n}$, then an appropriate state merger is $\oplus(U) = (\boldsymbol{n}, \bigcup_{u \in U} P(u), \bigcup_{u \in U} S(u))$. Since we restrict the state merger to nodes with the same associated matching, the possibilities to reduce the size of the relaxed MDD are also limited. However, since $|M|$ is at most the product $|s_1| \, |s_2|$ of the lengths of the two input strings s_1 and s_2, the size of each layer is still polynomially bounded by $O(|s_1| \, |s_2|)$.

Let $\mathrm{mat}(\mathcal{D}_M) = \{\boldsymbol{m}(u) \mid u \in V(\mathcal{D}_M) \setminus \{\mathrm{r}\}\} \subseteq M$ be the set of matchings derived from $\mathcal{D}_M$. To see that $\mathrm{mat}(\mathcal{D}_M)$ is indeed a feasible set of matchings to solve the model $\mathrm{ILP}(\mathrm{mat}(\mathcal{D}_M))$ from Sect. 4, remember that each path from r in $\mathcal{D}_M$ encodes a feasible CS. Hence, each such path can also be described as a sequence of matchings from M. In particular this is true for the matchings of a RFLCS, which must be therefore also contained in $\mathrm{mat}(\mathcal{D}_M)$.

Problem Specific Upper Bounds. To reduce $\mathrm{mat}(\mathcal{D}_M)$ further we filter arcs and nodes based on sub-optimality. The idea is to compute for each node u an upper bound $Z^{\mathrm{ub}}(u)$ on the number of characters that can appear in a common subsequence after the character $c(\boldsymbol{m}(u))$ of matching $\boldsymbol{m}(u)$. Then we can prune each

---

Algorithm 1. Incremental Refinement

1: **Input:** set of matchings M, lower bound lb, maximum width threshold W
2: $s^{\text{best}} \leftarrow \varepsilon$
3: construct initial relaxed decision diagram $\mathcal{D}_M$
4: **repeat**
5:　　filter-bottom-up($\mathcal{D}_M$, max(lb, $|s^{\text{best}}|$))
6:　　$s^{\text{rfcs}} \leftarrow$ derive-primal-solution($\mathcal{D}_M$) and update s^{best} **if** $|s^{\text{best}}| < |s^{\text{rfcs}}|$
7: **until** no new best solution s^{best} found
8: determine priority ranking $a_1^*, \ldots, a_{|\Sigma|}^*$ of all characters
9: **repeat**
10:　　$M \leftarrow \text{mat}(\mathcal{D}_M)$
11:　　**for** $i \leftarrow 1$ to $|\Sigma| + 1$ **do**
12:　　　　refine(L_i, $a_1^*, \ldots, a_{|\Sigma|}^*$, W)
13:　　　　filter arcs between L_i and L_{i+1}
14:　　**end for**
15:　　**repeat**
16:　　　　filter-bottom-up($\mathcal{D}_M$, max(lb, $|s^{\text{best}}|$))
17:　　　　$s^{\text{rfcs}} \leftarrow$ derive-primal-solution($\mathcal{D}_M$) and update s^{best} **if** $|s^{\text{best}}| < |s^{\text{rfcs}}|$
18:　　**until** no new best solution s^{best} found
19: **until** $|M| < |\text{mat}(\mathcal{D}_M)|$
20: **return** ($\mathcal{D}_M$, s^{best})

---

node u in the relaxed MDD where $Z^{\text{lp}}(u) + Z^{\text{ub}}(u) < lb$ holds, where lb is a known lower bound on the length of the RFLCS and $Z^{\text{lp}}(u)$ is the length of the longest path from $\mathbf{r}$ to u. We compute the upper bound for each node u by

$$Z^{\text{ub}}(u) = \min\{|P(u)|, \text{UB}^{\text{lcs}}(\boldsymbol{m}(u)), \max_{\alpha=(w,u)} \{Z^{\text{ub}}(w) - 1\}, Z^{\text{lp}\uparrow}(u)\}. \quad (3)$$

The first term takes the number of characters into account that can still be appended to the CS after matching $\boldsymbol{m}(u)$. The second term $\text{UB}^{\text{lcs}}(\boldsymbol{m}(u)) = \text{LCS}(m_1, m_2)$ is based on DP and computes the length of the longest common subsequence from matching $\boldsymbol{m}(u)$ onward. Note that this bound can be obtained in constant time by using a data structure known as scoring matrix, which can be computed during preprocessing for two input strings in $O(|s_1||s_2|)$ time [6]. The third term takes the upper bounds from the parent nodes of u into account. Finally, the last term corresponds to the length of the longest path from u to any sink node in the relaxed MDD. Note that this term is only available if the whole relaxed MDD is already compiled.

6 Incremental Refinement

Our approach to compile a relaxed MDD $\mathcal{D}_M$ w.r.t. matching set M for the RFLCS problem is based on the *incremental refinement* (IR) algorithm from Cire and van Hoeve [12] for sequencing problems. Since $\mathcal{D}_M$ considers the CS constraints exactly and only relaxes the RF constraint, paths in $\mathcal{D}_M$ originating from $\mathbf{r}$ will correspond to CSs where characters may appear more than once. We use the ideas from [12] to ensure at least for some characters that they occur at most once at each path for refining $\mathcal{D}_M$. Cire and van Hoeve [12] showed that the size of a given relaxed MDD will be at most doubled to establish this property for one more character.

The algorithm applies repeatedly two major steps—filtering and refinement—until some termination condition is fulfilled. Let $a_1^*, a_2^*, \ldots, a_{|\Sigma|}^*$ be a ranking of the characters in Σ s.t. a_1^* is the most important character to appear at most once at each path in $\mathcal{D}_M$ to get a strong relaxation. The following refinement step is applied layer by layer starting with L_1: For each character $a^* = a_1^*, a_2^*, \ldots, a_{|\Sigma|}^*$ we identify nodes u s.t. $a^* \in P(u) \cap S(u)$ and split them into two new nodes u_1 and u_2 where an incoming arc $\alpha = (v, u)$ is redirected to u_1 if $a^* \in P(u) \setminus \{c(\alpha)\}$ and to u_2 otherwise. All outgoing arcs are replicated for both nodes u_1 and u_2. We do this as long as the size of the layer is below a maximum width threshold W. For more details and a correctness proof in the context of sequencing problems see [12]. Due to the splitting of nodes the corresponding states may be changed and some of the outgoing arcs from the current layer to the next layer may become infeasible. Those arcs are filtered for each layer after the refinement step finishes. Algorithm 1 shows this at lines 12 and 13. The algorithm terminates if set $\mathrm{mat}(\mathcal{D}_M)$ could not be further reduced by the previously applied refinement/filtering round. The other main parts of the algorithm are:

Initial Relaxed MDD. The IR algorithm starts with an initial relaxed MDD. Usually, this initial relaxed MDD is a naive one of width one, i.e., a relaxed MDD with just a single node at each layer. However, in our case we want to respect the CS constraints and only superimpose states that correspond to the same matching. Therefore we compile the initial $\mathcal{D}_M$ layer-by-layer in a top-down approach. At each layer L_i, $i \geq 1$, we expand all nodes using the transition function (2), thus creating for each feasible transition a corresponding node in L_{i+1} and adding the corresponding arc if the node is not sub-optimal according to Eq. 3. Then all nodes in L_{i+1} with the same corresponding matching are merged. Since no feasible CS can be longer than the upper bound $Z^{\mathrm{ub}}(\mathbf{r})$, the compilation of $\mathcal{D}_M$ stops at the $(Z^{\mathrm{ub}}(\mathbf{r}) + 1)$-th layer.

Character Ranking for Refinement. To determine priorities for the characters we use some structural information obtained from the initial MDD. For this purpose let $\mathrm{All}^{\uparrow}(u)$ for each node $u \in V(\mathcal{D}_M)$ be the set of characters that appear on all paths from node u to a sink node. Note that set $\mathrm{All}^{\uparrow}(u)$ can be efficiently computed in a recursive way by a single bottom-up pass. If there exists a node v with an incoming arc $\alpha = (u, v)$ s.t. $c(\alpha) \in \mathrm{All}^{\uparrow}(v)$ holds, then each path originating from **r** and leading to any sink node will be infeasible if the path traverses α since character $c(\alpha)$ will appear more than once in a corresponding CS, i.e., the RF constraint will be violated. In [12] such arcs could be safely removed without also removing any feasible solution from the relaxed MDD. In our case this is not possible since solutions have arbitrary length and the path from **r** to v could still correspond to a complete feasible solution. However, we can use these violations to determine for which character it is most important to appear on all paths at most once to get a strong relaxation. Hence, we count for each character how often such a violation occurs in $\mathcal{D}_M$ and sort the characters according to non-increasing numbers of violations. Ties are resolved by preferring characters that appear in more matchings.

Filtering and Deriving New Primal Solutions. Lines 4–7 and 15–18 perform the following steps. First the function filter-bottom-up performs a single bottom up pass where for each node u the length of the longest path $Z^{\mathrm{lp}\uparrow}(u)$ from u to any sink node is computed and the upper bound $Z^{\mathrm{ub}}(u)$ is updated accordingly. If $Z^{\mathrm{lp}}(u) + Z^{\mathrm{ub}}(u) < lb$ then node u and all incident arcs are removed from $\mathcal{D}_M$.

After filtering we try to derive from $\mathcal{D}_M$ a new best heuristic solution. Since each path in $\mathcal{D}_M$ originating from **r** corresponds to a CS, we can derive a RFCS by removing duplicate letters. This is done in two steps. First, a bottom-up pass is performed where primal bounds are computed: For each node $u \in \mathcal{D}_M$ we recursively determine set $B^{\uparrow}(u) = B^{\uparrow}(v) \cup \{c(\alpha')\}$ where outgoing arc $\alpha' = (u,v)$ maximizes $\alpha' = \arg\max_{\alpha=(u,v)} |B^{\uparrow}(v) \cup \{c(\alpha)\}|$. Ties are resolved by sticking at the first arc that maximizes the expression. If u has no outgoing arcs then $B^{\uparrow}(u) = \emptyset$. Note that $|B^{\uparrow}(\mathbf{r})|$ is a valid primal bound on the RFLCS problem, since only the union is taken to compute $B^{\uparrow}(.)$. To improve this bound further, the second step performs a top-down pass where set $B^{\downarrow}(v)$ is recursively computed for each node v using the information of the precisely computed set $B^{\uparrow}(v)$. Hence, $B^{\downarrow}(v) = B^{\downarrow}(u) \cup \{c(\alpha')\}$ where incoming arc $\alpha' = (u,v)$ maximizes $\alpha' = \arg\max_{\alpha=(u,v)} |B^{\downarrow}(u) \cup \{c(\alpha)\} \cup B^{\uparrow}(v)|$ using $|B^{\downarrow}(v) \cup \{c(\alpha)\}|$ as tie breaking criterion. A sink node v' that maximizes $|B^{\downarrow}(v')|$ then provides the strongest primal bound. A respective RFCS is derived by going from v' backwards to **r**, skipping any character that already occurred along the path.

If a new best heuristic solution could be obtained in this way then the filter-bottom-up step is repeated and we try again to obtain a new best heuristic solution.

Main Procedure: Algorithm 2 shows the main procedure to solve an instance of the RFLCS problem. As input the algorithm takes the set of input strings S, a possibly known lower bound on the RFLCS length or zero, and the maximum width threshold W for the relaxed MDDs. The original set of matchings M reduced by performing iteratively the following steps. The first step processes the input strings s_1 and s_2 by removing characters that have no associated matching in M and characters that appear immediately one after the other in the input strings. For example, if character $a \in \Sigma$ appears in an input string at both position i and $i+1$ then a can be removed from $i+1$ without removing

Algorithm 2. Main Procedure for solving the RFLCS problem

1: **Input:** input strings S, lower bound lb, maximum width threshold W
2: $s^{\text{best}} \leftarrow \varepsilon$
3: derive original M w.r.t. S
4: **repeat**
5:     process S w.r.t. M
6:     $M \leftarrow \{m \in M \mid \text{UB}^{\text{BLUM}}(m) \geq \max(lb, |s^{\text{best}}|)\}$
7:     $(\mathcal{D}_M, s^{\text{rfcs}}) \leftarrow \text{IR}(M, \max(lb, |s^{\text{best}}|), W)$
8:     $M \leftarrow \text{mat}(\mathcal{D}_M)$ and update s^{best} if $|s^{\text{rfcs}}| > |s^{\text{best}}|$
9:     **return** s^{best} if $s^{\text{best}} = Z^{\text{lp}\uparrow}(\mathbf{r})$
10: **until** no characters can be removed from input strings
11: $s^{\text{ilp}} \leftarrow$ solve ILP(M)
12: update s^{best} if $|s^{\text{ilp}}| > |s^{\text{best}}|$
13: **return** s^{best}

---

any feasible solution. Furthermore, if the pattern $abab$ with $a, b \in \Sigma$ has been discovered in one of the input strings then the last b can be removed from the input string due to the RF constraint. Next, M is reduced by removing matchings $m \in M$ where the upper bound used in [6], denoted by $\text{UB}^{\text{BLUM}}(m)$ is lower than our currently best primal bound. This upper bound is based on the first two terms in Eq. (3), i.e., on the number of characters that can appear in a RFCS that contains $m \in M$ and on the length of the LCS that contains m. Note that the difference to Eq. (3) is that $\text{UB}^{\text{BLUM}}(m)$ is an upper bound on the length of a complete RFCS containing m wheres Eq. (3) describes an upper bound on the remaining part from m onward. With this reduced set M we compile a relaxed MDD $\mathcal{D}_M$. If the length of the hereby derived RFCS s^{rfcs} is equal to the longest path in $\mathcal{D}_M$ then s^{rfcs} is an optimal solution and the algorithm terminates. Otherwise, if due to the reduced set $\text{mat}(\mathcal{D}_M)$ further characters can be removed from s_1 and s_2 then we repeat the procedure until no further characters can be removed. Note that since the size of the input strings are reduced at each iteration also $\text{UB}^{\text{BLUM}}(m)$ changes, which may further reduce set M. Finally the ILP model from Sect. 4 is solved for set M.

7   Experimental Results

To test and compare our approach we used two benchmark sets from [4]. The first set, SET1, consists of 1680 randomly generated instances. For each combination of the input string lengths $n \in \{32, 64, 128, 256, 512, 1024, 2048, 4096\}$ and the alphabet sizes $|\Sigma| \in \{\frac{n}{8}, \frac{n}{4}, \frac{3n}{8}, \frac{n}{2}, \frac{5n}{8}, \frac{3n}{4}, \frac{7n}{8}\}$ there are 30 instances. The second set, SET2, consists of 30 randomly generated instances for each combination of the alphabet size $|\Sigma| \in \{4, 8, 16, 32, 64, 128, 256, 512\}$ and the maximal repetition of each character, reps $\in \{3, 4, 5, 6, 7, 8\}$. This set has a total of 1440 instances.

The algorithms were implemented using GNU C++ 5.4.1. All tests were executed on a single core of an Intel Xeon E5649 with 2.53 GHz and 16 GB RAM. The ILP model from Sect. 4 was solved with CPLEX 12.7 with a CPU-time limit of 3600 s. For Algorithm 2, henceforth denoted as MDD+CPLEX, the

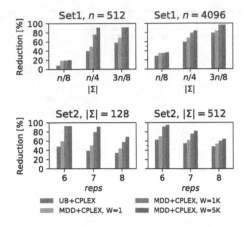

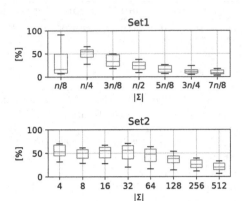

Fig. 2. Average reduction of matchings obtained from UB+CPLEX and MDD+CPLEX with different maximum width thresholds W.

Fig. 3. Average difference between the obtained reduction rates of UB+CPLEX and MDD+CPLEX with $W = 5000$.

maximum width threshold was set to $W = 5000$. This value was determined in preliminary experiments s.t. set M could be reduced as much as possible and as many instances as possible can be solved to optimality within the memory limit of 16 GB. MDD+CPLEX is compared to the approach from Blum et al. [6], henceforth denoted as UB+CPLEX, where the ILP(M') model is solved with the reduced set of matchings $M' = \{m \in M \mid \mathrm{UB}^{\mathrm{BLUM}}(m) < lb\}$. Both approaches use the lengths of the currently best known solutions from the literature as initial lower bound lb. Note that the compiled relaxed MDDs from Algorithm 1 are not strictly limited to W since the initial relaxed MDD could already contain layers that contain more nodes than W. However, such layers are not further refined during the compilation.

The average reduction rate of matchings from M is shown in Fig. 2 for 12 instance classes by means of bar plots. The first bar corresponds always to the UB+CPLEX approach whereas the next three bars corresponds to the MDD+CPLEX approach with different values for $W \in \{1, 1000, 5000\}$. Note that $W = 1$ means that only the initial relaxed MDD is compiled and no further refinement will take place. As expected, the obtained reduction rate increases with W. The boxplots in Fig. 3 report the average difference $\mathrm{red_{MDD}} - \mathrm{red_{UB}}$ between the average reduction rate $\mathrm{red_{UB}}$ obtained from UB+CPLEX and $\mathrm{red_{MDD}}$ obtained from MDD+CPLEX in percentage points aggregated over the ratio between n and $|\Sigma|$ in case of SET1 and over $|\Sigma|$ in case of SET2. On average the MDD+CPLEX approach is able to reduce the original set of matchings by more than 25.79% and 41.28% as UB+CPLEX does, for SET1 and SET2, respectively.

Detailed aggregated results are presented in Tables 1 and 2 where the first two columns show the instance characteristics and the third column shows the average length of the so far best known solution from the literature. Columns obj report for each tested approach the average length of the best obtained solutions. In case of MDD+CLPEX these solutions are either those obtained from the ILP model or the ones found during the compilation of the relaxed MDDs. In case of UB+CPLEX a "-" symbol indicates that CPLEX was not able at all to derive a primal solution within the time and memory limits. Average optimality gaps, shown in columns gap, are calculated by $100\% \times (ub - obj)/ub$ where ub is for each approach the best obtained upper bound. In case of MDD+CPLEX this is either the upper bound obtained from the ILP model or the length of the longest path from a compiled relaxed MDD. Columns t_{prep} list average preprocessing times in CPU seconds including the computation of the reduced set of matchings M (see Algorithm 2, Line 10). Columns t_{tot} list average total computation times in CPU seconds until the algorithm terminates including t_{prep} plus the time CPLEX needs and columns #opt report the total numbers of instances solved to optimality. In case of MDD+CPLEX the second number corresponds to the number of instances where optimality could already be proven by the compiled relaxed MDD at Line 9 in Algorithm 2. Hence, the number of times where it was not required to solve the ILP model at all. Average reduction rates of the original set of matchings are reported by columns red.

Regarding the number of instances solved to proven optimality, note that already UB+CPLEX was quite successful with a total of 90.26%. More precisely, 1489 out of 1680 instances from SET1 and 1327 out of 1440 instances of SET2 could be solved to proven optimality by UB+CPLEX. Nevertheless, MDD+CPLEX is able to solve significantly more instances to proven optimality: 1541 instances from SET1 and 1381 instances of SET2, and thus a total of 93.65%. Moreover, in 90.90% of all instances it was not necessary to solve the ILP model at all, since Algorithm 2 terminated early at Line 9. Hence, the obtained upper bound from the compiled relaxed MDD was equal to the length of the currently best found solution in these cases. Concerning the computation times, the UB+CPLEX approach was on average in only two cases faster then the MDD+CPLEX approach regarding benchmark set SET1 and only in one case regarding benchmark set SET2. Finally, the MDD+CPLEX approach is able to obtain in 135 cases better results than the currently best-known-solutions from the literature. For each considered problem class, MDD+CPLEX is able to provide on average better results than UB+CPLEX.

Table 1. Results on SET1 instances.

$|\Sigma|$	n	so far best.	UB+CPLEX obj	gap [%]	t [s]	#opt	red [%]	MDD+CPLEX obj	gap [%]	t_{prep} [s]	t_{tot} [s]	#opt	red [%]
$n/8$	32	4.00	4.00	0.00	0.16	30	3.72	4.00	0.00	<0.01	<0.01	30/30	94.94
	64	8.00	8.00	0.00	1.69	30	4.67	8.00	0.00	<0.01	<0.01	30/30	78.23
	128	16.00	16.00	0.00	11.46	30	4.80	16.00	0.00	0.04	0.04	30/30	44.45
	256	31.97	31.97	0.00	298.59	30	5.47	31.97	0.00	0.42	0.42	30/30	25.47
	512	63.90	32.30	49.50	3615.63	0	8.16	62.80	1.82	60.82	3434.53	2/2	20.29
	1024	116.30	0.00	100.00	3676.46	0	15.89	113.47	11.22	271.15	3946.52	0/0	23.34
	2048	185.07	-	-	0.48	0	22.45	186.23	16.29	1134.30	4135.18	0/0	28.49
	4096	284.80	-	-	2.04	0	28.71	292.03	11.05	4297.07	4297.07	0/0	36.59
$n/4$	32	7.83	7.83	0.00	0.03	30	22.13	7.83	0.00	<0.01	<0.01	30/30	79.54
	64	14.67	14.67	0.00	0.21	30	21.08	14.67	0.00	<0.01	<0.01	30/30	84.06
	128	25.93	25.93	0.00	4.82	30	21.76	25.93	0.00	0.47	0.47	30/30	87.34
	256	43.97	43.97	0.00	22.93	30	33.17	43.97	0.00	4.61	4.61	30/30	90.67
	512	68.57	68.57	0.00	389.04	30	39.54	68.57	0.00	28.43	38.93	30/28	91.01
	1024	105.07	104.97	0.96	2064.94	21	48.42	105.07	0.00	122.81	188.27	30/24	91.56
	2048	155.73	120.47	26.52	3314.28	4	56.72	156.87	0.37	576.39	1571.46	24/15	85.91
	4096	227.23	12.77	95.05	3650.80	0	59.48	230.37	0.73	1996.26	4036.48	15/6	83.99
$3n/8$	32	8.77	8.77	0.00	0.02	30	30.44	8.77	0.00	<0.01	<0.01	30/30	78.03
	64	15.53	15.53	0.00	0.06	30	32.30	15.53	0.00	<0.01	<0.01	30/30	81.16
	128	24.90	24.90	0.00	0.70	30	36.58	24.90	0.00	0.06	0.06	30/30	86.44
	256	39.97	39.97	0.00	1.52	30	55.50	39.97	0.00	0.50	0.50	30/30	89.20
	512	59.97	59.97	0.00	17.62	30	57.92	59.97	0.00	4.26	4.26	30/30	91.16
	1024	90.73	90.73	0.00	29.12	30	70.02	90.73	0.00	13.59	13.59	30/30	93.59
	2048	131.13	131.17	0.05	476.47	29	72.92	131.17	0.00	50.39	50.39	30/30	94.79
	4096	193.20	192.77	0.50	1030.16	25	78.79	193.37	0.00	163.90	163.99	30/29	96.65
$n/2$	32	8.87	8.87	0.00	0.01	30	37.51	8.87	0.00	<0.01	<0.01	30/30	75.08
	64	14.80	14.80	0.00	0.02	30	46.60	14.80	0.00	<0.01	<0.01	30/30	81.22
	128	22.93	22.93	0.00	0.08	30	52.04	22.93	0.00	0.01	0.01	30/30	82.57
	256	35.20	35.20	0.00	0.46	30	60.22	35.20	0.00	0.07	0.07	30/30	87.37
	512	53.13	53.13	0.00	2.72	30	69.40	53.13	0.00	0.92	0.92	30/30	90.49
	1024	79.13	79.13	0.00	7.38	30	75.93	79.13	0.00	3.45	3.45	30/30	93.05
	2048	115.70	115.70	0.00	21.32	30	80.40	115.70	0.00	19.65	19.65	30/30	94.59
	4096	167.97	167.97	0.00	93.68	30	86.74	167.97	0.00	26.89	26.89	30/30	95.84
$5n/8$	32	8.60	8.60	0.00	0.01	30	46.29	8.60	0.00	<0.01	<0.01	30/30	72.45
	64	13.30	13.30	0.00	0.01	30	52.75	13.30	0.00	<0.01	<0.01	30/30	78.35
	128	21.20	21.20	0.00	0.03	30	59.95	21.20	0.00	<0.01	<0.01	30/30	83.47
	256	32.53	32.53	0.00	0.11	30	67.54	32.53	0.00	0.02	0.02	30/30	86.36
	512	47.83	47.83	0.00	0.61	30	74.69	47.83	0.00	0.10	0.10	30/30	88.68
	1024	70.20	70.20	0.00	1.12	30	81.45	70.20	0.00	0.76	0.76	30/30	91.35
	2048	103.97	103.97	0.00	4.66	30	84.96	103.97	0.00	3.27	3.27	30/30	93.98
	4096	150.57	150.57	0.00	16.11	30	88.65	150.57	0.00	10.26	10.26	30/30	95.77
$3n/4$	32	8.17	8.17	0.00	0.01	30	47.56	8.17	0.00	<0.01	<0.01	30/30	71.83
	64	12.53	12.53	0.00	0.01	30	53.92	12.53	0.00	<0.01	<0.01	30/30	71.72
	128	19.70	19.70	0.00	0.02	30	65.68	19.70	0.00	<0.01	<0.01	30/30	79.50
	256	29.97	29.97	0.00	0.04	30	72.89	29.97	0.00	0.01	0.01	30/30	84.41
	512	44.57	44.57	0.00	0.26	30	77.45	44.57	0.00	0.03	0.03	30/30	88.28
	1024	65.20	65.20	0.00	0.53	30	83.86	65.20	0.00	0.26	0.26	30/30	92.07
	2048	94.67	94.67	0.00	1.45	30	88.57	94.67	0.00	0.51	0.51	30/30	94.18
	4096	136.77	136.77	0.00	6.06	30	90.14	136.77	0.00	4.19	4.19	30/30	95.22
$7n/8$	32	7.67	7.67	0.00	0.01	30	47.37	7.67	0.00	<0.01	<0.01	30/30	64.92
	64	11.57	11.57	0.00	0.01	30	56.15	11.57	0.00	<0.01	<0.01	30/30	73.63
	128	18.40	18.40	0.00	0.02	30	63.40	18.40	0.00	<0.01	<0.01	30/30	76.54
	256	27.80	27.80	0.00	0.03	30	74.05	27.80	0.00	0.01	0.01	30/30	84.25
	512	40.60	40.60	0.00	0.09	30	80.25	40.60	0.00	0.03	0.03	30/30	87.14
	1024	60.57	60.57	0.00	0.47	30	85.41	60.57	0.00	0.11	0.11	30/30	91.16
	2048	88.00	88.00	0.00	2.54	30	85.93	88.00	0.00	0.63	0.63	30/30	91.89
	4096	127.37	127.37	0.00	4.76	30	91.37	127.37	0.00	2.17	2.17	30/30	94.71

Table 2. Results on SET2 instances.

$\|\Sigma\|$	reps	so far best	UB+CPLEX obj	gap [%]	t [s]	#opt	red [%]	MDD+CPLEX obj	gap [%]	t_{prep} [s]	t_{tot} [s]	#opt	red [%]
4	3	3.47	3.47	0.00	<0.01	30	34.16	3.47	0.00	<0.01	<0.01	30/30	65.78
	4	3.77	3.77	0.00	<0.01	30	32.24	3.77	0.00	<0.01	<0.01	30/30	76.59
	5	3.83	3.83	0.00	<0.01	30	35.31	3.83	0.00	<0.01	<0.01	30/30	81.44
	6	3.90	3.90	0.00	0.01	30	25.62	3.90	0.00	<0.01	<0.01	30/30	85.92
	7	3.97	3.97	0.00	0.02	30	18.34	3.97	0.00	<0.01	<0.01	30/30	88.59
	8	3.97	3.97	0.00	0.02	30	19.01	3.97	0.00	<0.01	<0.01	30/30	89.54
8	3	6.23	6.23	0.00	<0.01	30	38.40	6.23	0.00	<0.01	<0.01	30/30	66.94
	4	6.87	6.87	0.00	0.01	30	34.77	6.87	0.00	<0.01	<0.01	30/30	71.70
	5	7.40	7.40	0.00	0.02	30	33.12	7.40	0.00	<0.01	<0.01	30/30	80.10
	6	7.53	7.53	0.00	0.02	30	25.51	7.53	0.00	<0.01	<0.01	30/30	79.22
	7	7.70	7.70	0.00	0.03	30	21.51	7.70	0.00	<0.01	<0.01	30/30	80.36
	8	7.77	7.77	0.00	0.07	30	19.09	7.77	0.00	<0.01	<0.01	30/30	80.65
16	3	9.70	9.70	0.00	0.01	30	43.91	9.70	0.00	<0.01	<0.01	30/30	71.78
	4	11.57	11.57	0.00	0.02	30	42.73	11.57	0.00	<0.01	<0.01	30/30	79.44
	5	12.93	12.93	0.00	0.04	30	28.84	12.93	0.00	<0.01	<0.01	30/30	79.97
	6	14.00	14.00	0.00	0.12	30	23.82	14.00	0.00	<0.01	<0.01	30/30	83.96
	7	14.93	14.93	0.00	0.29	30	21.32	14.93	0.00	0.01	0.01	30/30	84.81
	8	14.80	14.80	0.00	0.46	30	19.26	14.80	0.00	0.01	0.01	30/30	86.09
32	3	16.13	16.13	0.00	0.02	30	57.94	16.13	0.00	<0.01	<0.01	30/30	78.26
	4	19.00	19.00	0.00	0.05	30	48.26	19.00	0.00	<0.01	<0.01	30/30	81.46
	5	21.63	21.63	0.00	0.52	30	33.72	21.63	0.00	0.04	0.04	30/30	85.76
	6	23.73	23.73	0.00	1.39	30	25.52	23.73	0.00	0.10	0.10	30/30	85.91
	7	25.57	25.57	0.00	3.65	30	21.18	25.57	0.00	0.61	0.61	30/30	89.21
	8	27.50	27.50	0.00	5.19	30	18.22	27.50	0.00	0.99	0.99	30/30	88.80
64	3	25.43	25.43	0.00	0.04	30	65.65	25.43	0.00	<0.01	<0.01	30/30	82.34
	4	30.37	30.37	0.00	0.22	30	57.79	30.37	0.00	0.04	0.04	30/30	86.45
	5	34.93	34.93	0.00	3.36	30	44.23	34.93	0.00	0.74	0.74	30/30	86.26
	6	39.13	39.13	0.00	12.93	30	37.10	39.13	0.00	2.03	2.03	30/30	90.35
	7	43.63	43.63	0.00	28.43	30	29.84	43.63	0.00	7.51	7.92	30/29	89.54
	8	45.53	45.53	0.00	84.45	30	24.83	45.53	0.00	14.14	27.62	30/25	84.73
128	3	36.77	36.77	0.00	0.25	30	70.96	36.77	0.00	0.02	0.02	30/30	85.18
	4	45.03	45.03	0.00	3.20	30	60.94	45.03	0.00	0.37	0.37	30/30	87.78
	5	53.43	53.43	0.00	13.48	30	54.18	53.43	0.00	3.30	3.30	30/30	90.42
	6	61.53	61.53	0.00	47.21	30	48.04	61.53	0.00	8.17	8.17	30/30	92.79
	7	68.47	68.47	0.00	337.66	30	39.01	68.47	0.00	29.22	36.33	30/28	91.40
	8	74.60	74.43	1.43	1941.38	20	33.90	74.60	0.11	74.85	1010.01	28/12	68.99
256	3	55.03	55.03	0.00	0.66	30	77.45	55.03	0.00	0.07	0.07	30/30	89.31
	4	68.93	68.93	0.00	4.99	30	74.05	68.93	0.00	1.82	1.82	30/30	92.27
	5	81.43	81.43	0.00	41.75	30	63.63	81.43	0.00	12.95	12.95	30/30	93.83
	6	93.60	93.60	0.00	406.78	30	56.99	93.60	0.00	45.63	48.11	30/28	93.00
	7	104.50	104.40	0.80	1764.49	24	52.06	104.50	0.00	129.27	369.43	30/20	87.91
	8	115.07	110.77	8.91	3562.90	1	43.12	115.03	2.70	375.68	3167.78	10/2	62.22
512	3	81.63	81.63	0.00	0.83	30	86.31	81.63	0.00	0.58	0.58	30/30	92.84
	4	101.13	101.13	0.00	10.56	30	80.23	101.13	0.00	9.01	9.01	30/30	93.71
	5	121.03	121.03	0.00	162.42	30	72.44	121.03	0.00	37.06	37.06	30/30	95.16
	6	138.40	137.13	1.81	2040.66	21	62.80	138.60	0.00	143.51	148.66	30/27	94.86
	7	155.17	126.53	23.97	3570.00	1	55.24	156.00	0.70	679.41	2115.36	20/10	82.00
	8	173.07	19.67	90.03	3633.19	0	47.95	174.23	3.25	1221.36	4499.14	3/1	64.24

8 Conclusions

In this work we approached the RFLCS problem by transforming an instance into a maximum independent set (MIS) problem instance as this is done by Blum et al. [6]. The MIS problem is subsequently solved by the ILP solver CPLEX. Our major contribution is to heavily reduce the conflict graph of the MIS problem by means of relaxed MDDs. This has multiple advantages: (1) reducing the conflict graph leads to an improved performance of CPLEX s.t. more instances could be solved faster to proven optimality, (2) the compiled relaxed MDDs present a discrete relaxation of the RFLCS problem meaning that upper bounds can be additionally derived and (3) it is also possible to quickly derive heuristic solutions from the MDDs. In many cases it was not necessary anymore to solve the ILP for the MIS problem since the upper bound from the MDD corresponded to the length of the derived heuristic solution and thus optimality was already proven. Overall, for many benchmark instances new state-of-the-art results could be obtained.

In the literature there are works where relaxed decision diagrams are successfully embedded into a branch-and-bound algorithm s.t. branching is done over nodes in the relaxed decision diagram. Since relaxed MDDs provide also strong upper bonds for the RFLCS problem it may be promising future work to develop such a branch-and-bound algorithm for the RFLCS problem to solve even larger instances to optimality. Moreover, it seems promising to apply relaxed decision diagrams also on other LCS-related problems.

References

1. Adi, S.S., et al.: Repetition-free longest common subsequence. Discret. Appl. Math. **158**(12), 1315–1324 (2010)
2. Aho, A., Hopcroft, J., Ullman, J.: Data Structures and Algorithms. Addison-Wesley, Boston (1983)
3. Bergman, D., Cire, A.A., van Hoeve, W.J., Hooker, J.N.: Decision diagrams for optimization. In: OŚullivan, B., Wooldridge, M. (eds.) Artificial Intelligence: Foundations, Theory, and Algorithms. Springer, Cham (2016). https://doi.org/10.1007/978-3-319-42849-9
4. Blum, C., Blesa, M.J.: Construct, merge, solve and adapt: application to the repetition-free longest common subsequence problem. In: Chicano, F., Hu, B., García-Sánchez, P. (eds.) EvoCOP 2016. LNCS, vol. 9595, pp. 46–57. Springer, Cham (2016). https://doi.org/10.1007/978-3-319-30698-8_4
5. Blum, C., Blesa, M.J.: A comprehensive comparison of metaheuristics for the repetition-free longest common subsequence problem. J. Heuristics **24**(3), 551–579 (2017). https://doi.org/10.1007/s10732-017-9329-x
6. Blum, C., et al.: Solving longest common subsequence problems via a transformation to the maximum clique problem. Technical report. AC-TR-20-003 (2020). Submitted to Computers & Operations Research
7. Bonizzoni, P., Vedova, G.D., Dondi, R., Pirola, Y.: Variants of constrained longest common subsequence. Inf. Process. Lett. **110**(20), 877–881 (2010)

8. Brisk, P., Kaplan, A., Sarrafzadeh, M.: Area-efficient instruction set synthesis for reconfigurable system-on-chip designs. In: Proceedings of DAC 2004 - The 41st Annual Design Automation Conference. pp. 395–400. IEEE Press (2004)

9. Castelli, M., Beretta, S., Vanneschi, L.: A hybrid genetic algorithm for the repetition free longest common subsequence problem. Oper. Res. Lett. **41**(6), 644–649 (2013)

10. Chen, Y.C., Chao, K.M.: On the generalized constrained longest common subsequence problems. J. Comb. Optim. **21**(3), 383–392 (2011)

11. Chowdhury, S.R., Hasan, M.M., Iqbal, S., Rahman, M.S.: Computing a longest common palindromic subsequence. Fundamenta Informaticae **129**(4), 329–340 (2014)

12. Cire, A.A., Hoeve, W.V.: Multivalued decision diagrams for sequencing problems. Oper. Res. **61**(6), 1411–1428 (2013)

13. Gusfield, D.: Algorithms on Strings, Trees, and Sequences: Computer Science and Computational Biology. Cambridge University Press, Cambridge (1997)

14. Hooker, J.N.: Decision diagrams and dynamic programming. In: Gomes, C., Sellmann, M. (eds.) CPAIOR 2013. LNCS, vol. 7874, pp. 94–110. Springer, Heidelberg (2013). https://doi.org/10.1007/978-3-642-38171-3_7

15. Jiang, T., Lin, G.-H., Ma, B., Zhang, K.: The longest common subsequence problem for arc-annotated sequences. In: Giancarlo, R., Sankoff, D. (eds.) CPM 2000. LNCS, vol. 1848, pp. 154–165. Springer, Heidelberg (2000). https://doi.org/10.1007/3-540-45123-4_15

16. Jiang, T., Lin, G., Ma, B., Zhang, K.: A general edit distance between RNA structures. J. Comput. Biol. **9**(2), 371–388 (2002)

17. Kruskal, J.B.: An overview of sequence comparison: time warps, string edits, and macromolecules. SIAM Rev. **25**(2), 201–237 (1983)

18. Smith, T., Waterman, M.: Identification of common molecular subsequences. J. Mol. Biol. **147**(1), 195–197 (1981)

19. Storer, J.A.: Data Compression: Methods and Theory. Computer Science Press, Inc., Rockville (1987)

20. Tsai, Y.T.: The constrained longest common subsequence problem. Inf. Process. Lett. **88**(4), 173–176 (2003)

Distributing Battery Swapping Stations for Electric Scooters in an Urban Area

Thomas Jatschka[1(✉)], Fabio F. Oberweger[1], Tobias Rodemann[2], and Gunther R. Raidl[1]

[1] Institute of Logic and Computation, TU Wien, Vienna, Austria
{tjatschk,raidl}@ac.tuwien.ac.at, e1551139@student.tuwien.ac.at
[2] Honda Research Institute Europe, Offenbach, Germany
tobias.rodemann@honda-ri.de

Abstract. We investigate the problem of setting up battery swapping stations for electric scooters in an urban area from a computational optimization point of view. For the considered electric scooters batteries can be swapped quickly in a few simple steps. Depleted batteries are recharged at these swapping stations and provided again to customers once fully charged. Our goal is to identify optimal battery swapping station locations as well as to determine their capacities appropriately in order to cover a specified level of assumed demand at minimum cost. We propose a Mixed Integer Linear Programming (MILP) formulation that models the customer demand over time in a discretized fashion and also considers battery charging times. Moreover, we propose a Large Neighborhood Search (LNS) heuristic for addressing larger problem instances for which the MILP model cannot practically be solved anymore. Prototype implementations are experimentally evaluated on artificial benchmark scenarios. Moreover, we also consider an instance derived from real-world taxi and bus stop shelter data of Manhattan. With the MILP model, instances with up to 1000 potential station locations and up to 2000 origin/destination demand pairs can be solved to near optimality, while for larger instances the LNS is a highly promising choice.

Keywords: Facility location problem · E-mobility · Battery swapping stations · Mixed integer linear programming · Large Neighborhood Search

1 Introduction

Recharging the batteries of electric vehicles is usually a time-consuming process that hinders the large-scale adoption of such vehicles, especially when their range without reloading is too limited. An alternative possibility is to build electric vehicles in which the batteries can be replaced with charged ones. Batteries for

Thomas Jatschka acknowledges the financial support from Honda Research Institute Europe. We thank Honda R&D Co., Ltd. for technical insights.

N. Olenev et al. (Eds.): OPTIMA 2020, LNCS 12422, pp. 150–165, 2020.
https://doi.org/10.1007/978-3-030-62867-3_12

electric scooters are compact enough to be replaced directly by any customer in a few simple steps. Replacement batteries are provided in exchange for the used ones at swapping stations. Returned batteries are recharged at these stations, and once fully charged, they are again provided for exchange.

We aim at investigating how to best distribute such battery swapping stations in a given urban area and how many battery slots and corresponding batteries are required at each station. Our optimization goal is to minimize the setup costs for stations in dependence of their numbers of slots and required batteries in order to cover a specified amount of user demand over multiple consecutive time periods. It is assumed that customers who want to change batteries specify their trip data (origin, destination, approximate time) online and are automatically assigned to an appropriate station for the exchange (if one exists). This way, a better utilization of the swapping stations can be achieved. However, such an automated assignment also needs to consider a certain customer dropout as not every customer is willing to travel to a predestined station if the detour is long. We assume that all scooters in our system are homogeneous and therefore require the same batteries and have the same range. Moreover, since the scooters are operating in an urban area, it is safe to assume that a scooter's range is usually larger than the length of a customer's single trip. Hence, we do not consider multiple battery swapping stops for a single trip. In fact, a scooter battery is typically exchanged after multiple trips only. We model this problem as a mixed integer linear program (MILP). Smaller problem instances can be solved by directly applying a state-of-the-art MILP solver. To address the aspect of scalability to larger instances, where the MILP solver does not yield satisfactory solutions anymore, a *Large Neighborhood Search* (LNS) heuristic is proposed. The approaches are experimentally evaluated on artificial benchmark scenarios as well as one instance derived from real-world yellow taxi trip data and bus stop shelter station data of Manhattan.

Section 2 reviews relevant related work. Section 3 presents the problem formalization in the form of a MILP. The LNS heuristic is described in Sect. 4. Section 5 explains how the benchmark scenarios are generated. Experimental results of the proposed solution methods are given in Sect. 6. Finally, Sect. 7 concludes this article and gives an outlook on future work.

2   Related Work

In general, our problem can be classified as a location-allocation optimization problem [1]. Specifically, our problem is closely related to the capacitated multiple allocation Fixed Charge Facility Location Problem (FLP) [2] in which customers need to be assigned to facilities in order to satisfy their demand while minimizing costs for building facilities and serving customers. Moreover, the customer demand can be split arbitrarily between multiple facilities. When allocating customers to facilities from the perspective of the facility provider without considering the customers' preferences, one frequently has to expect a certain amount of customer dropout which we model with the help of a decay function as done in, e.g., [3–5]. Facility location problems with time dependent parameters are also referred to as multi-period FLPs [2]. One example for a multi-period FLP can be found in [6], where the dynamic maximal covering problem is considered.

Moreover, our problem exhibits similarities with the Capacitated Deviation-Flow Refueling Location Model (CDFRLM) introduced in [7], which is an extension of the Flow Refueling Location Model (FRLM) introduced by Kuby and Lim [8]. The FRLM aims to locate a fixed amount of refueling stations to maximize the total flow volume refueled. Several extensions of the FRLM have been proposed in the last years, such as the capacitated FRLM [9] in which the demand a station can satisfy is limited. The Deviation Flow Refueling Location Model (DFRLM) [5] relaxes the FRLM by allowing customers to deviate from their shortest O/D pair paths in order to go to a refueling station. Moreover, it is assumed that the number of customers willing to take a deviation from the shortest path is exponentially decreasing with the length of the deviation. In [7], the Capacitated Deviation-Flow Refueling Location Model (CDFRLM) is presented which also introduces station capacities to the DFRLM,

While there already exists work for setting up a system of battery swapping stations, e.g., [10,11], to the best of our knowledge, there is no previous work that considers specifically the aspect of recharging and reusing returned batteries and its implications concerning station capacities when optimizing station locations and configurations.

3   The Multi-period Battery Swapping Station Location Problem

In this section we formalize the problem of setting up battery swapping stations for electric scooters in an urban area. The *Multi-Period Battery Swapping Station Location Problem* (MBSSLP), as we call it, minimizes the costs for setting up battery swapping stations to satisfy a requested expected total demand over a whole day. To be able to consider battery charging times, we consider a day in a discretized fashion as a set of equally long consecutive time intervals given as a set of the start times $\mathcal{T}$ of the intervals; w.l.o.g., we assume $\mathcal{T} = \{1, \ldots, t_{\max}\}$. We make the simplifying assumption that charging any battery always takes the same time and only completely recharged batteries are provided to customers again. Moreover, as trips in an urban environment are usually rather short, we further assume that trips start and end in the same time interval.

Let $G = (V, A, w)$ be a weighted directed graph with node set V corresponding to all relevant geographic locations, arc set $A \subseteq V \times V$, corresponding to shortest paths between locations, and arc weights $w : A \to \mathbb{R}^+$ representing the respective travel times. We assume battery swapping stations can be set up at a subset of locations $L = \{1, \ldots, n\} \subseteq V$. Moreover, each location $l \in L$ has associated a maximal number of possible battery charging slots $s_l \geq 0$, fixed setup cost c_l for setting up a station at this location, and building costs per slot $c_l^s \geq 0$. Customer travel demands are given by origin-destination (O/D) pairs $Q \subseteq V \times V$; let $m = |Q|$. The expected number of users that need to change batteries on trip $q \in Q$ during a time interval $t \in T$ is denoted as d_q^t. The minimal amount of expected total customer demand that shall be satisfied over all time intervals in T is denoted by $d_{\min}$. Moreover, we are given a maximum detour length $w_{\max}^{\text{detour}}$ by which a feasible path including a battery swap for some $q \in Q$ may be longer than a shortest path from the origin to the destination of q. Finally, the number of time intervals required for completely recharging a battery is referred to as t^c.

It is assumed that customers would always take a shortest possible path p_q for an O/D pair $q = (u, v) \in Q$, except when they have to make a detour for swapping batteries. Let the set of arcs of a shortest path p_{uv} from node $u \in V$ to node $v \in V$ be $A(p_{uv}) \subseteq A$ and its length $w(p_{uv}) = \sum_{e \in A(p_{uv})} w(e)$. Moreover, we consider for an O/D pair $q = (uv) \in Q$ a shortest path that includes a certain location $l \in L$ as intermediate stop and denote it by p_q^l. The combination of a shortest path from u to l and a shortest path from l to v forms such a shortest path p_q^l, and its length is $w(p_q^l) = w(p_{ul}) + w(p_{lv})$. Let L_q be the set of locations $l \in L$ for which $w(p_q^l) \leq w(p_q) + w_{\max}^{\text{detour}}$ for $q \in Q$, i.e., the locations that may be used for battery swaps for O/D pair q.

A solution to the MBSSLP is primarily given by a pair of vectors $x = (x_l)_{l \in L} \in \{0, 1\}^n$ and $y = (y_l)_{l \in L}$ with $y_l \in \{0, \ldots, s_l\}$, where $x_l = 1$ indicates that a swapping station is to be established at location l and y_l is the respective number of battery slots. Moreover, let a_{ql}^t denote the part of the expected demand of O/D pair $q \in Q$ which we assign to a location $l \in L_q$ during time period $t \in T$. Similarly to [5], we consider the loss of users in dependence of the detour length by applying a penalty coefficient $g(q, l)$ to a_{ql}^t in order to obtain the actually expected satisfied demand $\tilde{a}_{ql}^t$ of O/D pair q at location l. As suggested in [5,12] we use the sigmoid function for this penalty coefficient, i.e., $g(q, l) = 1/(1 + \alpha e^{\beta(w(p_q^l) - w(p_q)) - \delta})$, where $w(p_q^l) - w(p_q)$ is the detour distance for going to the swapping station, δ_q is a reference distance, and α and β are parameters determine the shape of the function.

Based on the variables x, y, a, and $\tilde{a}$ the MBSSLP can be expressed as the following MILP.

$$\min \sum_{l \in L} (c_l x_l + c_l^s y_l) \tag{1}$$

$$x_l \cdot s_l \geq y_l \qquad\qquad \forall l \in L \quad (2)$$

$$\tilde{a}_{ql}^t = g(q, l) \cdot a_{ql}^t \qquad\qquad \forall t \in T,\ q \in Q,\ l \in L_q \quad (3)$$

$$\sum_{l \in L_q} a_{ql}^t \leq d_q^t \qquad\qquad \forall t \in T,\ q \in Q \quad (4)$$

$$\sum_{t'=\max(1, t-t^c)}^{t} \sum_{q \in Q | l \in L_q} \tilde{a}_{ql}^{t'} \leq y_l \qquad\qquad \forall t \in T,\ l \in L \quad (5)$$

$$\sum_{t=1}^{t_{\max}} \sum_{q \in Q} \sum_{l \in L_q} \tilde{a}_{ql}^t \geq d_{\min} \tag{6}$$

$$x_l \in \{0, 1\} \qquad\qquad \forall l \in L \qquad (7)$$
$$y_l \in \{0, \ldots, s_l\} \qquad\qquad \forall l \in L \qquad (8)$$
$$0 \le a_{ql}^t, \tilde{a}_{ql}^t \le s_l \qquad\qquad \forall t \in \mathcal{T},\, q \in Q,\, l \in L_q \qquad (9)$$

The goal of the objective function (1) is to find a feasible solution that minimizes the setup costs for stations and their battery slots. Inequalities (2) ensure that battery slots can only be allocated to a location $l \in L$ if a station is opened there. For better readability equalities (3) introduce variables $\tilde{a}_{ql}^t$ by applying the penalty coefficients $g(q, l)$ to variables a_{ql}^t. Constraints (4) enforce that the total demand assigned from an O/D pair q to locations does not exceed d_q^t for all $t \in \mathcal{T}$. Inequalities (5) ensure the required capacity y_l at all locations over all time intervals. Note that by using $\tilde{a}_{ql}^t$ instead of a_{ql}^t in (5), we "overbook" stations to consider the expected case, similarly as in [13]. Inequalities (5) also model that swapped batteries can be reused after t^c time intervals. The minimal satisfied demand to be fulfilled over all time intervals is expressed by inequality (6). Finally, the domains of the variables are given in (7)–(9).

4   Large Neighborhood Search

Large Neighborhood Search (LNS) [14] is a prominent metaheuristic for addressing difficult combinatorial optimization problems, which builds upon effective lower-level heuristics. A basic LNS in essence follows a classical local search framework, but usually much larger neighborhoods are considered in each iteration. The key-idea is to search these neighborhoods not in a naive enumerative way but to apply some "more clever" problem-specific procedure to solve the subproblem induced by each neighborhood in order to obtain the best or a promising heuristic solution from the neighborhood. Frequently, LNS follows a destroy and recreate scheme: A current incumbent solution is partially destroyed, typically by freeing a subset of the decision variables and fixing the others to their current values, and then repaired again by finding best or at least promising values for the freed variables.

We first show how to construct an initial solution in a fast greedy way. Afterwards, the search and destroy operators of our LNS are described.

4.1   Greedy Construction Heuristic

The construction heuristic generates a solution station-wise. In each iteration of the algorithm a new station is opened and demand is allocated to it. In order to decide at which location to open a station next, we first calculate how much additional demand a new station at each so far unused location could satisfy w.r.t. the already opened stations. The location with the highest ratio of additionally satisfied demand to corresponding building costs is then chosen for opening the next station.

To calculate the amount of demand a station $l \in L$ can satisfy, demand is assigned from each $q \in Q \mid l \in L_q$ for all time periods $t \in T$ to l until either the station's maximum capacity is exhausted or all demand has been assigned. The iteration order of Q is hereby decided by the decay function g such that O/D pairs with lower decay value w.r.t. l are considered first.

The construction algorithm terminates when one of the following conditions is met: at least $d_{\min}$ demand is satisfied, stations are opened at all possible locations, or no more demand can be assigned to a station anymore.

4.2 Destroy and Repair Operators

Let (x, y, a) be a solution to the MBSSLP. Moreover, let $L(x) \subseteq L$ be the set of locations for which $x_l = 1$. In a first step we create an undirected graph $G^L = (V, E)$ where $(u, v) \in E$ for $u, v \in V$ if and only if $\{u, v\} \subseteq L_q$ for at least one O/D pair $q \in Q$.

We then derive a set of locations L_{repair} that are considered for repairing via an (r, k)-*repair operator*. The operator iteratively adds k random node sets to L_{repair} where each node set is generated by choosing a random vertex $v \in V$ as well as r random neighbors of v in G^L (less if the degree of v is less than r). Afterwards, k random locations from $L(x)$ are added to L_{repair}. Should, during the generation of L_{repair}, a randomly selected vertex already be in L_{repair} the repair operator chooses a new random vertex if possible. From L_{repair} we derive the set $L_{\text{destroy}} = L_{\text{repair}} \cap L(x)$, and close all stations at these locations.

When repairing the solution, one needs to consider how much more demand needs to be satisfied in order to make the solution feasible again and how much demand from which O/D pairs is still available to be assigned to a station. For this purpose, let $D' = (d'^t_q)_{t \in T, q \in Q}$ be the demand not yet assigned to any opened location in the destroyed solution, i.e., $d'^t_q = d^t_q - \sum_{l \in L(x) \backslash L_{\text{destroy}}} a^t_{ql}$. Moreover, let d_{sat} be the amount of total demand satisfied in the partially destroyed solution, i.e., $d_{\text{sat}} = \sum_{l \in L(x) \backslash L_{\text{destroy}}} \sum_{t=1}^{t_{\max}} \sum_{q \in Q} \tilde{a}^t_{ql}$. Hence, the goal of the repair function is to assign at least $d'_{\min} = d_{\min} - d_{\text{sat}}$ demand from D' to the locations $L' = L_{\text{destroy}} \cup L_{\text{repair}}$. For this purpose, let $I(L', D', d'_{\min})$ be the residual MBSSLP instance in which L, $D = (d^t_q)_{t \in T, q \in Q}$, and $d_{\min}$ are replaced with L', D', and $d'_{\min}$. We determine a promising heuristic solution to $I(L', D', d'_{\min})$ using a relaxation of the MILP (1)–(9): Allowing the y_l variables to be continuous, i.e., replacing (8) by $0 \leq y_l \leq s_l, \forall l \in L$, while still keeping the x_l variables integral significantly speeds up the solving of the MILP. Obtained fractional values for y_l are finally rounded up to obtain a feasible solution to the original MBSSLP again, assuming one exists.

Note that the described solving of the relaxation of the MILP followed by rounding can also be used as a standalone heuristic for the original MBSSLP, which is applicable as long as the instance is not too large. We refer to this approach as y-*Relaxed MILP Heuristic* (RMH$_y$). Additionally, we also considered solving the full linear relaxation of the original MILP, i.e., the linear program in which all x_l as well as y_l variables are continuous, and rounding up obtained fractional x_l as well as y_l values to the next integers; we call this heuristic *Linear Programming Heuristic* (LPH). In Sect. 6 we compare these approaches to each other, showing that the RMH_y heuristic is a better choice for repairing solutions than the LPH heuristic.

5   Test Instances

As no real problem instances are available to us we created artificial test instances with characteristics that might be expected in real scenarios. The creation of this instances is described next. Moreover, we derived one problem instance from real-world taxi trip and bus stop data of Manhattan as described in Sect. 5.2.

5.1   Random Instances for the MBSSLP

The instances are simplified scenarios modeled after a typical work day where people go to work in the morning and return home in the evening. Battery swapping stations as well as origin and destination locations of customers are located within a square of length $\lceil \xi \sqrt{n} \rceil$ with $\xi = 800$. We generate a network graph $G = (V, E)$ following a similar procedure as used in [7,15] by first sampling $|V| = 5n$ random points from the square and then constructing an euclidean spanning tree w.r.t. V. Afterwards, n additional randomly chosen edges $(u, v) \in V \times V$ are added to E.

The set of potential battery swapping station locations L is generated by choosing n random nodes from V. Costs for building a station are chosen uniformly at random from $\{50, \ldots, 70\}$ for each station. Costs for adding a battery slot to a station are set to 40. Each battery swapping station can have at most 70 battery slots.

Origin and destination locations are chosen from a random subset $V' \subseteq V$ with $|V'| = \min(\frac{m}{2}, 5n)$. To each $v \in V'$ a random weight γ_v is assigned according to a log-normal distribution with mean $\mu = \ln(100)$ and standard deviation $\sigma = 0.5$. Moreover, we also assign weights γ_q to each OD-pair $q = (u, v) \in V' \times V'$ such that γ_q corresponds to $f_{\text{PDF}}(w(p_q), \mu, \sigma)$ with f_{PDF} being the probability density function of a lognormal distribution with mean $\mu = \ln(5000)$ and standard deviation $\sigma = 0.2$. The total demand d_q^{total} of an O/D-pair $q = (u, v)$ is then calculated as $d_q^{\text{total}} = \gamma_u \cdot \gamma_v \cdot \gamma_q$. We then set Q to be the set of m O/D-pairs q of $V' \times V'$ for which d_q^{total} is highest.

The swapping demand of each O/D-pair is distributed over 24 time periods, $\mathcal{T} = \{1, \ldots, 24\}$ and recharging a battery requires one time period, i.e., $t^c = 1$. We assume each customer to travel twice on his corresponding path, once in the morning to get to work and once in the evening to travel back home, and we assume that customers need to swap batteries once per trip counted here as demand. The demand during each time period $t \in \mathcal{T}$ is determined by two normal distributions $\mathcal{N}_{\text{morning}}(8, 1)$ and $\mathcal{N}_{\text{evening}}(18, 2)$, respectively. From each distribution 100 samples t are generated and transformed to valid integral values by $t := (\lceil t \rceil \mod t_{\max}) + 1$. Afterwards, d_q^{total} is distributed over $\mathcal{T}$ according to the frequency in which the time periods $t \in \mathcal{T}$ appear in the generated samples.

The maximal deviation distance of the users, $w_{\max}^{\text{detour}}$, is set to $\xi/2$ and the parameters of the distance decay function are set to $\alpha = 100$, $\beta = 0.1$, and $\delta_q = w_{\max}^{\text{detour}}/10$ for all $q \in Q$. Fig. 1 shows the decay value $g(q, l)$ in dependence of the deviation distance $w(p_q^l) - w(p_q)$ with the chosen parameterization.

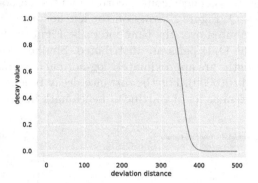

Fig. 1. Decay $g(q, l)$ in dependence of the deviation distance $w(p_q^l) - w(p_q)$.

Eight groups of test instances for different combinations of n and m have been generated as described in Sect. 5, and each group consists of thirty instances. In Sect. 6 we evaluate the instances with $d_{\min}$ being set either to 30% or to 80% of the total swapping demand.

5.2   Manhattan Instance

Next to artificial benchmark instances we also derived an instance from real-world yellow taxi trip data and bus stop shelter data of Manhattan, which we call here Manhattan instance. The underlying street network of the instance corresponds to the street network graph of Manhattan provided by the Python package OSMNX[1]. Origin/Destination pairs of our instance correspond to trips between the taxi zones[2] of Manhattan. The partitioning of Manhattan into taxi zones is shown in Fig. 3. For each taxi zone one random origin and one random destination location were chosen from the set of nodes of the network graph that are associated with the corresponding taxi zone.

The set of O/D-pairs and their corresponding demands have been derived from the 2016 Yellow Taxi Trip Data[3]. The taxi data set was first preprocessed and all trips with invalid data as well as trips made on a weekend have been removed from the data set. Furthermore, we have also removed all trips which do not start and end in Manhattan. From the preprocessed data set we then

---

[1] https://github.com/gboeing/osmnx.
[2] https://data.cityofnewyork.us/Transportation/NYC-Taxi-Zones/d3c5-ddgc.
[3] https://data.cityofnewyork.us/Transportation/2016-Yellow-Taxi-Trip-Data/k67s-dv2t.

extracted for each trip the pickup time, the pickup zone, the drop-off zone, as well as the passenger count. Each pickup time was rounded down to the nearest hour and afterwards an average daily passenger count for each triple (pickup hour, pickup zone, drop-off zone) was calculated. In total, the final table contains 4498 unique pickup/drop-off zone pairs which also constitute the instance's set of O/D pairs Q. These passenger counts correspond to the hourly demands d_q^t of the O/D pairs $q \in Q$. Figure 2 shows on the left how the total demand over all O/D pairs is distributed over the time intervals. Figure 2 shows on the right how the lengths of the O/D pairs are distributed. Similarly to our benchmark instances, the trip lengths are approximately log-normal distributed with a mean between $\ln(5000)$ and $ln(6000)$. For the distance decay function and $w_{\max}^{\text{detour}}$ we use the same parameters as for the artificial benchmark instances.

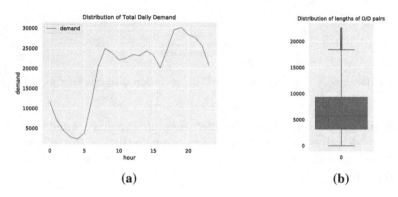

(a)   (b)

Fig. 2. Distributions of (a) demand and (b) trip length of the O/D pairs from the real-world data based instance.

The set of potential battery swapping station locations L is derived from the bus stop shelters[4] of Manhattan by selecting 500 locations randomly. Figure 3 shows the distribution of the stations.

As shown in Fig. 2 left the demand at each hour is quite high. Therefore we choose a capacity limit of 200 for each battery swapping station, The costs for building a station as well for adding a battery charging slot are chosen as for the artificial instances.

6   Computational Results

All algorithms were implemented in Julia[5] 1.4.2. All test runs have been executed on an Intel Xeon E5-2640 v4 2.40 GHz machine in single-threaded mode with a time limit of thirty minutes. Gurobi[6] 8.1.0 was used for solving the MILPs.

---

[4] https://data.cityofnewyork.us/Transportation/Bus-Stop-Shelters/qafz-7myz.
[5] https://julialang.org/.
[6] https://www.gurobi.com/.

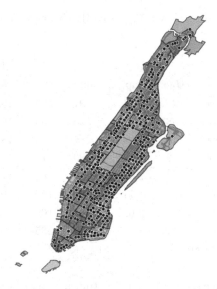

Fig. 3. Taxi zones of Manhattan and potential locations for swapping stations.

First, we investigate the performance of the standalone MILP model given by Eqs. (1)–(9) as well as the standalone RMH_y and the LPH approach. Afterwards, the results of the LNS are discussed. Finally, in Sect. 6.3 we present the results on the instance derived from real-world data for the LNS approach as well as the MILP models. All instances are evaluated with d_{min} being set either to 30% or to 80% of the total swapping demand. Hence, let $d_{min}[\%]$ refer to d_{min} as percentage of the total swapping demand.

6.1 MILP Approaches

All MILP models were solved with Gurobi 8.1.0. In case no optimal solution was found within the time limit, the solver returned the best found feasible solution if it exists.

Table 1 shows a summary of the performance of the exact MILP approach, RMH_y and LPH for each instance group in our benchmark set. Column "gap[%]" shows the average optimality gaps for each instance group, the median computation times are shown in column "time[s]", and column "$|L(x)|$" lists the average number of opened stations in the solutions. Note that the gaps listed for RMH_y and LPH are determined also w.r.t. the lower bounds obtained by the original MILP.

Overall, with the exact MILP solving was aborted due to the time limit for almost all instances. However, for each instance at least one feasible solution was found. Instances with up to 1000 potential battery swapping stations and 2000 O/D-pairs can be solved by the MILP almost to optimality with a gap of less than 1%. For larger instances the optimality gaps deteriorate. Compared to the results of the original MILP model, RMH_y yields in general better average optimality

Table 1. Results of the original MILP, the RMH_y heuristic, and the LPH heuristic.

(a) MILP results for $d_{\min}[\%] = 30$.

		MILP			RMH_y			LPH		
		gap[%]	time[s]	$|L(x)|$	gap[%]	time[s]	$|L(x)|$	gap[%]	time[s]	$|L(x)|$
250	500	0.05	1800	25	2.61	91	25	18.62	2	81
	1000	0.02	1800	38	1.59	125	38	10.38	4	103
500	1000	0.03	1800	46	2.54	287	46	18.12	5	149
	2000	0.08	1800	72	1.60	686	71	10.06	12	190
1000	2000	0.24	1800	89	2.54	1295	88	17.95	20	279
	4000	2.69	1800	192	1.77	1800	129	9.78	47	346
2000	4000	9.09	1800	382	3.67	1800	166	18.01	81	532
	8000	6.78	1800	531	8.60	1800	535	10.92	238	660

(b) Results for $d_{\min}[\%] = 80$.

		MILP			RMH_y			LPH		
		gap[%]	time[s]	$|L(x)|$	gap[%]	time[s]	$|L(x)|$	gap[%]	time[s]	$|L(x)|$
250	500	0.03	1800	47	1.09	47	47	4.98	2	86
	1000	0.02	1800	72	0.32	536	72	2.47	5	121
500	1000	0.02	1800	84	1.01	464	84	4.85	7	158
	2000	0.08	1800	138	0.31	1800	137	2.37	18	226
1000	2000	0.12	1800	160	1.04	1800	159	4.78	25	294
	4000	1.92	1800	305	0.35	1800	260	2.33	64	425
2000	4000	3.64	1800	488	1.40	1800	316	4.81	95	559
	8000	29.54	1800	1248	0.49	1800	515	2.31	236	815

gaps for the three largest instance groups. The LPH approach was able to solve
all instances to optimality w.r.t. the linear relaxation of the original MILP in
less than 5 min on average. However, the derived feasible MBSSLP solutions
are significantly worse than the solutions generated by RMH_y especially for
$d_{\min}[\%] = 30$. For instances nearly solved to optimally, we can also observe that
the number of opened stations in the solutions are as expected. RMH_y solutions
require a marginally smaller number of opened stations than the MILP solutions.
Solutions generated from the LPH approach, on the other hand, require a much
higher number of opened stations than the other approaches. Hence, LPH does
not seem to be a good choice as repair procedure for the LNS.

Figure 4 provides a more detailed comparison of the optimality gaps of the
MILP, RMH_y and LPH solutions. The figure shows boxplots of the optimality
gaps for each instance group and approach and confirms our previous observations. Note that for a better comparison between the approaches Fig. 4b is cut
off and only shows optimality gaps up to 7% since solutions to the instances
with $n = 1000, m = 4000$ as well as $n = 2000, m = 8000$ generated by the MILP

feature optimality gaps up to 45%. For the largest instances with $n \geq 1000$ and $m \geq 4000$, RMH_y starts to produce better results than the MILP while LPH does not seem to be able to compete with RMH_y for any instance group. However, since RMH_y requires solving a large MILP as well, this approach also has its limits concerning scalability. Therefore, in the next section we investigate the LNS that uses in each iteration RMH_y to (re-)optimize only a comparably small part of a solution.

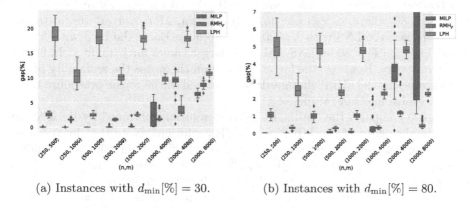

(a) Instances with $d_{\min}[\%] = 30$.   (b) Instances with $d_{\min}[\%] = 80$.

Fig. 4. Optimality gaps of the MILP, RMH_y and LPH solutions.

6.2 Large Neighborhood Search

For the size parameters of the repair operator we consider here, after preliminary tests $r = 4$ and $k \in \{4, 14, 20\}$. These values are promising as the MILPs corresponding to the repair subproblems can usually be solved to a small remaining optimality gap within seconds. As the LNS is a heuristic approach, it also does not make much sense to solve the MILPs always to proven optimality; instead

Table 2. Results of the LNS.

n	m	$d_{\min}[\%] = 30$						$d_{\min}[\%] = 80$					
		k = 4		k = 14		k = 20		k = 4		k = 14		k = 20	
		gap[%]	iter	gap[%]	iter	gap[%]	iter	gap[%]	iter	gap[%]	iter	gap[%]	iter
250	500	<u>1.05</u>	2549	1.62	385	1.70	217	<u>0.57</u>	3222	0.75	520	0.79	270
	1000	<u>0.83</u>	1465	1.14	207	1.21	117	0.32	1843	<u>0.23</u>	263	0.25	117
500	1000	<u>1.31</u>	2094	1.64	418	1.83	207	<u>0.72</u>	2305	0.77	559	0.81	299
	2000	<u>1.06</u>	982	1.22	230	1.29	132	0.48	1115	0.33	282	<u>0.31</u>	156
1000	2000	<u>1.72</u>	1177	1.95	399	2.05	241	<u>1.00</u>	1375	1.02	482	1.03	317
	4000	<u>1.41</u>	604	1.44	203	1.45	132	0.78	606	0.46	214	<u>0.42</u>	128
2000	4000	<u>2.58</u>	698	2.64	292	2.69	211	1.59	720	1.46	331	<u>1.39</u>	251
	8000	3.28	306	3.10	128	<u>3.06</u>	93	2.00	280	1.11	128	<u>1.06</u>	87

we terminated the MILP solver when a solution with an optimality gap of at most 0.0005% has been reached. Each LNS run was terminated after 30 min. The results of the LNS are shown in Table 2. For each considered minimum demand coverage d_{min} and each neighborhood size parameter k, the average number of iterations "iter" and the average optimality gap "gap[%]" (w.r.t. the lower bounds obtained by the original MILP).

The table shows that, naturally, the LNS can perform less iterations the larger k is. For instances with $d_{min}[\%] = 30$ we can see that the solutions tend to deteriorate as k is increasing. However, this is not the case for instances with $d_{min}[\%] = 80$ where we can see no such pattern. Moreover, as the instances become larger, the LNS with $k = 20$ starts to outperform the LNS with $k = 4$. Hence, for $d_{min}[\%] = 80$ an LNS with even larger values for k might yield better results in theory. However, the larger k is chosen the worse the scalability of the LNS becomes as the MILP that needs to be solved in the repair procedure takes longer to solve for larger values of k.

Figure 5 compares the optimality gaps of solutions obtained by the LNS to the optimality gaps of the MILP and RMH_y solutions. Note that for a better comparison between the approaches Fig. 5b is cut off and only shows optimality gaps up to 7%. For instances with $d_{min}[\%] = 30$ we can see that the LNS is on average on all instance groups able to produce better solutions than RMH_y. This particularly holds for the largest instance group, where the gap of RMH_y deteriorates to over 8% but the LNS' gaps are still within 4%. For instances with $d_{min}[\%] = 80$, both, the LNS as well as RMH_y, perform quite well with gaps usually less than 2%. The LNS solutions are here slightly worse than the RMH_y solutions for larger instances.

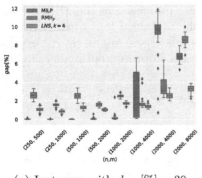

(a) Instances with $d_{min}[\%] = 30$.

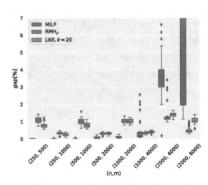

(b) Instances with $d_{min}[\%] = 80$.

Fig. 5. Comparison of the optimality gaps of the LNS solutions to the solutions of the other approaches.

Overall, we can say that the LNS works reasonably well over all considered benchmark instances, and it is reasonable to expect it to scale much better to even larger instances than RMH_y or solving the original MILP directly.

6.3   Results on the Manhattan Instance

In this section we show how well the MILP approaches as well as the LNS were able to deal with the real-world data based Manhattan instance. While the size of n and m is similar to some of our benchmark instances, the Manhattan instance is much harder to solve than our benchmark instances due to the shape of Manhattan as well as the instance's geographic distribution of demand.

Tables 3 and 4 show respective results. Each solution approach was applied to the instance six times with different values for $d_{min}[\%]$. For each approach the tables lists the total costs of the solutions, the corresponding optimality gaps (always w.r.t. the lower bounds obtained from the linear relaxation of the original MILP), and the computation times in seconds. The direct MILP approach was only able to find (non optimal) solutions for the lowest levels of $d_{min}[\%]$. RMH_y and LPH could obtain feasible solutions for all cases except with $d_{min}[\%] = 60$. Concerning RMH_y and LPH, one can see that, as one might expect, gaps of LPH are usually significantly larger than those of RMH_y, but LPH is much faster and is, in contrast to RMH_y, also able to yield a feasible solution for $d_{min}[\%] = 50$.

Table 4 shows the results obtained by the LNS with $r = 3$ and different values for k. Listed are total costs of the solutions, the corresponding optimality gaps (if a lower bound is known from the MILP), and the number of destroy and repair iterations. Most importantly, in contrast to the above MILP/LP approaches, the LNS could also find a feasible solution for $d_{min}[\%] = 60$. Moreover, except for the lowest level of $d_{min}[\%] = 10$, the LNS was able to find the best solutions. The number of performed destroy and repair iterations stays approximately the same for increasing levels of $d_{min}[\%]$. However, as expected, the number of iterations decreases the larger the value for k.

Table 3. LPH, RMH_y, and MILP results for the Manhattan instance.

$d_{min}[\%]$	LPH			RMH_y			MILP		
	costs	gap[%]	time[s]	costs	gap[%]	time[s]	costs	gap[%]	time[s]
10	155797	1.27	179	_153886_	0.04	1801	_153886_	0.04	1801
20	325775	2.90	140	321773	1.69	1801	320168	1.20	1801
40	692976	1.06	196	_689600_	0.57	1801	-	-	-
50	_892035_	0.77	704	-	-	-	-	-	-
60	-	-	-	-	-	-	-	-	-

Table 4. LNS results for the Manhattan instance.

$d_{min}[\%]$	$k = 4$			$k = 7$			$k = 14$		
	costs	gap[%]	iter	costs	gap[%]	iter	costs	gap[%]	iter
10	153900	0.05	92	153890	0.05	19	154025	0.13	2
20	319769	1.07	87	319334	0.94	43	318939	0.82	19
40	688298	0.39	87	687769	0.31	42	687983	0.34	16
50	890049	0.55	83	888920	0.43	44	887926	0.32	24
60	1095190	-	89	1093898	-	43	1095097	-	15

7 Conclusions and Future Work

We presented the new Multi-Period Battery Swapping Station Location Problem (MBSSLP) for distributing battery swapping stations in an urban area. On our benchmark instances, directly solving the proposed MILP model is reasonable for instances with up to 1000 stations and $2000\,\Omega$/D-pairs, where solutions with small gaps could be obtained. For larger instances solving the MILP model becomes quickly infeasible and heuristics need to be employed to find approximate solutions. Relaxing the y variables and rounding obtained fractional values, i.e., our RMH_y, is a viable approach by which significantly larger instances can be solved reasonably well, nevertheless it also has its limits. We therefore proposed an LNS that effectively utilizes RMH_y and provides better scalability. This can in particular be seen in the results for the real-world data based Manhattan instance.

We remark that the proposed LNS still has room for improvement. For example, different strategies for selecting the nodes to be removed or considered for addition may be investigated. Moreover, adaptive mechanisms for choosing among different destroy and re-create methods may be useful. Last but not least, there are also alternative ways to address the scalability issue, for example by approaches based on (hierarchical) clustering and iterative refinement.

In future work the MBSSLP model should also be further refined to reflect real-world aspects in a more realistic way. For example, battery swapping stations are usually not extended slot by slot but by modules which consist of multiple new battery slots. So far, we also have not yet considered a pricing model for customers or costs for maintaining the battery swapping stations and the batteries.

References

1. Cooper, L.: Location-allocation problems. Oper. Res. **11**(3), 331–343 (1963)
2. Laporte, G., Nickel, S., da Gama, F.S. (eds.): Location Science. Springer, Switzerland (2015). https://doi.org/10.1007/978-3-319-13111-5
3. Verter, V., Lapierre, S.D.: Location of preventive health care facilities. Ann. Oper. Res. **110**(1), 123–132 (2002)

4. Berman, O., Larson, R.C., Fouska, N.: Optimal location of discretionary service facilities. Transp. Sci. **26**(3), 201–211 (1992)
5. Kim, J.G., Kuby, M.: The deviation-flow refueling location model for optimizing a network of refueling stations. Int. J. Hydrogen Energy **37**(6), 5406–5420 (2012)
6. Zarandi, M.H.F., Davari, S., Sisakht, S.A.H.: The large-scale dynamic maximal covering location problem. Math. Comput. Modell. **57**(3), 710–719 (2013)
7. Hosseini, M., MirHassani, S., Hooshmand, F.: Deviation-flow refueling location problem with capacitated facilities: model and algorithm. Transp. Res. Part D Transp. Environ. **54**, 269–281 (2017)
8. Kuby, M., Lim, S.: The flow-refueling location problem for alternative-fuel vehicles. Socio-Econ. Plan. Sci. **39**(2), 125–145 (2005)
9. Upchurch, C., Kuby, M., Lim, S.: A model for location of capacitated alternative-fuel stations. Geogr. Anal. **41**(1), 85–106 (2009)
10. Mak, H.Y., Rong, Y., Shen, Z.J.M.: Infrastructure planning for electric vehicles with battery swapping. Manage. Sci. **59**(7), 1557–1575 (2013)
11. Zeng, M., Pan, Y., Zhang, D., Lu, Z., Li, Y.: Data-driven location selection for battery swapping stations. IEEE Access **7**, 133760–133771 (2019)
12. Kuby, M.J., Kelley, S.B., Schoenemann, J.: Spatial refueling patterns of alternative-fuel and gasoline vehicle drivers in Los Angeles. Transp. Res. Part D Transp. Environ. **25**, 84–92 (2013)
13. Murali, P., Ordñez, F., Dessouky, M.M.: Facility location under demand uncertainty: response to a large-scale bio-terror attack. Socio-Econ. Plan. Sci. **46**(1), 78–87 (2012). Special Issue: Disaster Planning and Logistics: Part 1
14. Gendreau, M., Potvin, J.Y., et al.: Handbook of Metaheuristics, vol. 3. Springer, Cham (2019)
15. Capar, I., Kuby, M., Leon, V.J., Tsai, Y.J.: An arc cover-path-cover formulation and strategic analysis of alternative-fuel station locations. Eur. J. Oper. Res. **227**(1), 142–151 (2013)

Optimal Combination of Tensor Optimization Methods

Dmitry Kamzolov[1]($\boxtimes$) (iD), Alexander Gasnikov[1,3] (iD),
and Pavel Dvurechensky[2,3] (iD)

[1] Moscow Institute of Physics and Technology, Moscow, Russia
dkamzolov@yandex.ru
[2] Weierstrass Institute for Applied Analysis and Stochastics, Berlin, Germany
[3] Institute for Information Transmission Problems RAS, Moscow, Russia

Abstract. We consider the minimization problem of a sum of a number of functions having Lipshitz p-th order derivatives with different Lipschitz constants. In this case, to accelerate optimization, we propose a general framework allowing to obtain near-optimal oracle complexity for each function in the sum separately, meaning, in particular, that the oracle for a function with lower Lipschitz constant is called a smaller number of times. As a building block, we extend the current theory of tensor methods and show how to generalize near-optimal tensor methods to work with inexact tensor step. Further, we investigate the situation when the functions in the sum have Lipschitz derivatives of a different order. For this situation, we propose a generic way to separate the oracle complexity between the parts of the sum. Our method is not optimal, which leads to an open problem of the optimal combination of oracles of a different order.

Keywords: Tensor method · Inexact method · Second-order method · Complexity

1 Introduction

Higher-order (tensor) methods, which use the derivatives of the objective up to order p, recently have become an area of intensive research effort in optimization, despite the idea is quite old and goes back to the works of P. Chebyshev and L. Kantorovich ([5,18]). One of the reasons is that the lower complexity bounds

The work of D. Kamzolov in Sects. 1–4 is funded by RFBR, project number 19-31-27001. The work of A. Gasnikov and P. Dvurechensky in Sects. 1–4 of the paper is supported by RFBR grant 18-29-03071 mk. The work in Sect. 5 is supported by the Ministry of Science and Higher Education of the Russian Federation (Goszadaniye) No. 075-00337-20-03, project No. 0714-2020-0005.

N. Olenev et al. (Eds.): OPTIMA 2020, LNCS 12422, pp. 166–183, 2020.
https://doi.org/10.1007/978-3-030-62867-3_13

were obtained in [1,2,26], which opened a question of optimal methods, and it was shown in [26] that Taylor expansion of a convex function can be made convex by appropriate regularization, leading to tractable tensor step implementable in practice. Recently nearly optimal methods were obtained in [13,26], and extensions for Hölder continuous higher-order derivatives were proposed in [16,28]. In this paper, we consider an interesting question that is still open in the theory of tensor methods. *Namely, if a tensor method minimizes a function f up to accuracy ε in $N_f(\varepsilon)$ oracle calls and possibly another tensor method minimizes a function g in $N_g(\varepsilon)$ oracle calls, is it possible to combine these two methods to minimize f + g up to accuracy ε in $\tilde{O}(N_f(\varepsilon))$ oracle calls for f and $\tilde{O}(N_g(\varepsilon))$ oracle calls for g?* To say more, we would like to have a generic approach which can take as an input different particular algorithms for each component. For simplicity, we consider a sum of two functions, but we believe that the approach can be generalized for an arbitrary number of functions. Note that in the last few years, the answer to this question plays a crucial role in the development of optimal algorithms for convex decentralized distributed optimization [3,9,14,19,21,27].

Some results in this direction are known for the first-order methods $p = 1$ [3,10,19,20,22] and for the case of the sum of two functions with the second being so simple that it can be incorporated directly in the tensor step [17] like in composite first-order methods [24]. Yet, the general theory on how to combine different methods to obtain optimal complexity for tensor methods is not yet developed for $p \geq 2$.

First, we consider uniformly convex sum of two functions $f + g$ each having Lipschitz derivatives of the same order p. Our approach is based on the recent framework of near-optimal tensor methods [13], which extends the algorithm of [23] to tensor methods. Our idea is to apply the near-optimal tensor method to the sum, considering g as a composite and including it into the tensor step without its Taylor approximation. Then each tensor step requires to solve properly regularized uniformly convex auxiliary problem. This is again done by the nearly optimal tensor method. Since the auxiliary problem turns out to be very well conditioned, it is possible to solve it very fast, and we only need to call the oracle for g. The careful analysis allows to separate the oracle complexity as we call the oracle for f only on outer iterations and oracle for g only on the inner, resulting in the optimal number of oracle calls for f and for g separately. As a building block, we explain how to extend near-optimal tensor methods to work with inexact tensor step, extending the current theory since existing near-optimal methods assume that the tensor step is exact. If the function is not uniformly convex, one can use a standard regularization technique with a small regularization parameter.

Note, there exist number of accelerated envelopes that allows to accelerate tensor methods: Monteiro–Svaiter envelop [4,12,17,23,25], Doikov–Nesterov envelope [7]. Further we will use Monteiro–Svaiter envelope. Note that it seems that Doikov–Nesterov envelop and standard direct Nesterov's tensor acceleration [26] doesn't well suited for our purposes. Note also, that for all envelops for the moment it's not known with what accuracy we should solve auxiliary problem. In Monteiro–Svaiter envelop we working on this in Appendix B. Among different variants of Monteiro–Svaiter envelop we preferred variant from [4], but we generalize (see Appendixes) [4] on composite case [17] and on uniformly convex problem target functions [12].

Second, we consider the case when f and g has Lipshitz derivatives of different order p_f and p_g respectively. We apply a similar technique as above, but using non-accelerated tensor methods as building blocks. We demonstrate that in this case, complexities can also be separated, but they turn out to be not optimal. This states an open problem of an optimal combination of optimal methods that use oracles of a different order. As far as we know for the moment there exists only one optimal result concerns the methods of different orders. This is the result from [3], where authors considered sliding of optimal 0-order and 1-order methods.

2   Problem Statement and Preliminaries

In what follows, we work in a finite-dimensional linear vector space E. Its dual space, the space of all linear functions on E, is denoted by E^*. For $x \in E$ and $s \in E^*$, we denote by $\langle s,x \rangle$ the value of a linear function s at x. For the (primal) space E, we introduce a norm $\| \cdot \|_E$. Then the dual norm is defined in the standard way:

$$\|s\|_{E^*} = \max_{x \in E} \left\{ \langle s,x \rangle : \|x\|_E \leq 1 \right\}.$$

Finally, for a convex function $f : \mathbf{dom}\, f \to R$ with $\mathbf{dom}\, f \subseteq E$ we denote by $\nabla f(x) \in E^*$ one of its subgradients.

We consider the following convex optimization problem:

$$\min_{x \in E} F(x) = f(x) + g(x), \tag{1}$$

where $f(x)$ and $g(x)$ are convex functions with Lipschitz p-th derivative, it means that

$$\|D^p f(x) - D^p f(y)\| \leq L_{p,f} \|x - y\|. \tag{2}$$

Then Taylor approximation of function $f(x)$ can be written as follows:

$$\Omega_p(f,x;y) = f(x) + \sum_{k=1}^{p} \frac{1}{k!} D^k f(x) \left[y - x\right]^k, y \in E.$$

By the standard arguments from [26], we get from (2) the following inequality

$$|f(y) - \Omega_p(f,x;y)| \leq \frac{L_{p,f}}{(p+1)!} \|y - x\|^{p+1}. \tag{3}$$

Now we introduce an additional condition for the functions.

Definition 1. *Function $F(x)$ is r-uniformly convex $(p + 1 \geq r \geq 2)$ if*

$$F(y) \geq F(x) + \langle \nabla F(x), y - x \rangle + \frac{\sigma_r}{r} \|y - x\|^r, \quad \forall x, y \in E$$

with constant σ_r.

One of the main examples of r-uniformly convex functions is $\frac{1}{r}\|x\|^r$ from Lemma 5 [7].

Lemma 1. *For fixed $r \geq 2$, consider the following function:*

$$f_r(x) = \frac{1}{r}\|x\|^r, \quad x \in \mathbb{E}.$$

Function $f_r(x)$ is uniformly convex of degree r with $\sigma_r = 2^{2-r}$.

Problem (1) can be solved by tensor methods [26] or its accelerated versions [4,13,17,25]. All these methods have a common subproblem to solve:

$$T_H(x) = \underset{y}{\operatorname{argmin}} \left\{ \Omega_p(f + g, x; y) + \frac{H_{p,f} + H_{p,g}}{p!} \|y - x\|^{p+1} \right\}.$$

For $H_{p,f} \geq L_{p,f}, H_{p,g} \geq L_{p,g}$ this subproblem is convex and hence can be effectively solved by some sub-solver, like a first-order method. Note that this method does not use information about sum type problem and compute their derivatives the same number of times. We want to separate computation complexity of high-order derivatives for sum of two functions. In next section we will describe this idea in more details.

As an accelerated optimal method, we introduce Accelerated Taylor Descent (ATD) from [4]. But for our paper we need to get a Composite Accelerated Taylor Descent(CATD).

Algorithm 1 is a generalization of ATD from [4] for composite optimization problem. It means that we try to minimize sum of two functions $F(x) = f(x) + g(x)$, where $g(x)$ is a proper closed convex function and subproblem (4) with $g(x)$ is easy to solve. Note that if $g(x)$ smooth and has a gradient, so $g'(y_{k+1}) = \nabla g(y_{k+1})$, but if $g(x)$ has only subgradient, we should introduce $g'(y_{k+1})$. Similarly to (2.9) from [6] by using optimality condition for (4) we define

$$g'(y_{k+1}) = -\nabla \Omega_p(f, \tilde{x}_k; y_{k+1}) - \frac{(p+1)H_{p,f}}{p!} \|y_{k+1} - \tilde{x}_k\|^{p-1}(y_{k+1} - \tilde{x}_k).$$

Theorem 1. *Let $F(x) = f(x) + g(x)$, where f denote a convex function whose p^{th} derivative is L_p-Lipschitz, $g(x)$ is a proper closed convex function and let x_* denote a minimizer of F. Then CATD satisfies, with $c_p = 2^{p-1}(p+1)^{\frac{3p+1}{2}}/(p-1)!$,*

$$F(y_k) - F(x_*) \leq \frac{c_p L_p R^{p+1}}{k^{\frac{3p+1}{2}}}, \tag{5}$$

Algorithm 1. Composite Accelerated Taylor Descent

1: **Input:** convex function $f : \mathbb{R}^d \to \mathbb{R}$ such that $\nabla^p f$ is L_p-Lipschitz, proper closed convex g : $\mathbb{R}^d \to \mathbb{R}$.
2: Set $A_0 = 0$, $x_0 = y_0$
3: **for** $k = 0$ **to** $k = K - 1$ **do**
4:     Compute a pair $\lambda_{k+1} > 0$ and $y_{k+1} \in \mathbb{R}^d$ such that

$$\frac{1}{2} \le \lambda_{k+1} \frac{H_{p,f} \cdot \|y_{k+1} - \tilde{x}_k\|^{p-1}}{(p-1)!} \le \frac{p}{p+1},$$

where

$$y_{k+1} = \underset{y}{\operatorname{argmin}} \left\{ \Omega_p(f, \tilde{x}_k; y) + \frac{H_{p,f}}{p!} \|y - \tilde{x}_k\|^{p+1} + g(y) \right\}, \tag{4}$$

and

$$a_{k+1} = \frac{\lambda_{k+1} + \sqrt{\lambda_{k+1}^2 + 4\lambda_{k+1} A_k}}{2}, \ A_{k+1} = A_k + a_{k+1}, \text{ and } \tilde{x}_k = \frac{A_k}{A_{k+1}} y_k + \frac{a_{k+1}}{A_{k+1}} x_k.$$

5:     Update $x_{k+1} := x_k - a_{k+1}\nabla f(y_{k+1}) - a_{k+1}g'(y_{k+1})$.
6: **return** y_K

---

where
$$R = \|x_0 - x^*\| \tag{6}$$
is the maximal radius of the initial set. Furthermore each iteration of ATD can be implemented in $\tilde{O}(1)$ calls to a pth-order Taylor expansion oracle, where $\tilde{O}$ means up to logarithmic factors.

We prove this theorem similarly to the proof of [4] in Appendix A.

Now we assume that function $F(x)$ is additionally r-uniformly convex, hence we may get a speed up by using restarts. Next, we introduce CATD with restarts and its convergence theorem.

Algorithm 2. CATD with restarts

1: **Input:** r-uniformly convex function $F : \mathbb{R}^d \to \mathbb{R}$ with constant σ_r and CATD conditions.
2: Set $z_0 = x_0 = 0$ and $R_0 = \|z_0 - x_*\|$.
3: **for** $k = 0$, **to** K **do**
4:     Set $R_k = R_0 \cdot 2^{-k}$ and

$$N_k = \max \left\{ \left\lceil \left(\frac{rc_p L_p 2^r}{\sigma_r} R_k^{p+1-r} \right)^{\frac{2}{3p+1}} \right\rceil, 1 \right\}. \tag{7}$$

5:     Set $z_{k+1} := y_{N_k}$ as the output of CATD started from z_k and run for N_k steps.
6: **return** z_K

Theorem 2. *CATD with restarts for r-uniformly convex function F with constant σ_r converges with N_r steps of CATD per restart and with N_F total number of CATD steps, where*

$$N_F = \tilde{O}\left[\left(\frac{L_{p,f}R^{p+1-r}}{\sigma_r}\right)^{\frac{2}{3p+1}}\right].$$

We prove this theorem similarly to [12] in Appendix C.

3   Uniformly Convex Functions

We consider similar to (1) problem.

$$\min F(x) = f(x) + g(x), \tag{8}$$

where additionally $F(x)$ is r-uniformly convex function. We also assume, that $p + 1 \geq r$.

Algorithm 2 applied to the problem (8) converges with the next convergence speed. To reach $F(x_N) - F(x^*) \leq \varepsilon$, we need $N_f + N_g$ iterations, where

$$N_f = \tilde{O}\left[\left(\frac{L_{p,f}R^{p+1-r}}{\sigma_r}\right)^{\frac{2}{3p+1}}\right], \tag{9}$$

$$N_g = \tilde{O}\left[\left(\frac{L_{p,g}R^{p+1-r}}{\sigma_r}\right)^{\frac{2}{3p+1}}\right]. \tag{10}$$

Note that for this method we compute $N_f + N_g$ derivatives for both $f(x)$ and $g(x)$ functions. We want to separate this computations and compute N_f derivatives for the function f and N_g derivatives for the function g.

Next we will describe the our framework. We assume that $L_{p,f} < L_{p,g}$, it means that $N_f < N_g$. For that case we consider problem 8 as a composite problem with $g(x)$ as a composite part. We solve this problem by Algorithm 2. In this algorithm we have tensor subproblem (4). To solve this subproblem we run another Algorithm 2 with objective function $\Omega_p(f, \tilde{x}_k; y) + \frac{H_{p,f}}{p!}\|y - \tilde{x}_k\|^{p+1} + g(y)$ up to the desired accuracy. As we will prove next, this subproblem may be solved linearly by the desired accuracy, so we should not worry too much about the level of the desired accuracy. We write more details about the correctness of this part and the more precise level of desired accuracy in Appendix B. As a result we get Algorithm 3.

Now we prove that this framework split computation's complexities.

Theorem 3. *Assume $F(x)$ is r-uniformly convex function ($r \geq 2$), $f(x)$ and $g(x)$ are convex functions with Lipshitz p-th derivative ($p \geq 1$, $p + 1 \geq r$) and $L_{p,f} < L_{p,g}$. Then by using our framework with $H_{p,f} = 2L_{p,f}$, method converges to $F(x_N) - F(x^*) \leq \varepsilon$ with N_f as (9) computations of derivatives $f(x)$ and N_g as (10) computation of derivatives $g(x)$.*

Algorithm 3. Tensor Methods Combination

1: **Input:** r-uniformly convex function $F(x) = f(x) + g(x)$ with constant σ_r, convex functions $f(x)$ and $g(x)$ such that $\nabla^p f$ is $L_{p,f}$-Lipschitz and $\nabla^p g$ is $L_{p,g}$-Lipschitz.

2: Set $z_0 = y_0 = x_0$
3: **for** $k = 0$, **to** $K - 1$ **do**
4:     Run Algorithm 2 for problem $f(x) + g(x)$, where $g(x)$ is a composite part.
5:     **for** $m = 0$, **to** $M - 1$ **do**
6:         Run Algorithm 2 up to desired accuracy for subproblem

$$\min_y \left(\Omega_p(f, \tilde{x}_k; y) + \frac{H_{p,f}}{p!} \|y - \tilde{x}_k\|^{p+1} + g(y) \right)$$

7: **return** z_K

Proof. As we prove in 2 for the outer composite method with constant $H_{p,f} = 2L_{p,f}$ we need to make

$$N_{out} = \tilde{O}\left[\left(\frac{2pL_{p,f}R^{p+1-r}}{\sigma_r} \right)^{\frac{2}{3p+1}} \right]$$

outer steps, it means that we need to compute $N_{out} = N_f$ derivatives of $f(x)$. Now we compute how much steps of inner method we need. Note that inner function has Lipshitz p-th derivative $H_{p,f} + L_g$. Also it is $(p+1)$-uniformly convex with σ_{p+1}. To compute σ_{p+1} we need to split $H_{p,f}$ into two parts $H_{p,f} = H_1 + H_2$, where the first part needs to make $\Omega_p(f, x; y) + \frac{H_1}{p!} \|y - x\|^{p+1}$ a convex function and the second part needs to make $\frac{H_2}{p!} \|y - x\|^{p+1}$ a uniformly convex term. Hence, from Lemma 1 we have $\sigma_{p+1} = \frac{H_2(p+1)2^{2-p}}{p!}$. We take $H_1 = H_2 = L_{p,f}$. As a result, the number of inner iterations equal to

$$N_{inn} = \tilde{O}\left[\left(\frac{2L_{p,f} + L_{p,g}}{\frac{(p+1)L_{p,f}2^{2-p}}{p!}} \right)^{\frac{2}{3p+1}} \log\left(\frac{F(x_0) - F(x^*) + H_{p,f}R^{p+1}}{\varepsilon} \right) \right]$$

$$= \tilde{O}\left[\left(\frac{2L_{p,f} + L_{p,g}}{\frac{(p+1)L_{p,f}2^{2-p}}{p!}} \right)^{\frac{2}{3p+1}} \right] \overset{L_{p,f} \leq L_{p,g}}{=} \tilde{O}\left[\left(\frac{L_{p,g}}{L_{p,f}} \right)^{\frac{2}{3p+1}} \right].$$

(11)

Hence the total number of inner iterations and total number of derivative's computations of $g(x)$ is

$$N_g = N_{out} \cdot N_{inn} = \tilde{O}\left[\left(\frac{L_{p,f}R^{p+1-r}}{\sigma_r} \right)^{\frac{2}{3p+1}} \right] \cdot \tilde{O}\left[\left(\frac{L_{p,g}}{L_{p,f}} \right)^{\frac{2}{3p+1}} \right]$$

$$= \tilde{O}\left[\left(\frac{L_{p,g}R^{p+1-r}}{\sigma_r} \right)^{\frac{2}{3p+1}} \right].$$

So we prove the theorem and split computation complexities.

Note, that this framework also easily adapts to methods without accelerating like [6,26]. But, unfortunately, it is much harder to adapt for other acceleration schemes. As we know, it is possible to adapt this framework for speed ups from [12,17] for $p \geq 2$, but for $p = 1$ it may arise some troubles because of adaptive inner regularisation and hence hard subproblem. As for [26] acceleration it also hard to adapt, because the inner subproblem is much harder with increasing complexity.

Also note that this framework can be generalized to the problem of the sum of m functions.

4   General Convex Functions

We consider (1) problem for convex functions.

If we will use Algorithm 1 for problem (1) we get next convergence speed. To reach $F(x_N) - F(x^*) \leq \varepsilon$, we need $N_f + N_g$ iterations, where

$$N_f = \tilde{O}\left[\left(\frac{L_{p,f} R^{p+1}}{\varepsilon}\right)^{\frac{2}{3p+1}}\right],\tag{12}$$

$$N_g = \tilde{O}\left[\left(\frac{L_{p,g} R^{p+1}}{\varepsilon}\right)^{\frac{2}{3p+1}}\right].\tag{13}$$

Now we prove that the our framework split computation's complexities for convex functions.

Theorem 4. *Assume $f(x)$ and $g(x)$ are convex functions with Lipshitz p-th derivative ($p \geq 1$, $p + 1 \geq q$) and $L_{p,f} < L_{p,g}$. Then by using our framework with $H_{p,f} = 2L_{p,f}$, method converges to $F(x_N) - F(x^*) \leq \varepsilon$ with N_f as (12) computations of derivatives $f(x)$ and N_g as (13) computation of derivatives $g(x)$.*

Proof. For the outer method 1 with constant $H_{p,f} = 2L_{p,f}$, we make

$$N_{out} = \tilde{O}\left[\left(\frac{2L_{p,f} R^{p+1}}{\varepsilon}\right)^{\frac{2}{3p+1}}\right]$$

outer steps, it means that we need to compute $N_{out} = N_f$ derivatives of $f(x)$. For inner method 1 to solve subproblem (4) similarly we has the same rate as (11) Hence the total number of inner iterations and total number of derivative's computations of $g(x)$ is

$$N_g = N_{out} \cdot N_{inn} = \tilde{O}\left[\left(\frac{2L_{p,f} R^{p+1}}{\varepsilon}\right)^{\frac{2}{3p+1}}\right] \cdot \tilde{O}\left[\left(\frac{L_{p,g}}{L_{p,f}}\right)^{\frac{2}{3p+1}}\right]$$

$$= \tilde{O}\left[\left(\frac{L_{p,g} R^{p+1}}{\varepsilon}\right)^{\frac{2}{3p+1}}\right].$$

So for convex function computation complexities are also splitting.

5   Multi-composite Tensor Method

The natural generalization of framework 3 is to use for the sum of two functions with different smoothness and hence different order of methods. But as we know, in the literature there is no method that works with the sum of two functions with different smoothness. We need to use tensor methods for the smallest order. To improve this situation we introduce the new type of problem, where $f(x)$ and $g(x)$ have different smoothness order. Similar idea for the first and second order was in the paper [8]. Next we propose a tensor method to solve such problem with splitting the complexities.

We introduce a multi-composite tensor optimization problem.

$$F(x) = f(x) + g(x) + h(x), \tag{14}$$

where $h(x)$ is a simple proper closed convex function, $f(x)$ is a convex functions with Lipschitz q-th derivative and $g(x)$ is a convex functions with Lipschitz p-th derivative. By using Theorem 1 from [26] we can get for $f(x)$ if $H_{q,f} \geq qL_{q,f}$, that

$$\Omega_q(f,x;y) + \frac{H_{q,f}}{(q+1)!}\|y-x\|^{q+1}$$

is convex and

$$f(y) \leq \Omega_q(f,x;y) + \frac{H_{q,f}}{(q+1)!}\|y-x\|^{q+1}. \tag{15}$$

Now we propose our method

$$T_{H_{q,f},H_{p,g}}(x) \in \operatorname*{Argmin}_y \left\{ \Omega_q(f,x;y) + \frac{H_{q,f}}{(q+1)!}\|y-x\|^{q+1} \right. \tag{16}$$

$$\left. + \Omega_p(g,x;y) + \frac{H_{p,g}}{(p+1)!}\|y-x\|^{p+1} + h(y) \right\}, \tag{17}$$

then

$$x_{t+1} = T_{H_{q,f},H_{p,g}}(x_t). \tag{18}$$

One can see that our method based on method [26] and combine models of two functions. Next we start to prove, that our method converges and split the complexities.

We assume that exists at least one solution x_* of problem (1) and the level sets of F are bounded. By the first-order optimality condition for $T = T_{H_{q,f},H_{p,g}}(x)$ we get:

$$\nabla\Omega_q(f,x;T) + \frac{H_{q,f}(T-x)}{q!}\|T-x\|^{q-1}$$

$$+\nabla\Omega_p(g,x;T) + \frac{H_{p,g}(T-x)}{p!}\|T-x\|^{p-1} + \partial h(T) = 0. \tag{19}$$

For the proof we need next small lemma.

Lemma 2. *For any $x \in E$, $H_{q,f} \geq qL_{q,f}$ and $H_{p,g} \geq pL_{p,g}$, we have*

$$F(T_{H_{q,f},H_{p,g}}(x)) \leq \min_y \left\{ F(y) + \frac{H_{q,f} + L_{q,f}}{(q+1)!} \|y - x\|^{q+1} + \frac{H_{p,g} + L_{p,g}}{(p+1)!} \|y - x\|^{p+1} \right\}. \tag{20}$$

Proof.

$$F(T_{H_{q,f},H_{p,g}}(x)) \leq \min_y \left\{ \Omega_q(f,x;y) + \frac{H_{q,f}}{(q+1)!} \|y - x\|^{q+1} \right.$$

$$\left. + \Omega_p(g,x;y) + \frac{H_{p,g}}{(p+1)!} \|y - x\|^{p+1} + h(y) \right\}$$

$$\overset{(3)}{\leq} \min_y \left\{ F(y) + \frac{H_{q,f} + L_{q,f}}{(q+1)!} \|y - x\|^{q+1} + \frac{H_{p,g} + L_{p,g}}{(p+1)!} \|y - x\|^{p+1} \right\}.$$

This leads us to the main theorem, that proves the convergence speed of our method.

Theorem 5. *If $f_q(x)$ is convex functions with Lipshitz constant $L_{q,f}$ for q-th derivative, $f_p(x)$ is convex functions with Lipshitz constant $L_{p,g}$ for p-th derivative; $H_{q,f} \geq qL_{q,f}$ and $H_{p,g} \geq pL_{p,g}$. α_t is chosen such that $\alpha_0 = 1$ and $\alpha_t \in [0;1]$ $t \geq 1$, then for any $t \geq 0$ for method (18) we have*

$$F(x_{t+1}) - F(x_*) \leq A_t \sum_{i=0}^{t} \left[C_f \frac{\alpha_i^{q+1}}{A_i} \|x_i - x_*\|^{q+1} + C_g \frac{\alpha_i^{p+1}}{A_i} \|x_i - x_*\|^{p+1} \right], \tag{21}$$

where

$$C_f = \frac{H_{q,f} + L_{q,f}}{(q+1)!}, \quad C_g = \frac{H_{p,g} + L_{p,g}}{(p+1)!},$$

$$A_t = \begin{cases} 1, & t = 0 \\ \prod\limits_{i=1}^{t} (1 - \alpha_i), & t \geq 1. \end{cases} \tag{22}$$

Proof. From (20)

$$F(x_{t+1}) \leq \min_y \left\{ F(y) + \frac{H_{q,f} + L_{q,f}}{(q+1)!} \|y - x_t\|^{q+1} + \frac{H_{p,g} + L_{p,g}}{(p+1)!} \|y - x_t\|^{p+1} \right\}$$

$$\leq F(y) + C_f \|y - x_t\|^{q+1} + C_g \|y - x_t\|^{p+1}.$$

If we take $y = x_t + \alpha_t(x_* - x_t)$, then by convexity

$$F(x_{t+1}) \leq F(y) + C_f \alpha_t^{q+1} \|x_* - x_t\|^{q+1} + C_g \alpha_t^{p+1} \|x_* - x_t\|^{p+1}$$

$$\leq (1 - \alpha_t) F(x_t) + \alpha_t F(x_*) + C_f \alpha_t^{q+1} \|x_* - x_t\|^{q+1} + C_g \alpha_t^{p+1} \|x_* - x_t\|^{p+1}.$$

Hence

$$F(x_{t+1}) - F(x_*) \leq (1 - \alpha_t) \left(F(x_t) - F(x_*) \right)$$

$$+ C_f \alpha_t^{q+1} \|x_* - x_t\|^{q+1} + C_g \alpha_t^{p+1} \|x_* - x_t\|^{p+1}.$$

For $t = 0$ and $\alpha_0 = 1$ we get

$$F(x_1) - F(x_*) \leq C_f \|x_* - x_0\|^{q+1} + C_g \|x_* - x_0\|^{p+1}.$$

For $t > 0$ we divide both sides by A_t:

$$\frac{1}{A_t}(F(x_{t+1}) - F(x_*)) \leq \frac{(1 - \alpha_t)}{A_t}(F(x_t) - F(x_*))$$

$$+ C_f \frac{\alpha_t^{q+1}}{A_t} \|x_* - x_t\|^{q+1} + C_g \frac{\alpha_t^{p+1}}{A_t} \|x_* - x_t\|^{p+1}$$

$$\overset{(22)}{\leq} \frac{1}{A_{t-1}}(F(x_t) - F(x_*))$$

$$+ C_f \frac{\alpha_t^{q+1}}{A_t} \|x_* - x_t\|^{q+1} + C_g \frac{\alpha_t^{p+1}}{A_t} \|x_* - x_t\|^{p+1}.$$

By summarising both sides we obtain (20).

Next we can fix parameters of this theorem and get next corollary.

Corollary 1. *For method* (18) *and* $\alpha_t = \frac{p+1}{t+p+1}$ *we have*

$$F(x_{t+1}) - F(x_*) \leq E_q \frac{(H_{q,f} + L_{q,f})R^{q+1}}{(t+p+1)^q} + E_p \frac{(H_{p,g} + L_{p,g})R^{p+1}}{(t+p+1)^p}, \qquad (23)$$

where

$$E_k = \frac{(p+1)^{k+1}}{(k+1)!}, \quad k = \{q, p\}.$$

Proof. We use

$$F(x_{t+1}) - F(x_*) \leq A_t \sum_{i=0}^{t} \left[C_f \frac{\alpha_i^{q+1}}{A_i} \|x_i - x_*\|^{q+1} + C_g \frac{\alpha_i^{p+1}}{A_i} \|x_i - x_*\|^{p+1} \right]$$

$$\overset{(6)}{\leq} C_f R^{q+1} \sum_{i=0}^{t} \frac{A_t \alpha_i^{q+1}}{A_i} + C_g R^{p+1} \sum_{i=0}^{t} \frac{A_t \alpha_i^{p+1}}{A_i}.$$

Now we compute these sums for $\alpha_t = \frac{p+1}{t+p+1}$:

$$A_t = \prod_{i=1}^{t}(1 - \alpha_i) = \prod_{i=1}^{t} \frac{i}{i+p+1} = \frac{t!\,(p+1)!}{(t+p+1)!} = (p+1)! \prod_{i=1}^{p+1} \frac{1}{t+i}$$

$$\geq \frac{(p+1)!}{(t+1)^{p+1}}.$$

For the first sum we get

$$\sum_{i=1}^{t} \frac{A_t \alpha_i^{p+1}}{A_i} = \sum_{i=1}^{t} \frac{(p+1)^{p+1} \prod_{j=1}^{p+1}(i+j)}{(i+p+1)^{p+1}(p+1)!} \cdot (p+1)! \prod_{i=1}^{p+1} \frac{1}{t+i}$$

$$= (p+1)^{p+1} \sum_{i=1}^{t} \prod_{j=1}^{p+1} \frac{i+j}{i+p+1} \prod_{i=1}^{p+1} \frac{1}{t+i}$$

$$\leq \frac{(p+1)^{p+1}}{(t+p+1)^p}.$$

For the second sum we get

$$\sum_{i=1}^{t} \frac{A_t \alpha_i^{q+1}}{A_i} = \sum_{i=1}^{t} \frac{(p+1)^{q+1} \prod_{j=1}^{p+1}(i+j)}{(i+p+1)^{q+1}(p+1)!} \cdot (p+1)! \prod_{i=1}^{p+1} \frac{1}{t+i}$$

$$= (p+1)^{q+1} \sum_{i=1}^{t} \frac{\prod_{j=1}^{p+1}(i+j)}{(i+p+1)^{q+1}} \cdot \prod_{i=1}^{p+1} \frac{1}{t+i}$$

$$\leq \frac{(p+1)^{q+1}}{(t+p+1)^q}.$$

From this two formulas for sums we get (23).

Finally, we prove that our method converges with the desired speed and split the complexities. Note that this algorithm can be generalized for the sum of m functions.

6   Conclusion

In this paper, we consider the minimization of the sum of two functions $f + g$ each having Lipshitz p-th order derivatives with different Lipschitz constants. We propose a general framework to accelerate tensor methods by splitting computational complexities. As a result, we get near-optimal oracle complexity for each function in the sum separately for any $p \geq 1$, including the first-order methods. To be more precise, if the near optimal complexity to minimize f is $N_f(\varepsilon)$ iterations and to minimize g is $N_g(\varepsilon)$, then our method requires no more than $\tilde{O}(N_f(\varepsilon))$ oracle calls for f and $\tilde{O}(N_g(\varepsilon))$ oracle calls for g to minimize $f + g$. We prove, that our framework works with both convex and uniformly convex functions. To get this result, we additionally generalize near-optimal tensor methods for composite problems with inexact inner tensor step.

Further, we investigate the situation when the functions in the sum have Lipschitz derivatives of a different order. For this situation, we propose a generic way to separate the oracle complexity between the parts of the sum. It is the first tensor method that works with functions with different smoothness. Our method is not optimal, which leads to an open problem of the optimal combination of oracles of a different order.

Acknowledgements. We would like to thank Yu. Nesterov for fruitful discussions on inexact solution of tensor subproblem.

A   Proof of Composite Accelerated Taylor Descent

This section is a rewriting of proof from [4], with adding composite part into the proof. Next theorem based on Theorem 2.1 from [4]

Theorem 6. *Let $(y_k)_{k\geq 1}$ be a sequence of points in $\mathbb{R}^d$ and $(\lambda_k)_{k\geq 1}$ a sequence in $\mathbb{R}_+$. Define $(a_k)_{k\geq 1}$ such that $\lambda_k A_k = a_k^2$ where $A_k = \sum_{i=1}^k a_i$. Define also for any $k \geq 0$, $x_k = x_0 - \sum_{i=1}^k a_i(\nabla f(y_i) + g'(y_i))$ and $\tilde{x}_k := \frac{a_{k+1}}{A_{k+1}}x_k + \frac{A_k}{A_{k+1}}y_k$. Finally assume if for some $\sigma \in [0,1]$*

$$\|y_{k+1} - (\tilde{x}_k - \lambda_{k+1}\nabla f(y_{k+1}))\| \leq \sigma \cdot \|y_{k+1} - \tilde{x}_k\|, \tag{24}$$

then one has for any $x \in \mathbb{R}^d$,

$$F(y_k) - F(x) \leq \frac{2\|x\|^2}{\left(\sum_{i=1}^k \sqrt{\lambda_i}\right)^2}, \tag{25}$$

and

$$\sum_{i=1}^k \frac{A_i}{\lambda_i}\|y_i - \tilde{x}_{i-1}\|^2 \leq \frac{\|x^*\|^2}{1 - \sigma^2}. \tag{26}$$

To prove this theorem we introduce auxiliaries lemmas based on Lemmas 2.2–2.5 and 3.1, Lemmas 2.6 and 3.3 one can take directly from [4] without any changes.

Lemma 3. *Let $\psi_0(x) = \frac{1}{2}\|x - x_0\|^2$ and define by induction $\psi_k(x) = \psi_{k-1}(x) + a_k\Omega_1(F, y_k, x)$. Then $x_k = x_0 - \sum_{i=1}^k a_i(\nabla f(y_i) + g'(y_i))$ is the minimizer of ψ_k, and $\psi_k(x) \leq A_k F(x) + \frac{1}{2}\|x - x_0\|^2$ where $A_k = \sum_{i=1}^k a_i$.*

Lemma 4. *Let (z_k) be a sequence such that*

$$\psi_k(x_k) - A_k F(z_k) \geq 0. \tag{27}$$

Then one has for any x,

$$F(z_k) \leq F(x) + \frac{\|x - x_0\|^2}{2A_k}. \tag{28}$$

Proof. One has (recall Lemma 3):

$$A_k F(z_k) \leq \psi_k(x_k) \leq \psi_k(x) \leq A_k F(x) + \frac{1}{2}\|x - x_0\|^2.$$

Lemma 5. *One has for any x,*

$$\psi_{k+1}(x) - A_{k+1}F(y_{k+1}) - (\psi_k(x_k) - A_k F(z_k))$$

$$\geq A_{k+1}(\nabla f(y_{k+1}) + g'(y_{k+1})) \cdot \left(\frac{a_{k+1}}{A_{k+1}}x + \frac{A_k}{A_{k+1}}z_k - y_{k+1}\right) + \frac{1}{2}\|x - x_k\|^2.$$

Proof. Firstly, by simple calculation we note that:

$$\psi_k(x) = \psi_k(x_k) + \frac{1}{2}\|x - x_k\|^2, \text{ and } \psi_{k+1}(x) = \psi_k(x_k) + \frac{1}{2}\|x - x_k\|^2 + a_{k+1}\Omega_1(f, y_{k+1}, x),$$

so that

$$\psi_{k+1}(x) - \psi_k(x_k) = a_{k+1}\Omega_1(F, y_{k+1}, x) + \frac{1}{2}\|x - x_k\|^2. \tag{29}$$

Now we want to make appear the term $A_{k+1}F(z_{k+1}) - A_kF(z_k)$ as a lower bound on the right hand side of (29) when evaluated at $x = x_{k+1}$. Using the inequality $\Omega_1(F, y_{k+1}, z_k) \leq f(z_k)$ we have:

$$a_{k+1}\Omega_1(F, y_{k+1}, x) = A_{k+1}\Omega_1(F, y_{k+1}, x) - A_k\Omega_1(F, y_{k+1}, x)$$
$$= A_{k+1}\Omega_1(F, y_{k+1}, x) - A_k\nabla F(y_{k+1}) \cdot (x - z_k) - A_k\Omega_1(F, y_{k+1}, z_k)$$
$$= A_{k+1}\Omega_1\left(F, y_{k+1}, x - \frac{A_k}{A_{k+1}}(x - z_k)\right) - A_k\Omega_1(F, y_{k+1}, z_k)$$
$$\geq A_{k+1}F(y_{k+1}) - A_kF(z_k)$$
$$+ A_{k+1}(\nabla f(y_{k+1}) + g'(y_{k+1})) \cdot \left(\frac{a_{k+1}}{A_{k+1}}x + \frac{A_k}{A_{k+1}}z_k - y_{k+1}\right),$$

which concludes the proof.

Lemma 6. *Denoting* $\lambda_{k+1} := \frac{a_{k+1}^2}{A_{k+1}}$ *and* $\tilde{x}_k := \frac{a_{k+1}}{A_{k+1}}x_k + \frac{A_k}{A_{k+1}}y_k$ *one has:*

$$\psi_{k+1}(x_{k+1}) - \Lambda_{k+1}F(y_{k+1}) - (\psi_k(x_k) - A_kF(y_k))$$
$$\geq \frac{A_{k+1}}{2\lambda_{k+1}}\left(\|y_{k+1} - \tilde{x}_k\|^2 - \|y_{k+1} - (\tilde{x}_k - \lambda_{k+1}(\nabla f(y_{k+1})) + g'(y_{k+1}))\|^2\right).$$

In particular, we have in light of (24)

$$\psi_k(x_k) - A_kF(y_k) \geq \frac{1 - \sigma^2}{2}\sum_{i=1}^{k}\frac{A_i}{\lambda_i}\|y_i - \tilde{x}_{i-1}\|^2.$$

Proof. We apply Lemma 5 with $z_k = y_k$ and $x = x_{k+1}$, and note that (with $\tilde{x} := \frac{a_{k+1}}{A_{k+1}}x + \frac{A_k}{A_{k+1}}y_k$):

$$(\nabla f(y_{k+1}) + g'(y_{k+1})) \cdot \left(\frac{a_{k+1}}{A_{k+1}}x + \frac{A_k}{A_{k+1}}y_k - y_{k+1}\right) + \frac{1}{2A_{k+1}}\|x - x_k\|^2$$
$$= (\nabla f(y_{k+1}) + g'(y_{k+1})) \cdot (\tilde{x} - y_{k+1}) + \frac{1}{2A_{k+1}}\left\|\frac{A_{k+1}}{a_{k+1}}\left(\tilde{x} - \frac{A_k}{A_{k+1}}y_k\right) - x_k\right\|^2$$
$$= (\nabla f(y_{k+1}) + g'(y_{k+1})) \cdot (\tilde{x} - y_{k+1}) + \frac{A_{k+1}}{2a_{k+1}^2}\left\|\tilde{x} - \left(\frac{a_{k+1}}{A_k}x_k + \frac{A_k}{A_{k+1}}y_k\right)\right\|^2.$$

This yields:

$$\psi_{k+1}(x_{k+1}) - A_{k+1}F(y_{k+1}) - (\psi_k(x_k) - A_kF(y_k))$$
$$\geq A_{k+1} \cdot \min_{x \in \mathbb{R}^d}\left\{(\nabla f(y_{k+1}) + g'(y_{k+1})) \cdot (x - y_{k+1}) + \frac{1}{2\lambda_{k+1}}\|x - \tilde{x}_k\|^2\right\}.$$

The value of the minimum is easy to compute.

For the first conclusion in Theorem 6, it suffices to combine Lemma 6 with Lemma 4, and Lemma 2.5 from [4]. The second conclusion in Theorem 6 follows from Lemma 6 and Lemma 3.

The following lemma shows that minimizing the pth order Taylor expansion (4) can be viewed as an implicit gradient step for some "large" step size:

Lemma 7. *Equation* (24) *holds true with* $\sigma = 1/2$ *for* (4), *provided that one has:*

$$\frac{1}{2} \le \lambda_{k+1} \frac{L_p \cdot \|y_{k+1} - \tilde{x}_k\|^{p-1}}{(p-1)!} \le \frac{p}{p+1}. \tag{30}$$

Proof. Observe that the optimality condition gives:

$$\nabla_y f_p(y_{k+1}, \tilde{x}_k) + \frac{L_p \cdot (p+1)}{p!}(y_{k+1} - \tilde{x}_k)\|y_{k+1} - \tilde{x}_k\|^{p-1} + g'(y_{k+1}) = 0. \tag{31}$$

In particular we get:

$$y_{k+1} - (\tilde{x}_k - \lambda_{k+1}(\nabla f(y_{k+1}) + g'(y_{k+1}))) = \lambda_{k+1}(\nabla f(y_{k+1}) + g'(y_{k+1}))$$
$$- \frac{p!}{L_p \cdot (p+1) \cdot \|y_{k+1} - \tilde{x}_k\|^{p-1}}(\nabla_y f_p(y_{k+1}, \tilde{x}_k) + g'(y_{k+1})).$$

By doing a Taylor expansion of the gradient function one obtains:

$$\|\nabla f(y) - \nabla_y f_p(y, x)\| \le \frac{L_p}{p!}\|y - x\|^p,$$

so that we find:

$$\|y_{k+1} - (\tilde{x}_k - \lambda_{k+1}(\nabla f(y_{k+1}) + g'(y_{k+1})))\|$$
$$\le \lambda_{k+1}\frac{L_p}{p!}\|y_{k+1} - \tilde{x}_k\|^p + \left|\lambda_{k+1} - \frac{p!}{L_p \cdot (p+1) \cdot \|y_{k+1} - \tilde{x}_k\|^{p-1}}\right| \cdot \|\nabla_y f_p(y_{k+1}, \tilde{x}_k) + g'(y_{k+1})\|$$
$$\le \|y_{k+1} - \tilde{x}_k\| \left(\lambda_{k+1}\frac{L_p}{p!}\|y_{k+1} - \tilde{x}_k\|^{p-1} + \left|\lambda_{k+1}\frac{L_p \cdot (p+1) \cdot \|y_{k+1} - \tilde{x}_k\|^{p-1}}{p!} - 1\right|\right)$$
$$= \|y_{k+1} - \tilde{x}_k\| \left(\frac{\eta}{p} + \left|\eta \cdot \frac{p+1}{p} - 1\right|\right)$$

where we used (31) in the second last equation and we let $\eta :=$ $\lambda_{k+1}\frac{L_p \cdot \|y_{k+1} - \tilde{x}_k\|^{p-1}}{(p-1)!}$ in the last equation. The result follows from the assumption $1/2 \le \eta \le p/(p+1)$ in (30).

Finally, if we replace $\|x^*\|$ by $\|x_0 - x^*\|$ in Lemma 3.3 and use Lemma 3.4 from [4] we prove Theorem 6.

B   Inexact solution of the subproblem

Suppose that (4) can not be solved exactly. Assume that we can find only inexact solution $\tilde{y}_{k+1}$ satisfies

$$\left\|\nabla\left(f_p(\tilde{y}_{k+1}, \tilde{x}_k) + \frac{L_p}{p!}\|\tilde{y}_{k+1} - \tilde{x}_k\|^{p+1} + g(\tilde{y}_{k+1})\right)\right\| \le \frac{L_p}{2p!}\|\tilde{y}_{k+1} - \tilde{x}_k\|^p. \tag{32}$$

In this case Lemma 7 should be corrected.

Lemma 8. *Equation* (24) *holds true with* $\sigma = 3/4$ *for* (32), *provided that one has:*

$$\frac{1}{2} \leq \lambda_{k+1} \frac{L_p \cdot \|\tilde{y}_{k+1} - \tilde{x}_k\|^{p-1}}{(p-1)!} \leq \frac{p}{p+1}.$$

Proof. Let's introduce

$$\Xi_{k+1} = \nabla \left(f_p(\tilde{y}_{k+1}, \tilde{x}_k) + \frac{L_p}{p!} \|\tilde{y}_{k+1} - \tilde{x}_k\|^{p+1} + g(\tilde{y}_{k+1}) \right).$$

The main difference with the proof of Lemma 7 is in the following line

$$\|\tilde{y}_{k+1} - (\tilde{x}_k - \lambda_{k+1}(\nabla f(\tilde{y}_{k+1}) + g'(\tilde{y}_{k+1})))\|$$

$$\leq \lambda_{k+1} \frac{L_p}{p!} \|\tilde{y}_{k+1} - \tilde{x}_k\|^p +$$

$$\left| \lambda_{k+1} - \frac{p!}{L_p \cdot (p+1) \cdot \|\tilde{y}_{k+1} - \tilde{x}_k\|^{p-1}} \right| \cdot \|\nabla_y f_p(\tilde{y}_{k+1}, \tilde{x}_k) + g'(\tilde{y}_{k+1})\| + \lambda_{k+1} \Xi_{k+1}$$

$$\leq \|\tilde{y}_{k+1} - \tilde{x}_k\| \left(\lambda_{k+1} \frac{L_p}{p!} \|\tilde{y}_{k+1} - \tilde{x}_k\|^{p-1} + \left| \lambda_{k+1} \frac{L_p \cdot (p+1) \cdot \|\tilde{y}_{k+1} - \tilde{x}_k\|^{p-1}}{p!} - 1 \right| \right)$$

$$+ \|\tilde{y}_{k+1} - \tilde{x}_k\| \cdot \frac{1}{2p} \cdot \lambda_{k+1} \frac{L_p \cdot \|\tilde{y}_{k+1} - \tilde{x}_k\|^{p-1}}{(p-1)!}.$$

To complete the proof it's left to notice that due to the (32)

$$\|\Xi_{k+1}\| \leq \frac{L_p}{2p!} \|\tilde{y}_{k+1} - \tilde{x}_k\|^p.$$

Based on (32) we try to relate the accuracy $\tilde{\varepsilon}$ we need to solve auxiliary problem to the desired accuracy ε for the problem (1). For this we use Lemma 2.1 from [15]. This Lemma guarantee that if

$$\left\| \nabla \left(f_p(\tilde{y}_{k+1}, \tilde{x}_k) + \frac{L_p}{p!} \|\tilde{y}_{k+1} - \tilde{x}_k\|^{p+1} + g(\tilde{y}_{k+1}) \right) \right\| \leq \frac{1}{4p(p+1)} \|\nabla F(\tilde{y}_{k+1})\|, \tag{33}$$

then (32) holds true. So it's sufficient to solve auxiliary problem in terms of (33).

Assume that $F(x)$ is r-uniformly convex function with constant σ_r ($r \geq 2$, $\sigma_r > 0$, see Definition 1), then from Lemma 2 [7] we have

$$F(\tilde{y}_{k+1}) - \min_{x \in E} F(x) \leq \frac{r-1}{r} \left(\frac{1}{\sigma_r} \right)^{\frac{1}{r-1}} \|\nabla F(\tilde{y}_{k+1})\|^{\frac{r}{r-1}}. \tag{34}$$

Inequalities (33), (34) give us guarantees that it's sufficient to solve auxiliary problem with the accuracy

$$\tilde{\varepsilon} = O \left((\varepsilon^{r-1} \sigma_r)^{\frac{1}{r}} \right)$$

in terms of criteria (33). Since auxiliary problem is every time r-uniformly convex we can apply (34) to auxiliary problem to estimate the accuracy in terms of function discrepancy. Anyway we will have that there is no need to think about it since the dependence of this accuracy are logarithmic. The only restrictive assumption we made is that $F(x)$ is r-uniformly convex. If this is not a case, like in Sect. 4, we may use regularisation tricks [11]. This lead us to $\sigma_2 \sim \varepsilon$. So the dependence $\tilde{\varepsilon}$ becomes worthier, but this doesn't change the main conclusion about possibility to skip the details concern the accuracy of the solution of auxiliary problem.

C   CATD with restarts

The proof of the Theorem 2.

Proof. As F is r-uniformly convex function we get

$$R_{k+1} = \|z_{k+1} - x_*\| \overset{(5)}{\leq} \left(\frac{r \left(F(z_{k+1}) - F(x_*) \right)}{\sigma_r} \right)^{\frac{1}{r}} \leq \left(\frac{r \left(\frac{c_p L_p R_k^{p+1}}{N_k^{\frac{3p+1}{2}}} \right)}{\sigma_r} \right)^{\frac{1}{r}}$$

$$= \left(\frac{r c_p L_p R_k^{p+1}}{\sigma_r N_k^{\frac{3p+1}{2}}} \right)^{\frac{1}{r}} \overset{(7)}{\leq} \left(\frac{R_k^{p+1}}{2^r R_k^{p+1-r}} \right)^{\frac{1}{r}} = \frac{R_k}{2}.$$

Now we compute the total number of CATD steps.

$$\sum_{k=0}^{K} N_k \leq \sum_{k=0}^{K} \left(\frac{r c_p L_p 2^r}{\sigma_r} R_k^{p+1-r} \right)^{\frac{2}{3p+1}} + K = \sum_{k=0}^{K} \left(\frac{r c_p L_p 2^r}{\sigma_r} (R_0 2^{-k})^{p+1-r} \right)^{\frac{2}{3p+1}} + K$$

$$= \left(\frac{r c_p L_p 2^r R_0^{p+1-r}}{\sigma_r} \right)^{\frac{2}{3p+1}} \sum_{k=0}^{K} 2^{\frac{-2(p+1-r)k}{3p+1}} + K.$$

References

1. Agarwal, N., Hazan, E.: Lower bounds for higher-order convex optimization. In: Conference On Learning Theory. PMLR (2018)
2. Arjevani, Y., Shamir, O., Shiff, R.: Oracle complexity of second-order methods for smooth convex optimization. Math. Program. **178**(1–2), 327–360 (2019)
3. Beznosikov, A., Gorbunov, E., Gasnikov, A.: Derivative-free method for decentralized distributed non-smooth optimization. arXiv preprint arXiv:1911.10645 (2019)
4. Bubeck, S., Jiang, Q., Lee, Y.T., Li, Y., Sidford, A.: Near-optimal method for highly smooth convex optimization. In: Conference on Learning Theory, pp. 492–507 (2019)
5. Chebyshev, P.: Collected Works, vol. 5. Strelbytskyy Multimedia Publishing, Kyiv (2018)
6. Doikov, N., Nesterov, Y.: Local convergence of tensor methods. arXiv preprint arXiv:1912.02516 (2019)

7. Doikov, N., Nesterov, Y.: Minimizing uniformly convex functions by cubic regularization of newton method. arXiv preprint arXiv:1905.02671 (2019)
8. Doikov, N., Richtárik, P.: Randomized block cubic newton method. In: International Conference on Machine Learning, pp. 1290–1298 (2018)
9. Dvinskikh, D., Gasnikov, A.: Decentralized and parallelized primal and dual accelerated methods for stochastic convex programming problems. arXiv preprint arXiv:1904.09015 (2019)
10. Dvinskikh, D., Omelchenko, S., Tiurin, A., Gasnikov, A.: Accelerated gradient sliding and variance reduction. arXiv preprint arXiv:1912.11632 (2019)
11. Dvurechensky, P., Gasnikov, A., Ostroukhov, P., Uribe, C.A., Ivanova, A.: Near-optimal tensor methods for minimizing the gradient norm of convex function. arXiv preprint arXiv:1912.03381 (2019)
12. Gasnikov, A., Dvurechensky, P., Gorbunov, E., Vorontsova, E., Selikhanovych, D., Uribe, C.A.: Optimal tensor methods in smooth convex and uniformly convex optimization. In: Conference on Learning Theory, pp. 1374–1391 (2019)
13. Gasnikov, A., et al.: Near optimal methods for minimizing convex functions with lipschitz p-th derivatives. In: Conference on Learning Theory, pp. 1392–1393 (2019)
14. Gorbunov, E., Dvinskikh, D., Gasnikov, A.: Optimal decentralized distributed algorithms for stochastic convex optimization. arXiv preprint arXiv:1911.07363 (2019)
15. Grapiglia, G.N., Nesterov, Y.: On inexact solution of auxiliary problems in tensor methods for convex optimization. arXiv preprint arXiv:1907.13023 (2019)
16. Grapiglia, G.N., Nesterov, Y.: Tensor methods for minimizing functions with Hölder continuous higher-order derivatives. arXiv preprint arXiv:1904.12559 (2019)
17. Jiang, B., Wang, H., Zhang, S.: An optimal high-order tensor method for convex optimization. In: Conference on Learning Theory, pp. 1799–1801 (2019)
18. Kantorovich, L.V.: On Newton's method. Trudy Matematicheskogo Instituta imeni VA Steklova **28**, 104–144 (1949)
19. Lan, G.: Lectures on optimization. Methods for machine learning. H. Milton Stewart School of Industrial and Systems Engineering, Georgia Institute of Technology, Atlanta, GA (2019)
20. Lan, G.: Gradient sliding for composite optimization. Math. Program. **159**(1–2), 201–235 (2016)
21. Lan, G., Lee, S., Zhou, Y.: Communication-efficient algorithms for decentralized and stochastic optimization. Math. Program. **180**(1), 237–284 (2018). https://doi.org/10.1007/s10107-018-1355-4
22. Lan, G., Ouyang, Y.: Accelerated gradient sliding for structured convex optimization. arXiv preprint arXiv:1609.04905 (2016)
23. Monteiro, R.D., Svaiter, B.F.: An accelerated hybrid proximal extragradient method for convex optimization and its implications to second-order methods. SIAM J. Optim. **23**(2), 1092–1125 (2013)
24. Nesterov, Y.: Gradient methods for minimizing composite functions. Math. Program. **140**(1), 125–161 (2013)
25. Nesterov, Y.: Lectures on Convex Optimization. SOIA, vol. 137. Springer, Cham (2018). https://doi.org/10.1007/978-3-319-91578-4
26. Nesterov, Y.: Implementable tensor methods in unconstrained convex optimization. Math. Program., 1–27 (2019). https://doi.org/10.1007/s10107-019-01449-1
27. Rogozin, A., Gasnikov, A.: Projected gradient method for decentralized optimization over time-varying networks. arXiv preprint arXiv:1911.08527 (2019)
28. Song, C., Ma, Y.: Towards unified acceleration of high-order algorithms under Hölder continuity and uniform convexity. arXiv preprint arXiv:1906.00582 (2019)

Nonlinear Least Squares Solver for Evaluating Canonical Tensor Decomposition

Igor Kaporin[(✉)] [iD]

Dorodnicyn Computing Center of FRC CSC RAS,
Vavilova street 40, Moscow, Russia
igorkaporin@mail.ru

Abstract. Nonlinear least squares iterative solver developed earlier by the author is applied for numerical solution of special system of multilinear equations arising in the problem of canonical tensor decomposition. The proposed algorithm is based on easily parallelizable computational kernels such as matrix-vector multiplications and elementary vector operations and therefore has a potential for a quite efficient implementation on modern high-performance computers. The results of numerical testing presented for certain examples of large scale dense and medium size sparse 3D tensors found in existing literature seem very competitive with respect to computational costs involved.

Keywords: Nonlinear least squares · Levenberg-Marquardt method · Preconditioned subspace descent · Canonical tensor decomposition

1 Introduction

Application areas of nonlinear least squares are numerous and include, for instance, acceleration of neural network learning processes using Levenberg - Marquardt type algorithms, pattern recognition, signal processing etc. This explains the need in further development of robust and efficient nonlinear least squares solvers.

In the present work, the problem of approximate evaluation of canonical decomposition of a three-way array will be considered. The canonical decomposition of multi-way arrays serves as an application tool for data analysis, where it has been used in a variety of fields including chemometrics, data mining, image compression, neuroscience and telecommunications.

The present paper is mainly based on the computational scheme developed in [10, 11] which has different options and parametrizations. Here we concretize the algorithm to make if more efficient for the problem of canonical tensor decomposition. In fact, the nonlinear solver described below is a general purpose one, and can be effectively applied to nonlinear problems other than tensor decomposition, especially in high-performance computing environment.

Supported by RFBR grant No.19-01-00666.

N. Olenev et al. (Eds.): OPTIMA 2020, LNCS 12422, pp. 184–195, 2020.
https://doi.org/10.1007/978-3-030-62867-3_14

2   General Description of Nonlinear LS Solver

A standard least squares problem is formulated as

$$x_* = \arg \min_{x \in R^n} \varphi(x), \tag{1}$$

where the function $\varphi : R^n \to R$ has the form

$$\varphi(x) = \frac{1}{2}\|f(x)\|^2 \equiv \frac{1}{2}f^T(x)f(x), \tag{2}$$

and $f(x)$ is a nonlinear mapping

$$f : R^n \to R^m, \qquad m \geq n, \tag{3}$$

Assuming sufficient smoothness of f, an iterative procedure is constructed to find the minimizer x_* numerically. Note that x_* satisfies the equation

$$\text{grad } \varphi(x_*) = 0, \tag{4}$$

where

$$\text{grad } \varphi(x) = J^T(x)f(x) \in R^n, \tag{5}$$

and

$$J(x) \equiv \frac{\partial f}{\partial x} \in R^{m \times n}, \tag{6}$$

is the Jacobian matrix of f at x.

2.1   Descent Along Normalized Direction

Let $x_0, x_1, \ldots, x_k, \ldots$ be the sequence of approximations to the stationary point x_* constructed in the course of iterations. Further on, we will use the notations

$$f_k = f(x_k), \quad J_k = J(x_k), \quad g_k = \text{grad}(x_k) = J_k^T f_k. \tag{7}$$

According to [6–11], the next approximation x_{k+1} to x_* is constructed as

$$x_{k+1} = x_k + \alpha_k p_k, \tag{8}$$

where p_k is a normalized direction vector satisfying the scaling condition

$$(J_k p_k)^T (f_k + J_k p_k) = 0, \tag{9}$$

and the stepsize parameter α_k satisfies

$$0 < \alpha_k < 2.$$

Next we consider sufficient conditions for the descent of $\varphi(x_k)$.

2.2   General Estimate for Residual Norm Reduction

Under rather mild conditions, see, e.g. [10], there exists the limiting stepsize $\widehat{\alpha}_k \in (0, 2)$ such that for all $0 < \alpha \le \widehat{\alpha}_k$ the estimate

$$\frac{\varphi(x_k + \alpha p_k)}{\varphi(x_k)} \le 1 - \left(\left(\alpha - \frac{\alpha^2}{2} \right) \vartheta_k^2 \right)^2 \tag{10}$$

is valid, where φ is defined in (2) and ϑ_k is determined as

$$\vartheta_k = \vartheta(f_k, J_k p_k) \equiv \frac{-(J_k p_k)^T f_k}{\|f_k\| \| J_k p_k \|}. \tag{11}$$

The latter is always nonnegative since by the normalization condition (9) it holds

$$\vartheta_k = \| J_k p_k \| / \| f_k \|. \tag{12}$$

Note that the quantity $\vartheta(f, Jp)$ represents the cosine of the Euclidean acute angle between m-vectors f and $(-Jp)$. Clearly, estimate (10) shows the importance of finding good directions p with values of $\vartheta(f, Jp)$ as large as possible.

Remark 1. The proof of (10) can be found in [6, 10, 11], where the limiting step-size $\widehat{\alpha} = \widehat{\alpha}(f, p)$ along a normalized direction p is defined as the maximum number such that the limiting stepsize condition

$$\| f(x + \alpha p) - f - \alpha Jp \| \le \left(\alpha - \frac{\alpha^2}{2} \right) \frac{\| Jp \|^2}{\| f \|} \tag{13}$$

is satisfied for all $0 < \alpha \le \widehat{\alpha}$.

Remark 2. It appears that $\widehat{\alpha}$ characterizes the nonlinearity of f in the neighborhood of x, while ϑ reflects the precision of approximate solution p of the "Newton equation" $f + Jp = 0$. Note that the latter may not (and often cannot) be solved exactly in the context of our considerations.

2.3   Choosing the Stepsize

Based on estimate (10) one can develop the following Armijo type procedure [2] for evaluating appropriate stepsize α_k providing for a certain decrease of the residual norm. Let $\widetilde{p}_k$ be a direction vector satisfying $\vartheta(f_k, J_k \widetilde{p}_k) = \vartheta_k > 0$ but, in general, not normalized. Therefore, we first normalize it using

$$p_k = \frac{-(J_k \widetilde{p}_k)^T f_k}{\| J_k \widetilde{p}_k \|^2} \widetilde{p}_k; \tag{14}$$

obviously, the normalization does not change the value of ϑ_k, and (9), (11) hold. Next we check the validity of estimate (10) for a decreasing sequence of trial values of $\alpha \in (0, 2)$; the standard choice is

$$\alpha^{(l)} = 2^{-l}, \qquad l = 0, 1, \ldots, l_{\max} - 1, \tag{15}$$

with $l_{\max} = 30$, which approximately corresponds to $\alpha^{(l)} > 2 \cdot 10^{-8}$. As soon as (10) be satisfied, one sets $\alpha_k = \alpha^{(l)}$. In numerical testing, the backtracking criterion (10) was often satisfied at once for $l = 0$ with the stepsize $\alpha_k = 1$.

2.4 Choosing Subspace Basis and Descent Direction

Let us choose the direction p_k as

$$p_k = V_k z_k, \tag{16}$$

where the columns of $V_k \in R^{n \times s}$ form the basis of the kth search subspace, and vector $z_k \in R^s$ contains the coefficients to be optimized. Here s is a small integer, typically $2 \le s \le 10$. We will restrict our attention to the choice of the subspace basis as

$$V_0 = [W_0 J_0^T f_0], \quad V_1 = [W_1 J_1^T f_1 \mid \widehat{p}_0], \quad V_2 = [W_2 J_2^T f_2 \mid W_2 J_2^T J_2 W_2 J_2^T f_2 \mid \widehat{p}_1]$$

(here we assumed that $s \ge 3$) and, in general,

$$V_k = [W_k J_k^T f_k \mid W_k J_k^T J_k W_k J_k^T f_k \mid \ \ldots \ \mid (W_k J_k^T J_k)^{s-2} W_k J_k^T f_k \mid \widehat{p}_{k-1}], \tag{17}$$

where $\widehat{p}_{k-1} = x_k - x_{k-1}$ is the previous direction (unnormalized) and symmetric positive definite matrix $W_k \in R^{n \times n}$ is a preconditioning matrix, see Sect. 2.5 below. Note that for $k = 0$ the last basis vector in (17) is omitted so that $V_0 \in R^{n \times 1}$. This construction is a particular case of the one proposed in [7,8].

A similar choice of the subspace basis was also discussed in [17], see also references therein. Closely related method for minimizing convex functions was considered in [12].

Our choice of z_k in (16) is subject to the condition of maximizing an appropriate lower bound for ϑ_k. From the definition (17) and (16) one readily has, denoting

$$U_k = J_k V_k \in R^{m \times s} \tag{18}$$

and introducing a (small) positive parameter ξ_k, the following estimate:

$$\max_{p=V_k z} \vartheta^2(f_k, J_k p) = \max_{p=V_k z} \frac{p^T J_k^T f_k f_k^T J_k p}{f_k^T f_k \ p^T J_k^T J_k p} \ge \max_{p=V_k z} \frac{p^T J_k^T f_k f_k^T J_k p}{f_k^T f_k \ p^T (J_k^T J_k + \xi_k I) p}$$

$$= \max_z \frac{z^T U_k^T f_k f_k^T U_k z}{f_k^T f_k \ z^T (U_k^T U_k + \xi_k V_k^T V_k) z}.$$

Finding the maximizer of the latter Rayleigh quotient as

$$z_k = (U_k^T U_k + \xi_k V_k^T V_k)^{-1} U_k^T f_k,$$

one obtains the following expression for the kth direction (defined up to the scalar multiple found by (14)):

$$\widetilde{p}_k = V_k (U_k^T U_k + \xi_k V_k^T V_k)^{-1} U_k^T f_k, \tag{19}$$

where the reasonable choice of the regularizing parameter is $\xi_k = O(\|f_k\|)$; see also [10,11] for some additional arguments supporting this formula. The resulting value of the squared cosine is

$$\vartheta_k^2 = f_k^T U_k (U_k^T U_k + \xi_k V_k^T V_k)^{-1} U_k^T f_k / f_k^T f_k.$$

In numerical tests, we simply used $\xi_k = \|f_k\|$, see [10] and Sect. 2.6 below.

Remark 3. Recalling the well-known Levenberg-Marquardt method (see, e.g., [3,4] and references cited therein), one can notice its coincidence with our construction for $s = 1$ (i.e., without adding $\widehat{p}_{k-1}$ to the basis) and $W_k = (J_k^T J_k + \xi_k I)^{-1}$, where $\xi_k > 0$ is a (small) regularizing parameter. Note also that the Levenberg-Marquardt direction $p_k^{(LM)} = (J_k^T J_k + \xi_k I)^{-1} J_k^T f_k$ is exactly the maximizer of the same as above Rayleigh quotient (but over the whole R^n):

$$p_k^{(LM)} = \arg\max_{p \in R^n} \vartheta^2(f_k, J_k p) = \arg\max_{p \in R^n} \frac{p^T J_k^T f_k f_k^T J_k p}{f_k^T f_k p^T (J_k^T J_k + \xi_k I) p}.$$

Hence, one can conclude that a good preconditioner W should satisfy a relation of the type $W_k^{-1} \approx J_k^T J_k + \xi_k I$ with some relatively small $\xi_k > 0$.

2.5   Using Explicit Preconditioning

As the preconditioner, a symmetric positive definite matrix $W_k = W_k(x) \in R^{n \times n}$ satisfying

$$W_k^{-1} \approx J_k^T J_k \tag{20}$$

is used. Clearly, the forming of W_k and multiplying it by a vector $q = W_k v$ must be as cheap as possible. Here we will consider preconditionings having a potential for a quite efficient implementation on modern high-performance computers. Instead of SSOR(ω) implicit preconditioning used earlier in [10,11], here we propose the simplest choice $W_k = D_k^{-1}$, where

$$D_k = \text{Diag}(J_k^T J_k). \tag{21}$$

is the diagonal part of $J_k^T J_k$. The case of a potentially more efficient factorized preconditioner $W_k = G_k^T G_k$ with sparse lower triangular G_k minimizing the K-condition number of $G_k J_k^T J_k G_k^T$, see, e.g., [9], will be considered elsewhere.

2.6   Description of Computational Algorithm

The above described preconditioned subspace descent algorithm can be summarized as follows. Note that indicating $J(x)$ and $f(x)$ as inputs means the availability of computational modules for the evaluation of vector $f(x)$ and matrix $J(x)$ for any given x.

Algorithm 1.
Key notations: $V_k = [v_1 | \ldots | v_{t_{\max}}]$, $U_k = [u_1 | \ldots | u_{t_{\max}}]$;
Input: $J(x) \in R^{m \times n}$, $f(x) \in R^m$, $x_0 \in R^n$;
Initialization:
$s = 5$, $\delta = 10^{-8}$,
$\zeta = 10^{-14}$, $\varepsilon = 10^{-12}$,
$\tau_{\min} = 10^{-8}$, $k_{\max} = 10000$,
$f_0 = f(x_0)$,
$\rho_0 = f_0^T f_0$,

Iterations:
for $k = 0, 1, \ldots, k_{\max} - 1$:
$\qquad J_k = J(x_k)$
$\qquad W_k = (\mathrm{Diag}(J_k^T J_k))^{-1}$
$\qquad v_1 := W_k J_k^T f_k$
$\qquad t_{\max} := \min(k + 1, s)$
$\qquad$ **for** $t = 2, \ldots, t_{\max} - 1$:
$\qquad\qquad v_t := W_k J_k^T J_k v_{t-1}$
$\qquad$ **end for**
$\qquad U_k = J_k V_k$
$\qquad S_k = U_k^T U_k + \sqrt{\rho_k} V_k^T V_k$
$\qquad S_k := S_k + \delta \mathrm{Diag}(S_k) + \zeta \mathrm{trace}(S_k) I$
$\qquad z_k = S_k^{-1}(U_k^T f_k)$
$\qquad p_k = \quad V_k z_k$
$\qquad v_{t_{\max}} := p_k$
$\qquad w_k = J_k p_k$
$\qquad \eta_k = w_k^T f_k$
$\qquad \zeta_k = w_k^T w_k$
$\qquad \beta_k = \eta_k / \zeta_k$
$\qquad p_k := \beta_k p_k$
$\qquad \theta_k = \eta_k^2 / (\rho_k \zeta_k)$
$\qquad \alpha^{(0)} = 1$
$\qquad$ **for** $l = 0, 1, \ldots, l_{\max} - 1$:
$\qquad\qquad x_k^{(l)} = x_k + \alpha^{(l)} p_k$
$\qquad\qquad f_k^{(l)} = f(x_k^{(l)})$
$\qquad\qquad \rho_k^{(l)} = (f_k^{(l)})^T f_k^{(l)}$
$\qquad\qquad \tau = \alpha^{(l)}(2 - \alpha^{(l)})\theta_k$
$\qquad\qquad$ **if** $(\tau < \tau_{\min})$ **return** x_k
$\qquad\qquad$ **if** $(\rho_k^{(l)}/\rho_k > 1 - (\tau/2)^2)$ **then**
$\qquad\qquad\qquad \alpha^{(l+1)} = \alpha^{(l)}/2$
$\qquad\qquad\qquad x_k^{(l+1)} = x_k + \alpha^{(l+1)} p_k$
$\qquad\qquad$ **else**
$\qquad\qquad\qquad$ **go to** NEXT
$\qquad\qquad$ **end if**
$\qquad$ **end for**
$\qquad$ NEXT: $x_{k+1} = x_k^{(l)}, \quad f_{k+1} = f_k^{(l)}, \quad \rho_{k+1} = \rho_k^{(l)}$;
$\qquad$ **if** $(\rho_{k+1} < \varepsilon^2 \rho_0)$ **or** $(\rho_{k+1} \geq \rho_k)$ **return** x_{k+1}
end for

3   Test Problems and Numerical Results

Below the results of application of Algorithm 1 to the problem of approximate canonical decomposition of 3D tensors are presented. For the test runs, one core of Pentium(R) Dual-Core CPU E6600 3.06 GHz, 3.25 Gbytes RAM desktop PC was used. We will consider different values of the subspace dimension ($2 \leq s \leq 9$). Since we are considering nonzero residual problems, the iterations typically terminate by the condition $\tau < \tau_{\min} = 10^{-8}$, see the corresponding line in Algorithm 1. Note that the iteration number k always coincides with the number of the Jacobian evaluations.

3.1   Test Problem Setting

The problem of approximate evaluation of Canonical Decomposition of 3D tensor was considered, e.g., in [1,13–16]. Given 3D array T, it is required to approximate it by the sum of 1-rank tensors:

$$t_{l_1,l_2,l_3} \approx \sum_{l=1}^{r} u(l_1, l)v(l_2, l)w(l_3, l);$$

if the equality is strict, then r is called tensor rank. By collecting the unknown entries of u, v, w into a single vector x, the residual function can be written as

$$f_{l_1+(l_2-1)m_1+(l_3-1)m_1m_2} = -t_{l_1,l_2,l_3} \tag{22}$$
$$+ \sum_{l=1}^{r} x_{(l-1)(m_1+m_2+m_3)+l_1} x_{lm_1+(l-1)(m_2+m_3)+l_2} x_{l(m_1+m_2)+(l-1)m_3+l_3},$$
$$1 \leq l_1 \leq m_1, \quad 1 \leq l_2 \leq m_2, \quad 1 \leq l_3 \leq m_3,$$

so that the number of equations and the number of unknowns are $m = m_1m_2m_3$ and $n = (m_1 + m_2 + m_3)r$, respectively. In this setting, there exist no isolated optimum solutions, at least due to the indeterminacy of scaling the multiplicands in each triple product $u(\cdot, l)v(\cdot, l)w(\cdot, l)$.

3.2   Test Case 1: Inverse 3D Distance Tensor

The first particular case we consider is (see, e.g., [16] and references cited therein)

$$t_{l_1,l_2,l_3} = \left(l_1^2 + l_2^2 + l_3^2\right)^{-1/2}, \tag{23}$$

with $m_1 = m_2 = m_3 = 100$ or $m_1 = m_2 = m_3 = 200$ and $r = 5$. The 3D array (23) arises from the numerical approximation of an integral equation with kernel $1/\|x - y\|$ acting on the unit cube and discretized on a uniform cubic grid. This is rather hard-to-solve nonzero residual problem; in particular, for any x the Jacobian $J(x)$ has rank deficiency.

Choosing Initial Guess It must be stressed that for such tensor decomposition problems, the choice of the initial guess is probably the most important tuning parameter. In test Case 1, we used a quasirandom sequence with elements in $\{-1, 0, 1\}$ given by the number-theoretic Moebius function $\mu(\cdot)$ (while in test Case 2 the so called logistic sequence was more effective, see Sect. 3.3 below). The initial guess was set as

$$\tilde{x}_j = \mu(q + j), \quad j = 1, \ldots, n, \tag{24}$$

where q is an arbitrary nonnegative number. Recall that the formal definition of μ is

$$1 = \left(\sum_{k=1}^{\infty} k^{-t} \right) \left(\sum_{k=1}^{\infty} \mu(k) k^{-t} \right)$$

i.e., $\mu(k)$ are the coefficients of the Dirichlet series representing the inverse Riemann zeta function. The integer sequence $\mu(j)$ for $1 \leq j \leq L$ was generated as follows: $\mu(1) = 1$,

$\mu(j) = 0, \ j = 2, \ldots, L$;
for $j = 1, \ldots, \lfloor L/2 \rfloor$:
    for $k = 2, \ldots, \lfloor L/j \rfloor$:
        $\mu(kj) := \mu(kj) - \mu(j)$
    end for
end for

This takes only $O(L \log L)$ subtractions and gives a seemingly chaotic sequence containing only the numbers $\{-1, 0, 1\}$.

Table 1. Performance of Algorithm 1 for problem (22), (23) with $m = 1\,000\,000$, $n = 1500$, $\text{nz}(J) = 15\,000\,000$, and the initial guess $\widetilde{x}_j = \mu(j + 2)$.

s	$\|x_*\|_\infty$	#J evaluations	#f evaluations	Total time, s
2	0.72	513	517	153
3	0.89	260	264	103
4	0.91	128	130	61
5	0.77	109	111	64
6	1.10	82	84	54
7	1.10	127	129	96
8	0.82	111	114	100
9	0.73	183	186	175

Results and Discussion Performance results obtained for $m_1 = m_2 = m_3 = 100$ and $m_1 = m_2 = m_3 = 200$ are given in Tables 1 and 2, respectively. In both cases, the initial guess (24) taken with $q = 2$ was used. All the test runs were terminated succesfully with the resulting optimal values $\|f(x_*)\| \approx 0.07388815$ and $\|f(x_*)\| \approx 0.16391815$ for the smaller and the larger problems, respectively (whereas the initial value of $\|f(x_0)\|$ was of the order 10^3). An important quality measure for the obtained solution x_* is its norm $\|x_*\|_\infty$ (the smaller, the better), and it is shown in the second column. Note also that in all cases the number of backtracking steps (which equals to the difference between the numbers in the fourth and third columns) was quite negligible, which confirms the efficiency of the developed stepsize selection procedure (Tables 1, 2, 3, 4 and 5).

Table 2. Performance of Algorithm 1 for problem (22), (23) with $m = 1\,000\,000$, $n = 1500$, $\mathrm{nz}(J) = 15\,000\,000$, and the initial guess $\tilde{x}_j = \frac{1}{2}\mu(j+2)$.

s	$\|x_*\|_\infty$	#J evaluations	#f evaluations	Total time, s
2	0.75	549	553	170
3	0.73	266	270	103
4	0.74	191	195	98
5	0.73	99	103	57
6	0.74	86	90	57
7	0.74	81	85	65
8	0.74	84	88	72
9	0.75	134	139	141

Table 3. Performance of Algorithm 1 for problem (22), (23) with $m = 8\,000\,000$, $n = 3000$, $\mathrm{nz}(J) = 120\,000\,000$, and the initial guess $\tilde{x}_j = \mu(j+2)$.

s	$\|x_*\|_\infty$	#J evaluations	#f evaluations	Total time, s
2	0.80	348	352	904
3	0.76	182	185	602
4	0.75	139	142	573
5	0.75	144	147	685
6	0.75	116	120	661
7	0.74	115	119	754

Table 4. Performance of Algorithm 1 for problem (22), (23) with $m = 8\,000\,000$, $n = 3000$, $\mathrm{nz}(J) = 120\,000\,000$, and the initial guess $\tilde{x}_j = \frac{1}{2}\mu(j+2)$.

s	$\|x_*\|_\infty$	#J evaluations	#f evaluations	Total time, s
2	0.75	358	360	947
3	0.85	193	195	652
4	0.74	159	161	668
5	0.76	191	195	936
6	0.75	124	128	711
7	0.73	116	121	813

Comparing the results in Table 1 with the similar data presented in Table 5 of [11], one can observe more than 2 times acceleration in total solution time for the same problem and initial guess. Obviously, this effect is due to the different construction of the search subspaces.

Further, comparing the results in Tables 1 and 2 with the corresponding data in Table 6.4 of [16], one can see that our method is comparable in performance with the Adaptive Algebraic Multigrid method (which takes 62 s and 441 s) and is several times faster than the Alternating Least Squares method (requiring 455 s and 2611 s). It must be stressed that the latter two methods are specialized procedures for approximating the Canonical Decomposition while our method represents a general purpose nonlinear least squares solver.

3.3 Test Case 2: Matrix Multiplication Tensor

The second particular case we consider (see, e.g., [11] and references cited therein) is known as "Brent Equations" [5] and arises in connection with the development of Fast Matrix Multiplication (FMM) algorithms. The problem setting is the same as in previous Subsection, except of the components of the 3D tensor, which are specified by

$$t_{l_1,l_2,l_3} = \delta(i_2 - j_1)\delta(j_2 - k_1)\delta(k_2 - i_1), \tag{25}$$

where

$$1 \le l_1 \le n_3 n_1 = m_1, \quad l_1 = i_1 + (i_2 - 1)n_3,$$

$$1 \le i_1 \le n_3, \quad 1 \le i_2 \le n_1;$$

$$1 \le l_2 \le n_1 n_2 = m_2, \quad l_2 = j_1 + (j_2 - 1)n_1,$$

$$1 \le j_1 \le n_1, \quad 1 \le j_2 \le n_2;$$

$$1 \le l_3 \le n_2 n_3 = m_3, \quad l_3 = k_1 + (k_2 - 1)n_2,$$

$$1 \le k_1 \le n_2, \quad 1 \le k_2 \le n_3.$$

Thus, we have $m = (n_1 n_2 n_3)^2$ and $n = (n_3 n_1 + n_1 n_2 + n_2 n_3)r$. The above described problem we will refer to as FMM $(n_1, n_2, n_3; r)$.

Choosing Initial Guess For each problem, 800 trials were run with iteration number limit 500 and with different initial guesses. The latter were determined by quasirandom sequences generated with the use of the so called logistic sequence (see, e.g. [18] and references cited therein):

$$\xi_0 \in \{0.1, 0.2, 0, 3, 0.4, 0.6, 0.7, 0, 8, 0.9\},$$

$$\xi_k = 1 - 2\xi_{k-1}^2, \qquad k = 1, 2, \dots, n,$$

$$x_0(j) = \frac{1}{4}\xi_{qj}, \qquad q = 1, 2, \dots, 100.$$

The best results are presented in Table 5.

Table 5. Performance of Algorithm 1 with $s = 7$ and residual norm reduction $\varepsilon = 10^{-14}$ for smaller FMM problems.

Problem	init.guess: q and ξ_0	$\|x_*\|_\infty$	#J evaluations	#f evaluations	Total time, s
(2,2,2;7)	5; 0.2	2.11	42	45	<0.01
(2,3,3;15)	76; 0.7	2.43	133	136	0.06
(2,3,4;20)	39; 0.6	2.37	362	364	0.34
(2,3,5;25)	19; 0.8	2.08	486	489	0.95
(3,3,3;23)	30; 0.9	1.32	182	185	0.27

Results and Discussion In Table 5, the performance data for Algorithm 1 obtained for canonical decomposition of FMM tensors parametrized as (2,2,2;7), (2,3,3;15), (2,3,4;20), (2,3,5;25), and (3,3,3;23) are given. The examples show more than two orders of magnitude acceleration of single solver application compared to [11]. This gives one an advantage in testing a much larger number of quasirandom initial guesses in a limited time.

As before, one can notice the almost minimum number of function evaluations per one Jacobian evaluation.

However, the fast solution of FMM's with larger $(n_1, n_2, n_3; r)$ (already for FMM (3,3,4,29)) seems still not possible for the current version of the nonlinear solver.

4   Concluding Remarks

In the present paper, a nonlinear least squares solver is developed which do not require second-order information, can efficiently use the sparsity of the Jacobian, and is formally applicable to all types of least squares problems. Moreover, the proposed version of the algorithm is well suited for an efficient implementation on modern high-performance computers. The results of numerical testing on problems related to canonical decomposition of 3D tensors have confirmed the efficiency and robustness of the Preconditioned Subspace Descent method in solving hard nonlinear problems.

References

1. Acar, E., Dunlavy, D.M., Kolda, T.G.: A scalable optimization approach for fitting canonical tensor decompositions. J. Chemometr. **25**(2), 67–86 (2011)
2. Armijo, L.: Minimization of functions having Lipschitz continuous first partial derivatives. Pac. J. Math. **16**(1), 1–3 (1966)
3. Bellavia, S., Gratton, S., Riccietti, E.: A Levenberg-Marquardt method for large nonlinear least-squares problems with dynamic accuracy in functions and gradients. Numer. Math. **140**(3), 791–825 (2018)
4. Bellavia, S., Morini, B.: Strong local convergence properties of adaptive regularized methods for nonlinear least squares. IMA J. Numer. Anal. **35**(2), 947–968 (2015)

5. Brent, R.P.: Algorithms for matrix multiplication (No. STAN-CS-70-157). Stanford University CA Department of Computer Science (1970)
6. Kaporin, I.E.: Esimating global convergence of inexact Newton methods via limiting stepsize along normalized direction. Rep. 9329, July 1993, Department of Mathematics, Catholic University of Nijmegen, Nijmegen, The Netherlands, 8p. (1993)
7. Kaporin, I.E.: The use of preconditioned Krylov subspaces in conjugate gradient type methods for the solution of nonlinear least square problems. (Russian) Vestnik Mosk. Univ., Ser. 15 (Computational Math. and Cybernetics) N 3, 26–31 (1995)
8. Kaporin, I.E., Axelsson, O.: On a class of nonlinear equation solvers based on the residual norm reduction over a sequence of affine subspaces. SIAM J. Sci. Comput. 16(1), 228–249 (1994)
9. Kaporin, I.: Using Chebyshev polynomials and approximate inverse triangular factorizations for preconditioning the conjugate gradient method. Comput. Math. Math. Phys. 52(2), 169–193 (2012)
10. Kaporin, I.: Preconditioned subspace descent methods for the solution of nonlinear systems of equations. In: Jacimovic, M., et al. (eds.) Proceedings of X International Conference on Optimization and Applications (OPTIMA 2019). Communications in Computer and Information Science (CCIS). Springer Nature Switzerland AG (2019)
11. Kaporin, I.: Preconditioned subspace descent method for nonlinear systems of equations. Open Comput. Sci. 10(1), 71–81 (2020)
12. Karimi, S., Vavasis, S.: Conjugate gradient with subspace optimization (2012). arXiv preprint arXiv:1202.1479v1
13. Kazeev, V.A., Tyrtyshnikov, E.E.: Structure of the Hessian matrix and an economical implementation of Newton's method in the problem of canonical approximation of tensors. Comput. Math. Math. Phys. 50(6), 927–945 (2010)
14. Oseledets, I.V., Savostyanov, D.V.: Minimization methods for approximating tensors and their comparison. Comput. Math. Math. Phys. 46(10), 1641–1650 (2006)
15. Sterck, H.D., Winlaw, M.: A nonlinearly preconditioned conjugate gradient algorithm for rank-R canonical tensor approximation. Numer. Linear Algebra Appl. 22(3), 410–432 (2015)
16. Sterck, H.D., Miller, K.: An adaptive algebraic multigrid algorithm for low-rank canonical tensor decomposition. SIAM J. Sci. Comput. 35(1), B1–B24 (2013)
17. Yuan, Y.X.: Recent advances in numerical methods for nonlinear equations and nonlinear least squares. Numer. Algebra Control Optim. 1(1), 15 34 (2011)
18. Yu, L., Barbot, J.P., Zheng, G., Sun, H.: Compressive sensing with chaotic sequence. IEEE Signal Process. Lett. 17(8), 731–734 (2010)

PCGLNS: A Heuristic Solver for the Precedence Constrained Generalized Traveling Salesman Problem

Michael Khachay[1,2,3]($\boxtimes$) (ID), Andrei Kudriavtsev[1,2]($\boxtimes$) (ID),
and Alexander Petunin[1,2]($\boxtimes$) (ID)

[1] Krasovsky Institute of Mathematics and Mechanics, Ekaterinburg, Russia
mkhachay@imm.uran.ru, andreikudrya1995@gmail.com, aapetunin@gmail.com
[2] Ural Federal University, Ekaterinburg, Russia
[3] Omsk State Technical University, Omsk, Russia

Abstract. The Precedence Constrained Generalized Traveling Salesman Problem (PCGTSP) is a specialized version of the well-known Generalized Traveling Salesman Problem (GTSP) having a lot of valuable applications in operations research. Despite the practical significance, results in the field of design, implementation, and numerical evaluation the algorithms for this problem remain still rare. In this paper, to the best of our knowledge, we propose the first heuristic solver for this problem augmented by numerical evaluation results of its performance against the public test instances library PCGTSPLIB. Our algorithm is an extension of the recent Large Neighborhood Search (GLNS) heuristic GTSP solver designed to take into account additional precedence constraints. Similarly to GLNS, the source code of all our algorithms is open, and the executables are freely accessible, which ensures the reproducibility of the reported numerical results.

Keywords: Generalized Traveling Salesman Problem · Precedence constraints · Heuristic solver · Large Neighborhood Search · TSPLIB

1 Introduction

The Precedence Constrained Generalized Traveling Salesman Problem (PCGTSP) is an extension of the well-known Generalized Traveling Salesman Problem (GTSP), which dates back to the seminal paper of G. Dantzig and J. Ramser [7] that proposed the first mathematical model for the supply problem for gas stations network from the dedicated terminal by a fleet of gasoline trucks.

As for the general GTSP, an instance of the PCGTSP (see, e.g. [24]) is defined by a finite point set V and some its partition $V_1, \ldots, V_m$ into distinct non-empty subsets also referred to as *clusters*. The goal is to find the shortest cyclic tour visiting each cluster exactly once. The main difference between the problems is that in PCGTSP the set of clusters is supposed to be partially ordered, such

© Springer Nature Switzerland AG 2020
N. Olenev et al. (Eds.): OPTIMA 2020, LNCS 12422, pp. 196–208, 2020.
https://doi.org/10.1007/978-3-030-62867-3_15

that each feasible tour departing from cluster V_1 should visit all other clusters in accordance with this predefined order.

The Precedence Constrained GTSP has several valuable industrial applications, among them are

- minimizing the total nonproductive *air* time spent by a tool during its movement from one part to another [2,25];
- efficient programming of Coordinate Measuring Machines (CMM) applied for industrial inspection processes [18,27];
- optimizing the processes of multi-hole drilling [8];
- optimal tool routing in metal sheet laser cutting [10,22].

For instance, in metal sheet cutting, we are given a cutting plan (similar to the presented at Fig. 1a), where each part to be cut off is defined by a closed contour. According to the traditional approach [6], each contour is discretized by pointing out so-called *piercing* points (Fig. 1b) producing the instance of the PCGTSP, where the partial order on the set of clusters is defined naturally by the nesting of the initial parts. Finally, the optimal solution of this instance presented in Fig. 1c provides a near-to-optimal tour of the laser cutter with respect to airtime minimization.

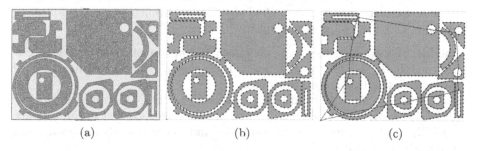

(a)   (b)   (c)

Fig. 1. Metal cutting: the initial cutting plan (a), the corresponding PCGTSP setting induced by piercing points located at the contours (b), and a near-to-optimal tour (c)

Related Work. Enclosing the classic Traveling Salesman Problem (TSP), the PCGTSP is strongly NP-hard even on the Euclidean plane [23] provided the number of clusters m belongs to the input. Similarly to the common GTSP, for any fixed m, the PCGTSP can be solved to optimality in polynomial time. Furthermore, the problem belongs to FPT with respect to the parameter m, since the well-known folk algorithm (see, e.g. [12]) that enumerates all the permutations of the given set of clusters can be easily adapted to the case of precedence constraints with naïve complexity bound $O((m-1)! \cdot m^2 n^3)$.

Although it is intuitively clear that the richer the predefined partial order, the simpler the appropriate PCGTSP instance[1], the dependence of theoretical

---

[1] Indeed, in the case of the linear order, the instance becomes trivial.

complexity bounds on the properties of the precedence constraints has not yet been established.

Perhaps, the only exception is two special types of the precedence constraints, for which polynomial time complexity of the PCGTSP is managed to be proven theoretically. The first type of such constraints was initially introduced by E. Balas [1] for the classic TSP. Efficient exact algorithms for the PCGTSP with precedence constraints of this type are proposed in recent papers [3,5]. Tours that fulfill constraints of the second type are referred to as quasi- and pseudo-pyramidal. Efficient parameterized algorithms for the PCGTSP with such precedence constraints are proposed in [19,20].

Despite its practical significance, in the field of algorithmic analysis, the PCGTSP remains weakly explored. To the best of our knowledge, publications reporting the algorithmic results for the problem are exhausted by several papers employing adapted versions of the classic dynamic programming (see, e.g., [4]), the recent paper [28] presenting an approach for construction branch-and-bound algorithms based on an original branching strategy, and a few papers [9,11,27] examining the applicability of heuristics and meta-heuristics to solving PCGTSP instances that come from the industrial applications.

Although the aforementioned results appear to be novel and promising, there are still many blank spots on the map of algorithmic analysis of the PCGTSP. In particular, from the point of numerical algorithms and heuristics, the following issues seem to be the most important

– the lack of open accessible (e.g., open-source) implementations similar to GK [16] or GLKH [17] for the common GTSP, which can be used in numerical experiments as well-known baselines;
– the absence of efficient Mixed Integer Linear Program (MILP) models[2] for the PCGTSP providing the opportunity to apply state-of-the-art optimizers like CPLEX or Gurobi for construction lower and upper bounds and examining the heuristic solutions.

In this paper, we try to bridge this gap.

Our contribution is two-fold:

(i) we propose a novel adaptive heuristic solver for the PCGTSP extending the recent Large Neighborhood Search algorithm GLNS designed in [31] for the common Generalized Traveling Salesman Problem;
(ii) employing our algorithm in combination with a novel MILP model for the PCGTSP and the state-of-the-art Gurobi optimizer [15], we obtain the first numerical results for the public library PCGTSPLIB of test instances proposed in [28]. All the source code, auxiliary scripts, and the test instances are freely available at [21].

The remaining part of the paper is organized as follows. In Sect. 2, we give a description of the Precedence Constrained Generalized Traveling Salesman

---

[2] To the best of our knowledge.

Problem. In Sect. 3, we explain the main idea of the proposed adaptive heuristic, which we call PCGLNS. Then, Sect. 4 and Sect. 5 provide a description of the basic removal and insertion heuristics adapted to the case of precedence constraints. Further, in Sect. 6 we propose our Mixed Integer Linear Programming (MILP) model for the PCGTSP and report the obtained numerical results. Finally, in Sect. 7 we come to the conclusions and overview some directions for future work.

2   Problem Statement

An instance of the Precedence Constrained Generalized Traveling Salesman Problem is given by an edge-weighted directed graph $G = (V, E, c)$ where V is a nodeset consisting of n nodes $v_1, \ldots, v_n$, a partition $V_1 \cup \ldots \cup V_m$ of its nodeset V into *clusters*, and an additional acyclic digraph $H = (I, \Pi), I = \{1, \ldots, m\}$, which encodes the *precedence constraints* defined on the set of clusters. Without loss of generality, we can assume that the graph G is complete and the weighting function c is non-negative.

A cyclic route (tour) $T = v_{i_1}, \ldots, v_{i_m}$ in the graph G is called *feasible*, if it fulfills the following conditions:

(i) the tour T departs from the cluster V_1, i.e. $v_{i_1} \in V_1$;
(ii) each cluster V_j is visited by T exactly once;
(iii) for any arc $(j, k) \in \Pi$, the cluster V_j should precede the cluster V_k in the tour T.

Hereinafter, we call a tour T a *partial tour*, if T fulfills points (i) and (iii) (but, maybe violates (ii)).

In PCGTSP, the goal is to find a feasible tour T of the minimum total cost

$$c(T) = c(v_{j_m}, v_{j_1}) + \sum_{j=1}^{m-1} c(v_{i_j}, v_{i_{j|1}}). \tag{1}$$

3   PCGLNS: The Main Idea

Our solver is an extension of the known GLNS algorithm proposed in [31] for the common GTSP, designed to take into account additional precedence constraints defined on a set of clusters. In turn, the GLNS appears to be an implementation of the Adaptive Large Neighborhood Search (ALNS) metaheuristic (see, e.g. [13]) combining the well-known *ruin and recreate* principle with online learning over a given set of basic removal and insertion heuristics.

The proposed algorithm follows the main scheme of the GLNS (presented in Algorithm 1). This scheme has several outer parameters. Among them are the predefined sets $\mathfrak{R}$ and $\mathfrak{J}$ of heuristics, acceptance and termination criteria, and the number of trials i_{max}. Acceptance and termination criteria and online learning methods were used by exactly the similar way as in the original paper. For the sake of brevity, we provide a short overview of them:

Algorithm 1. PCGLNS :: general framework

Input: an instance of the PCGTSP defined by a graph $G = (V, E, c)$, partition $V_1, \ldots, V_m = V$, and digraph $H = (I, \Pi)$
Output: an approximate solution $\bar{T}$

1: **for** $(0 < i \leq i_{\max})$ **do**
2:     construct an initial tour T
3:     $T_{best}(i) = T$
4:     **repeat**
5:         choose $H_{rem} \in \mathfrak{R}$ and $H_{ins} \in \mathfrak{I}$ from the predefined sets $\mathfrak{R}$ and $\mathfrak{I}$ of the removal and insertion heuristics, respectively, w.r.t. current distribution
6:         **ruin:** using H_{rem} sequentially, obtain a partial tour T' by exclusion a random number N_r of nodes from the tour T
7:         **repair:** applying H_{ins} exactly N_r times, build up a new tour T_{new} from T'
8:         optimize T_{new} locally
9:         **if** $c(T_{new}) < c(T_{best}(i))$ **then**
10:            $T_{best}(i) = T_{new}$
11:        **end if**
12:        **if** *acceptance_criterion* is met for T_{new} and T **then**
13:            $T = T_{new}$
14:            record the improvement made by H_{rem} and H_{ins}
15:        **end if**
16:    **until** *termination_criterion* is met
17:    update the distributions on $\mathfrak{R}$ and $\mathfrak{I}$ w.r.t. to improvements recorded
18: **end for**
19: **return** $\bar{T} = \arg\min\{T_{best}(i) : i \in \{1, \ldots, i_{\max}\}\}$

- as for the original GLNS solver [31], we use the standard simulated annealing *acceptance criterion* based on the *temperature* $\mathcal{T}$ decreasing successively along the iterations from a given initial value and affecting the acceptance probability $\mathsf{P}(T_{new})$ for newly found tours

$$\mathsf{P}(T_{new}) = \min\left\{e^{(c(T) - c(T_{new}))/\mathcal{T}}, 1\right\};$$

- *termination criterion* for the loop at Line 16 is met, if there are no improvements made to the current best-found solution T_{best} after a certain number of iterations, which can be given as the framework's input parameter;
- *online learning* is a tuning method for distibutions on the sets $\mathfrak{R}$ and $\mathfrak{I}$. At any warm restart (Line 5), we pick heuristics H_{rem} and H_{ins} to obtain new tour compared to the current best solution. Each heuristic gets a specific score based on current and best tour costs. The more scores have a heuristic, the more probable it will be picked on the next iteration.

The detailed description we postpone to the forthcoming paper and discuss only the modifications made to the used removal and insertion heuristics in order to adapt them to take into account the given precedence constraints.

4   Removal Heuristics

This section describes the predefined set $\mathfrak{R}$ consisting of three removal heuristics:

- *Worst removal* (see [14] for an example) is used to remove the node that maximizes the removal cost from the given tour;
- *Distance removal* is similar to the heuristic proposed [30] the pickup and delivery problem. The idea is to select a random seed node from the current tour, calculate distances from it to any other node in the tour, and remove the node, whose distance fulfils a given criterion, e.g. smallest, largest or random;
- *Segment removal* is used to simply remove a continuous segment of a given tour;

Since the modifications in all heuristics are very similar, we explain to them using the **unified worst removal** as an example.

Algorithm 2. Unified worst removal heuristic

Input: A partial tour $T = v_1, \ldots v_\ell$, $\lambda \in [0, \infty)$
Output: A new partial tour T_{new} shorter by one node

1: Randomly select $k \in \{2, ..., \ell\}$ with respect to the unnormalized probability mass function $[\lambda^0, \lambda^1, ..., \lambda^{\ell-1}]$, where $\ell = |V_T|$ (note that $k > 1$ since we should fix the first cluster)
2: For every v_j in T, find r_j by the formula

$$r_j = \begin{cases} c(v_{j-1}, v_j) + c(v_j, v_{j+1}) - c(v_{j-1}, v_{j+1}), & \text{if } j < \ell \\ c(v_{j-1}, v_j) + c(v_j, v_1) - c(v_{j-1}, v_1), & \text{otherwise} \end{cases}$$

3: Pick the vertex v_j from T with the k-th smallest value r_j
4: Remove v_j from T and connect v_{j-1} and v_{j+1} directly to obtain T_{new}
5: **return** T_{new}

Algorithm 2 presents the adapted version of worst removal heuristic, which is almost the same as in [31] except the only modification—the first node in current partial tour cannot be removed, since the cluster V_1 should be visited first to fulfil the precedence constrints.

5   Insertion Heuristics

The predefined set $\mathfrak{I}$ contains a set of the well-known heuristics proposed in [12]: nearest, farthest, random, and cheapest insertion. All of them share quite a common framework and differs in the way of picking the next cluster to insert:

- *Nearest insertion* picks the cluster, that contains some vertex at minimum distance to some vertex on the current partial tour;

- *Farthest insertion* picks the cluster, such that it's closest to the partial tour vertex is the farthest compared to closest vertices of other clusters;
- *Random insertion* uniformly randomly picks the cluster;
- *Cheapest insertion* picks the cluster, that minimizes the insertion cost.

In Algorithm 3, we illustrate our adaptation technique taking the **cheapest insertion** heuristic as an example.

Algorithm 3. Cheapest insertion heuristic

Input: A partial tour T, set of clusters $\mathcal{V}_{ins}$ not visited by T
Output: A partial tour T_{new} visiting one additional cluster

1: **for all** clusters $V_i \in \mathcal{V}_{ins}$ **do**
2:    In T find a maximal path $T_i' = v_{j_1}, \ldots, v_{j_p}$ which is valid for insertion a node from V_i
3:    For any $j_t \in T_i'$ and $v \in V_i$, define the insertion cost as follows

$$cost(j_t, v) = c(v_{j_t}, v) + c(v, v_{j_{t+1}}) - c(v_{j_t}, v_{j_{t+1}})$$

4: **end for**
5: Let

$$\bar{t}, \bar{v} = \arg\min\{cost(j_t, v) \colon V_i \in \mathcal{V}_{ins}, v \in V_i, j_t \in T_i'\}$$

6: Insert the node $\bar{v}$ into the edge $(v_{j_{\bar{t}}}, v_{j_{\bar{t}+1}})$

The precedence constraints are taking into account at Line 2. While, in the original version of the heuristic, new node can be inserted in any position of the initial tour T, in our modification, to fulfill the precedence constraints, after picking the cluster V_i, we restrict ourselves to the appropriate path T', where an insertion is possible. We are able to do this, since for every node's cluster in T we store its relation to the cluster V_i, i.e. ancestor, descendant or non-related.

6 Numerical Experiments

In this section, we present the numerical performance evaluation results of the proposed PCGLNS solver. As a testing ground, we employ the recent public library of benchmark PCGTSP instances extending the well-known TSPLIB and published in [28] and called PCGTSPLIB by its authors. For any instance, we (i) find a heuristic solution by the PCGLNS solver, after that, we (ii) use this solution as a baseline approximation for the branch-and-bound algorithm implemented in the state-of-the-art Gurobi optimizer 9 [15] to obtain the appropriate lower and upper bounds.

6.1 MILP Model

Our MILP model is a straightforward extension of the well-known L2ATSPxy model proposed in [29] for the Asymmetric TSP Problem with precedence constraints, the special case of the PCGTSP. We choose this particular model since it provides the best numerical performance among the known polynomial MILP formulations of the ATSP [26].

Similarly to [29], for any $i, j \in \{1, \ldots, n\}$, $i \neq j$, we introduce assignment binary variables x_{ij}, such that

$$x_{ij} = \begin{cases} 1, & \text{if the arc } (v_i, v_j) \text{ belongs to the tour } T, \\ 0, & \text{otherwise.} \end{cases}$$

Also, we define two sets of the auxiliary variables y_{pq} and u_{pq} for any $p, q \in \{1, \ldots, m\}$, $p \neq q$. Each of them is supposed to be continuous non-negative variable, but takes binary values, such that

$$u_{pq} = \begin{cases} 1, & \text{if the cluster } V_p \text{ precedes } V_q \text{ immediately in the tour } T, \\ 0, & \text{otherwise,} \end{cases}$$

$$y_{pq} = \begin{cases} 1, & \text{if the cluster } V_p \text{ can precede } V_q \text{ (not necessarily immediately),} \\ 0, & \text{otherwise.} \end{cases}$$

Then, our MILP model can be represented as follows

$$\min \sum_{i=1}^{n} \sum_{j=1, j \neq i}^{n} c_{ij} x_{ij} \tag{2}$$

subject to

$$\sum_{q=1, q \neq p}^{m} u_{pq} = 1 \ (p \in \{1, \ldots, m\}), \qquad \sum_{p=1, p \neq q}^{m} u_{pq} = 1 \ (q \in \{1, \ldots, m\}) \tag{3}$$

$$\sum_{i \in V_p} \sum_{j \in V_q} x_{ij} = u_{pq} \ (p, q \in \{1, \ldots, m\}) \tag{4}$$

$$\sum_{j=1, j \neq i}^{n} x_{ij} = \sum_{k=1, k \neq i}^{n} x_{ki} \ (i \in \{1, \ldots, n\}) \tag{5}$$

$$\sum_{q=1}^{m} u_{qq} = 0 \tag{6}$$

$$u_{pq} - y_{pq} \leq 0 \ (p \neq q \in \{2, \ldots, m\}) \tag{7}$$

$$y_{pq} + y_{qp} = 1 \ (p \neq q \in \{2, \ldots, m\}) \tag{8}$$

$$u_{pq} + y_{qr} + u_{rq} + y_{rp} + u_{pr} \leq 2 \ (p \neq q \neq r \in \{2, \ldots, m\}) \tag{9}$$

$$y_{pq} = 1 \quad (q \in \{2, \ldots, m\}, (p, q) \in \Pi) \tag{10}$$

$$u_{pr} = 0 \quad (r \in \{2, \ldots, m\}, (p, q) \in \Pi, (q, r) \in \Pi) \tag{11}$$

$$x_{ij} \in \{0, 1\}, u_{pq} \geq 0, y_{pq} \geq 0 \tag{12}$$

Here Eq. (3) ensure that each cluster is visited and only once. Equations (4) establish the dependence between variables x and u. Then, we include Eq. (5) to force each feasible tour to fulfill the Kirchhoff's first law. Equation (6) guarantees that, for any $q \in \{1, \ldots, m\}$ and any $i, j \in V_q$, we obtain $x_{ij} = 0$. Formulas (7)–(9) are included into the model for subtour elimination and enforcement the precedence constraints, which in turn are specified by equations (10). Finally, inclusion to the model equations (11) implies significant speed-up of the solution of the considered instances by Gurobi MILP optimizer.

6.2 Results and Discussion

The obtained numerical results are reported in Table 1 and organized as follows. The first group of columns with the common title 'Instance' presents the general information about each test instance including its number, identifier, and dimensions n and m. Then, the next column group gives information about approximate solutions found by our heuristic solver: approximate value App of the cost (2), running time (in seconds), and the appropriate relative error

$$\text{gap} = (\text{App} - \text{LB})/\text{App}.$$

Finally, the last column group collects the details concerning the further solution of test instances by the branch-and-bound algorithm implemented in Gurobi 9.0 with a warm start at the solution obtained by our solver: lower and upper bounds and the elapsed running time. We use time limits of 1 h and 12 h for the PCGLNS and for Gurobi, respectively.

As it follows from Table 1, our heuristic solver finds an optimal solution for 9 out of 40 instances of the PCGTSPLIB. Then, for 18 out the remaining instances, approximate solutions of relative error $< 10\%$ are found. All other instances (except #13 and #19, for which Gurobi was stopped due to time limit) are solved with relative error at most 50%. It is curious that high quality results are obtained for the largest instances in library ($rbgXXXx$).

At the moment, we have about 25% instances with not the highest quality solutions. We believe that, for almost all of them, the reason is application of the Branch-and-Bound algorithm implemented in Gurobi for accuracy validation of our heuristic solver.

In general, there are two groups of low-quality solutions. The first group considered failed because Gurobi ran out of time and could not even get a result of the initial relaxation in 12 h. Thus, we had no way to evaluate our approximate solution and decided to consider such results as low-quality. The second group consists of instances that our algorithm solved relatively fast, but Gurobi was not able to reduce the gap between the lower and upper bounds due to the lack of time, as well. Therefore, we resulted in relatively low-quality solutions

Table 1. Experimental results: optimal solutions are highlighted

Instance				PCGLNS			PCGLNS+Gurobi		
#	ID	nd # (n)	clr # (m)	App	time (sec)	gap (%)	LB	UB	time (sec)
1	ESC07	39	8	**1730**	2.17	**0.0**	**1730**		0.07
2	ESC12	65	13	**1390**	2.25	**0.0**	**1390**		0.52
3	ESC25	133	26	1418	2.92	2.5	**1383**		4.45
4	ESC47	244	48	1399	4.96	31.6	**1063**		52.01
5	ESC63	349	64	**62**	4.92	**0.0**	**62**		380.65
6	ESC78	414	79	14832	13.52	1.7	14581	14832	43200.01
7	br17.10	88	17	**43**	2.35	**0.0**	**43**		369.98
8	br17.12	92	17	**43**	2.35	**0.0**	**43**		107.28
9	ft53.1	281	53	6207	5.17	3.0	6022	6200	42099.00
10	ft53.2	274	53	6653	5.44	7.0	6184	6653	42137.00
11	ft53.3	281	53	8446	5.93	17.9	6936	8446	42194.00
12	ft53.4	275	53	**11822**	5.72	**0.0**	**11822**		23238.81
13	ft70.1	346	70	32848	8.77	n/a	Time limit		
14	ft70.2	351	70	33486	9.30	4.9	31840	33448	42021.00
15	ft70.3	347	70	35309	7.82	6.7	32944	35309	41173.00
16	ft70.4	353	70	44497	23.66	7.0	41378	44492	41827.00
17	kro124p.1	514	100	33320	14.47	10.2	29926	33320	34162.00
18	kro124p.2	524	100	35321	13.19	14.8	30101	35321	35379.00
19	kro124p.3	534	100	41340	20.74	n/a	Time limit		
20	kro124p.4	526	100	62818	29.95	25.7	46704	62818	41035.00
21	p43.1	203	43	22545	3.95	1.0	22327	22545	43150.00
22	p43.2	198	43	22841	4.52	2.0	22381	22839	43132.00
23	p43.3	211	43	23122	4.06	2.5	22540	23119	43058.00
24	p43.4	204	43	66857	4.73	32.1	45396	66848	43193.00
25	prob.100	510	99	1474	21.32	45.7	800	1442	38567.00
26	prob.42	208	41	232	4.09	14.9	**202**		1292.07
27	rbg048a	255	49	**282**	5.60	**0.0**	**282**		64.32
28	rbg050c	259	51	**378**	5.79	**0.0**	**378**		26.46
29	rbg109a	573	110	**848**	42.58	**0.0**	**848**		83.23
30	rbg150a	871	151	1415	144.90	0.1	**1414**		29094.62
31	rbg174a	962	175	1644	245.92	0.2	**1641**		5413.79
32	rbg253a	1389	254	2376	1192.06	0.3	2369	2372	43159.00
33	rbg323a	1825	324	2547	3599.60	0.5	**2533**		40499.30
34	rbg341a	1822	342	2101	3599.62	1.8	2064	2093	30687.00
35	rbg358a	1967	359	2080	3599.64	2.8	2021	2069	32215.00
36	rbg378a	1973	379	2307	3599.64	3.3	2231	2284	42279.00
37	ry48p.1	256	48	13135	5.06	7.7	12125	13135	43141.00
38	ry48p.2	250	48	13802	4.51	12.1	12130	13802	43033.00
39	ry48p.3	254	48	16540	5.16	20.8	13096	16540	43102.00
40	ry48p.4	249	48	25977	5.29	14.3	22266	25977	43057.00

by our algorithm. We believe that, in both cases, we can obtain better accuracy evaluations provided Gurobi will supplied with more running time or/and more powerful computation resources, e.g. cores, memory, and so on.

7   Conclusion

In this paper, we proposed the novel adaptive heuristic solver PCGLNS for the Precedence Constrained GTSP. The high performance of the solver was proved in numerical experiments with the novel MILP model and the state-of-the-art implementation of the branch-and-bound algorithm provided in Gurobi optimizer. The experiments were carried out against the library PCGTSPLIB of test instances published in recent paper [28]. To the best of our knowledge, solution results for these instances were published for the first time.

Although the proposed solver is fully implemented and is available for further numerical experiments [21], we are aware that both the algorithm itself and its implementation are far from perfect. For future work, we postponed the optimization of its general framework (which likely benefits from parallel implementation), incorporating other well-known heuristics including VNS and GA, and extending PCGTSPLIB by novel industry-inspired test instances.

Acknowledgements. This research was performed as part of research conducted in the Ural Mathematical Center and funded by the Russian Foundation for Basic Research, grants no. 19-07-01243 and 20-08-00873.

References

1. Balas, E.: New classes of efficiently solvable Generalized Traveling Salesman Problems. Ann. Oper. Res. **86**, 529–558 (1999). https://doi.org/10.1023/A:1018939709890
2. Castelino, K., D'Souza, R., Wright, P.K.: Toolpath optimization for minimizing airtime during machining. J. Manuf. Syst. **22**(3), 173–180 (2003). https://doi.org/10.1016/S0278-6125(03)90018-5
3. Chentsov, A.G., Khachai, M.Y., Khachai, D.M.: An exact algorithm with linear complexity for a problem of visiting megalopolises. Proc. Steklov Inst. Math. **295**(1), 38–46 (2016). https://doi.org/10.1134/S0081543816090054
4. Chentsov, A.: Problem of successive megalopolis traversal with the precedence conditions. Autom. Remote Control **75**(4), 728–744 (2014). https://doi.org/10.1134/S0005117914040122
5. Chentsov, A., Khachay, M., Khachay, D.: Linear time algorithm for precedence constrained asymmetric Generalized Traveling Salesman Problem. IFAC-PapersOnLine **49**(12), 651–655 (2016). https://doi.org/10.1016/j.ifacol.2016.07.767. 8th IFAC Conference on Manufacturing Modelling, Management and Control MIM 2016
6. Chentsov, A.G., Chentsov, P.A., Petunin, A.A., Sesekin, A.N.: Model of megalopolises in the tool path optimisation for CNC plate cutting machines. Int. J. Prod. Res. **56**(14), 4819–4830 (2018). https://doi.org/10.1080/00207543.2017.1421784
7. Dantzig, G.B., Ramser, J.H.: The truck dispatching problem. Manage. Sci. **6**(1), 80–91 (1959)
8. Dewil, R., Küçükoğlu, I., Luteyn, C., Cattrysse, D.: A critical review of multi-hole drilling path optimization. Arch. Comput. Meth. Eng. **26**(2), 449–459 (2019). https://doi.org/10.1007/s11831-018-9251-x

9. Dewil, R., Vansteenwegen, P., Cattrysse, D.: Construction heuristics for generating tool paths for laser cutters. Int. J. Prod. Res. **52**(20), 5965–5984 (2014). https://doi.org/10.1080/00207543.2014.895064

10. Dewil, R., Vansteenwegen, P., Cattrysse, D.: A review of cutting path algorithms for laser cutters. Int. J. Adv. Manuf. Technol. **87**(5), 1865–1884 (2016). https://doi.org/10.1007/s00170-016-8609-1

11. Dewil, R., Vansteenwegen, P., Cattrysse, D., Laguna, M., Vossen, T.: An improvement heuristic framework for the laser cutting tool path problem. Int. J. Prod. Res. **53**(6), 1761–1776 (2015). https://doi.org/10.1080/00207543.2014.959268

12. Fischetti, M., González, J.J.S., Toth, P.: A branch-and-cut algorithm for the symmetric Generalized Traveling Salesman Problem. Oper. Res. **45**(3), 378–394 (1997). https://doi.org/10.1287/opre.45.3.378

13. Gendreau, M., Potvin, J.Y.: Handbook of Metaheuristics. International Series in Operations Research & Management Science, 3rd edn., vol. 272. Springer International Publishing, Heidelberg (2019)

14. Ghilas, V., Demir, E., Van Woensel, T.: An adaptive large neighborhood search heuristic for the pickup and delivery problem with time windows and scheduled lines. Comput. Oper. Res. **72**(C), 12–30 (2016). https://doi.org/10.1016/j.cor.2016.01.018

15. Gurobi Optimization, LLC: Gurobi optimizer reference manual (2020). http://www.gurobi.com

16. Gutin, G., Karapetyan, D.: A memetic algorithm for the Generalized Traveling Salesman Problem. Nat. Comput. **9**(1), 47–60 (2010). https://doi.org/10.1016/j.cor.2009.05.004

17. Helsgaun, K.: Solving the equality Generalized Traveling Salesman Problem using the Lin-Kernighan-Helsgaun algorithm. Math. Program. Comput. **7**, 1–19 (2015). https://doi.org/10.1007/s12532-015-0080-8

18. Karuppusamy, N.S., Kang, B.Y.: Minimizing airtime by optimizing tool path in computer numerical control machine tools with application of A^* and genetic algorithms. Adv. Mech. Eng. **9**(12), 1687814017737448 (2017). https://doi.org/10.1177/1687814017737448

19. Khachay, M., Neznakhina, K.: Towards tractability of the Euclidean Generalized Traveling Salesman Problem in grid clusters defined by a grid of bounded height. In: Eremeev, A., Khachay, M., Kochetov, Y., Pardalos, P. (eds.) OPTA 2018. CCIS, vol. 871, pp. 68–77. Springer, Cham (2018). https://doi.org/10.1007/978-3-319-93800-4_6

20. Khachay, M., Neznakhina, K.: Complexity and approximability of the Euclidean Generalized Traveling Salesman Problem in grid clusters. Ann. Math. Artif. Intell. **88**(1), 53–69 (2020). https://doi.org/10.1007/s10472-019-09626-w

21. Kudriavtsev, A., Khachay, M.: PCGLNS: adaptive heuristic solver for the precedence constrained GTSP (2020). https://github.com/AndreiKud/PCGLNS/

22. Makarovskikh, T., Panyukov, A., Savitskiy, E.: Mathematical models and routing algorithms for economical cutting tool paths. Int. J. Prod. Res. **56**(3), 1171–1188 (2018). https://doi.org/10.1080/00207543.2017.1401746

23. Papadimitriou, C.: Euclidean TSP is NP-complete. Theoret. Comput. Sci. **4**, 237–244 (1977). https://doi.org/10.1016/0304-3975(77)90012-3

24. Punnen, A.P., Gutin, G., Punnen, A.P., (eds.): The Traveling Salesman Problem and Its Variations. Combinatorial Optimization, 1 edn., vol. 12. Springer, New York (2002)

25. Qudeiri, J.A., Yamamoto, H., Ramli, R.: Optimization of operation sequence in CNC machine tools using genetic algorithm. J. Adv. Mech. Des. Syst. Manuf. 1(2), 272–282 (2007). https://doi.org/10.1299/jamdsm.1.272

26. Roberti, R., Toth, P.: Models and algorithms for the asymmetric traveling salesman problem: an experimental comparison. EURO J. Transp. Logistics 1(1), 113–133 (2012). https://doi.org/10.1007/s13676-012-0010-0

27. Salman, R., Carlson, J.S., Ekstedt, F., Spensieri, D., Torstensson, J., Söderberg, R.: An industrially validated CMM inspection process with sequence constraints. Procedia CIRP 44, 138–143 (2016). https://doi.org/10.1016/j.procir.2016.02.136. 6th CIRP Conference on Assembly Technologies and Systems (CATS)

28. Salman, R., Ekstedt, F., Damaschke, P.: Branch-and-bound for the precedence constrained Generalized Traveling Salesman Problem. Oper. Res. Lett. 48(2), 163–166 (2020). https://doi.org/10.1016/j.orl.2020.01.009

29. Sarin, S.C., Sherali, H.D., Bhootra, A.: New tighter polynomial length formulations for the asymmetric traveling salesman problem with and without precedence constraints. Oper. Res. Lett. 33(1), 62–70 (2005). https://doi.org/10.1016/j.orl. 2004.03.007

30. Shaw, P.: A new local search algorithm providing high quality solutions to vehicle routing problems (1997)

31. Smith, S.L., Imeson, F.: GLNS: an effective large neighborhood search heuristic for the Generalized Traveling Salesman Problem. Comput. Oper. Res. 87, 1–19 (2017). https://doi.org/10.1016/j.cor.2017.05.010

Simultaneous Detection and Discrimination of Subsequences Which Are Nonlinearly Extended Elements of the Given Sequences Alphabet in a Quasiperiodic Sequence

Liudmila Mikhailova$^{(\boxtimes)}$ (ID) and Sergey Khamdullin (ID)

Sobolev Institute of Mathematics, 4 Koptyug Ave., 630090 Novosibirsk, Russia
{mikh,kham}@math.nsc.ru

Abstract. We consider a posteriori approach to the problem of noise-proof simultaneous detection and discrimination of subsequences-fragments having some given properties in a quasiperiodic sequence. The solution to the problem is stated for the case when the quantity of sought subsequences is unknown. We assume that 1) a finite alphabet of reference sequences is given; 2) a set of permissible deformations is defined for the alphabet, this set gathers all possible extensions of its elements (by duplicating their components); 3) every subsequence-fragment in the quasiperiodic sequence belongs to the set of permissible deformations; 4) subsequences-fragments do not intersect each other, and the difference between the initial positions of two neighboring fragments is limited from above by a given value.

We show that in the framework of a posteriori approach, the problem of simultaneous detection and discrimination reduces to solving an unexplored discrete optimization problem. A polynomial-time algorithm that guarantees the optimal solution to this optimization problem is proposed. The results of the numerical simulation are presented.

Keywords: Discrete optimisation problem · Quasiperiodic sequence · Detection · Discrimination · Polynomial-time solvability · Non-linear extension

1 Introduction

By a numerical quasiperiodic sequence, we mean any sequence that includes subsequences-fragments (subsequences formed by consecutive sequence elements) that have some predetermined characteristic properties with the following restrictions on their relative positions: 1) the fragments do not intersect each

The study was supported by the Russian Foundation for Basic Research, projects 19-07-00397 and 19-01-00308, by the Russian Academy of Science (the Program of basic research), project 0314-2019-0015.

N. Olenev et al. (Eds.): OPTIMA 2020, LNCS 12422, pp. 209–223, 2020.
https://doi.org/10.1007/978-3-030-62867-3_16

other; 2) they are included in the sequence as a whole; 3) the interval between the initial positions of two consecutive fragments is limited from above by the given constant. In this paper, a characteristic property of an individual fragment is the fragment coincidence with a sequence obtained as a nonlinear extension (by duplicating components) of some element of the given sequences set (alphabet). The problem of simultaneous detection and discrimination is to find all the subsequences-fragments and to identify each of them when processing the observed sequence being the sum of some unobservable quasiperiodic sequence and additive noise. In other words, we have to determine the initial index and to find the corresponding stretched alphabet element, for every fragment.

We can interpret numerical quasiperiodic sequences as time series obtained as a result of quasiperiodic pulse trains uniform sampling. In these pulse trains, both fluctuations in the interval between pulses and variations in the shape of the pulse relative to a given pattern are allowed. For example, such pulse trains arise when monitoring natural objects (underwater, aerospace, underground, biomedical, etc.) in the case of quasiperiodic repeatability of their states. This type of state repeatability implies that 1) the object can't be in two different states simultaneously; 2) the distance between two successive state repetitions is not greater than the given constant; 3) a typical state allows some fluctuations from one repetition to another. Below, in Sect. 6, we will illustrate the algorithm operation by examples of processing sequences, in the case when the alphabet includes ECG or PPG pulses. The shape and characteristic waves of these pulses are widely described in the literature (see, for example, [1–4]). Besides, it is the ECG and PPG signals that demonstrate a clear example of the described repeatability of states.

There are two different approaches to solving sequence processing problems: sequential and a posteriori. In the framework of the traditional, sequential approach, the problem is divided into several separate stages, each of which is implemented via well-studied methods and algorithms, such as: 1) optimal noise filtration [5,6], 2) change-point detection [7,8], 3) hypothesis testing and classification [9,10], etc. The advantages of this approach include the low complexity of the resulting algorithms, and the disadvantages—lack of global quality estimates. Even if each stage is implemented optimally, the resulting solution is not guaranteed to be optimal.

We consider a posteriori, non-traditional approach. Within this approach, there is no separation of the decision process into stages. This approach is less common because solving each new problem includes solving a unique, usually unexplored discrete optimization problem. Examples of applying a posteriori approach to problems of quasiperiodic sequence processing can be found in [11–15], and the works cited there.

The paper has the following structure. In Sect. 2, we present the data generation model (we give the formalization of the idea of a quasiperiodic sequence). In the same section, the detection and discrimination problem is formulated as an approximation one. The discrete optimization problem equivalent to the approximation problem is formulated in Sect. 3. There we also give a brief overview of its studied particular cases and problems close in the formulation. Section 4 is supporting, we introduce two auxiliary problems and suggest algorithms solving these problems. The main result is the exact polynomial-time algorithm presented and justified in Sect. 5. Finally, several examples of numerical simulation are given in Sect. 6.

This research continues the tradition of using a posteriori approach to problems of quasiperiodic sequences processing proposed and widely used by **Alexander V. Kel'manov**. The article became possible due to his numerous ideas. We dedicate it to his memory.

2   Data Generation Model and Approximation Problem

Assume that a finite set $V \subset \bigcup_{q=1}^{q_{\max}} \mathbb{R}^q$, $|V| = K$, of numerical sequences of length not exceeding $q_{\max}$ is given. We will call V the alphabet of reference sequences or the alphabet, and each sequence $U \in V$—the reference sequence. There is an example of the alphabet consisting of 8 reference sequences in Fig. 1.

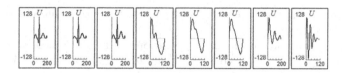

Fig. 1. Example of the alphabet

Define the set of all permissible nonlinear extensions of the sequence $U = (u_1, \ldots, u_q) \in V$ as the set consisting of all sequences of the form

$$(\underbrace{u_1, \ldots, u_1}_{k_1}, \underbrace{u_2, \ldots, u_2}_{k_2}, \ldots, \underbrace{u_q, \ldots, u_q}_{k_q}), \tag{1}$$

where $k_1, \ldots, k_q$ are positive integers. In the extended sequence (1), k_t, $t = 1, \ldots, q$, is the multiplicity of the element u_t, and

$$p = k_1 + \ldots + k_q \tag{2}$$

is the length of the extended sequence.

Assume that the possible length of the extended sequence (1) is limited from above by some constant ℓ, i.e.,

$$p \leq \ell. \tag{3}$$

Then formulas (1), (2), and (3) define the set of all permissible extensions of the sequence U into a sequence having the length not greater than ℓ.

For convenience, we give an alternative definition of the set of permissible extensions in terms of sequence elements indices. According to (1), (2), and (3), the contracting mapping $J : \{1, \ldots, p\} \longrightarrow \{1, \ldots, q\}$, where $p \in \{q, \ldots, \ell\}$, with the following properties

$$J(1) = 1, \; J(p) = q,$$
$$0 \le J(i+1) - J(i) \le 1, \; i = 1, \ldots, p-1,$$

(4)

defines a permissible mapping between the element indices in the reference and the extended sequences. With this mapping, the index sequence

$$\underbrace{1, \ldots, 1}_{k_1}, \underbrace{2, \ldots, 2}_{k_2}, \ldots, \underbrace{q, \ldots, q}_{k_q}$$

is the image of $\{1, \ldots, p\}$, where

$$k_t = \left| \left\{ i \,|\, J(i) = t, i \in \{1, \ldots, p\} \right\} \right|, \; t = 1, \ldots, q.$$

The set of all sequences of the form

$$u_{J(1)}, u_{J(2)}, \ldots, u_{J(i)}, \ldots, u_{J(p)},$$

where $p \in \{q, \ldots \ell\}$, and $J(i)$ satisfies (4), is the set of admissible extensions defined by (1), (2), and (3).

Let us set restrictions on the initial positions and durations of subsequences-fragments in a quasiperiodic sequence. Denote by $T_{\max}$ a positive integer that bounds from above the length of the interval between the beginnings of two adjacent subsequences. Recall that subsequences must not intersect and cross the bounds of the sequence.

Figure 2 depicts an example of a quasiperiodic sequence $X = (x_1, \ldots, x_N)$ of length N. This sequence includes subsequences coinciding with random extensions of the elements of the alphabet presented in Fig. 1.

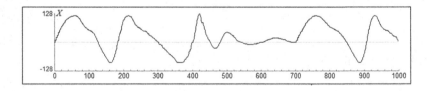

Fig. 2. Example of quasiperiodic sequence

Let the collection of subsequences is induced by a collection $\mathcal{U} = \{U^{(1)}, U^{(2)} \ldots\}$ of alphabet elements. Denote by $\mathcal{M} = \{n_1, n_2, \ldots\} \subseteq \{1, \ldots, N\}$ and $\mathcal{P} = \{p_1, p_2, \ldots\}$ the collections of initial indices and durations of the subsequences-fragments. We will assume that the collections $\mathcal{U}$, $\mathcal{M}$, and $\mathcal{P}$ have the same unknown size M which stands for the quantity of subsequences included in the quasiperiodic sequence X. The evident estimate $M \le M_{\max} \le N$ is valid for M, where $M_{\max}$ is the maximal possible quantity of subsequences. Let q_m, $m = 1, \ldots, M$, be the length of the sequence $U^{(m)}$. Then the elements of the collections $\mathcal{U}$, $\mathcal{M}$ and $\mathcal{P}$ are constrained by the following inequalities:

$$q_m \le p_m \le \ell \le T_{\max} \le N, \quad m = 1, \ldots, M, \tag{5}$$

$$p_{m-1} \le n_m - n_{m-1} \le T_{\max}, \quad m = 2, \ldots, M, \tag{6}$$

$$p_M \le N - n_M + 1. \tag{7}$$

Let $\mathcal{J} = \{J^{(1)}, \ldots, J^{(M)}\}$ be a collection of mappings of the form (4), in which $J^{(m)} : \{1, \ldots, p_m\} \longrightarrow \{1, \ldots, q_m\}$, $m = 1, \ldots, M$, corresponds to some extension of the sequence $U^{(m)}$ to length p_m.

From the above notations and definitions, it follows that the collections $\mathcal{U}$, $\mathcal{M}$, $\mathcal{P}$, and $\mathcal{J}$ uniquely determine the quasiperiodic sequence X, while the n-th element of the sequence can be written in the form

$$x_n = x_n(\mathcal{U}, \mathcal{M}, \mathcal{P}, \mathcal{J}) = \sum_{m=1}^{M} u^{(m)}_{J^{(m)}(n - n_m + 1)}, \quad n = 1, \ldots, N, \tag{8}$$

where

$$u^{(m)}_{J^{(m)}(i)} = 0, \quad m = 1, \ldots, M, \text{ if } i < 1 \text{ or } i > p_m. \tag{9}$$

It is easy to see that the right-hand side of (8) is the sum of M extended reference sequences from the alphabet V; these extended sequences are shifted from the initial serial number $n = 1$ and don't intersect.

The above model shows that if N, ℓ, and $T_{\max}$ are given, the alphabet V induces the set $\mathcal{X}$ of all admissible quasiperiodic sequences. At that, every element of this set is uniquely determined by the collections $\mathcal{U}$, $\mathcal{M}$, $\mathcal{P}$, and $\mathcal{J}$.

Assume that we observe a sequence $Y = (y_1, \ldots, y_N)$, which is the element-wise sum of an unobservable sequence $X \in \mathcal{X}$ and some sequence reflecting possible noise distortion. An example of the sequence Y is depicted in Fig. 3.

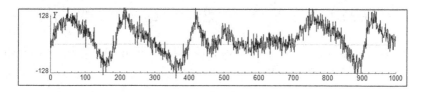

Fig. 3. Example of the sequence to be processed

Since the sequences X and Y can be considered as vectors in N-dimensional space, we can formulate the detection and discrimination problem (the problem of restoring an unobserved sequence) in the form of the following approximation problem:

$$\|Y - X\|^2 \longrightarrow \min_{X \in \mathcal{X}}. \tag{10}$$

It was previously shown [11], that even in the particular case of the problem, when $|V| = 1$, the cardinality of the set $\mathcal{X}$ grows exponentially as a function of N, so the brute-force searching through this set is hardly possible in a reasonable time.

Taking into account that $X = X(\mathcal{U}, \mathcal{M}, \mathcal{P}, \mathcal{J})$, the approximation problem (10) can be rewritten in the following equivalent form:

$$\|Y - X(\mathcal{U}, \mathcal{M}, \mathcal{P}, \mathcal{J})\|^2 \longrightarrow \min_{\mathcal{U}, \mathcal{M}, \mathcal{P}, \mathcal{J}}. \tag{11}$$

3 Optimization Problem and Related Problems

Expanding the squares of the norm in (11) and taking into account (8), (9) and constraints (5), (6), and (7), we obtain by simple calculations that

$$\|Y - X\|^2 = \sum_{n=1}^{N} y_n^2 + \sum_{m=1}^{M} \sum_{i=1}^{p_m} \{(u_{J^{(m)}(i)}^{(m)})^2 - 2y_{n_m+i-1}u_{J^{(m)}(i)}^{(m)}\}.$$

Here the first term in the right-hand side is constant and, hence, approximation problem (11) is equivalent to the following discrete optimization problem.

Problem 1. *Given:* the numerical sequence $Y = (y_1, \ldots, y_N)$, the positive integers $T_{\max}$, $q_{\max}$, and ℓ, such that $q_{\max} \le \ell$, the collection $V \subset \bigcup_{q=1}^{q_{\max}} \mathbb{R}^q$, $|V| = K$, of numerical sequences. *Find:* the collection $\mathcal{U} = (U^{(1)}, \ldots, U^{(m)}, \ldots)$ of numerical sequences, where $U^{(m)} = (u_1^{(m)}, \ldots, u_{q_m}^{(m)}) \in V$, the collection $\mathcal{M} = \{n_1, \ldots n_m, \ldots\}$ of indices of the sequence Y, the collection $\mathcal{P} = \{p_1, \ldots, p_m, \ldots\}$ of positive integers, the collection $\mathcal{J} = \{J^{(1)}, \ldots, J^{(m)}, \ldots\}$ of contracting mappings, where $J^{(m)} : \{1, \ldots, p_m\} \longrightarrow \{1, \ldots, q_m\}$, and the size M of these collections that minimize the objective function

$$F(\mathcal{U}, \mathcal{M}, \mathcal{P}, \mathcal{J}) = \sum_{m=1}^{M} \sum_{i=1}^{p_m} \{(u_{J^{(m)}(i)}^{(m)})^2 - 2y_{n_m+i-1}u_{J^{(m)}(i)}^{(m)}\},$$

under the constraints (5), (6), and (7) on the elements of the sought collections $\mathcal{M}$, $\mathcal{P}$, and under the constraints

$$J^{(m)}(1) = 1, \quad J^{(m)}(p_m) = q_m,$$
$$0 \le J^{(m)}(i) - J^{(m)}(i - 1) \le 1, \quad i = 2, \ldots, p_m,$$
$$m = 1, \ldots, M,$$

on the contracting mappings.

Problem 1 can be considered as the generalization of some previously studied problems. The particular case of Problem 1, when $|V| = 1$, was examined in [11]. The restriction $|V| = 1$ means that the quasiperiodic sequence includes fragments coinciding with the extensions of the same reference sequence, so problem [11] is a detection problem solely. In [11], the algorithm which finds the exact solution to this particular case in time $\mathcal{O}(T_{\max}^3 N)$ was suggested. Another particular case was examined in [12]. There, as in Problem 1, it was assumed that $|V| \geq 1$, but there were additional restrictions: $q_1 = \ldots = q_K = q$, $p_m = q$, $J^{(m)}(i) = i$, $m = 1, \ldots, M$. The modification of this particular case, when M is a part of the input (is given), was studied in [13]. In [12] and [13] the algorithms that allow obtaining optimal solutions in time $\mathcal{O}((T_{\max} + (K - 1)q)N)$ and $\mathcal{O}((T_{\max} + (K - 1)q)MN))$ respectively, were justified.

The problems of simultaneous detection and discrimination considered in [14] and [15] are close in formulation to Problem 1. In these problems, as in Problem 1, a quasiperiodic sequence includes fragments coinciding with admissibly distorted elements of the given sequence alphabet. However, in contrast to the problem considered in the current paper, it is assumed that the set of all permissible distortions in [14] and [15] consists of all subsequences obtained by removing a certain quantity of initial and final samples. In [14] the quantity of fragments is assumed to be unknown and in [15]—to be a part of the input. The exact polynomial algorithms solving these problems are presented in the cited papers.

4   Auxiliary Problems

In order to construct an algorithm solving Problem 1, we need to consider two following auxiliary problems.

Problem 2. *Given:* the array $\{w_{i,j}, i = 1, \ldots, p, j = 1, \ldots, q\}$ of real numbers. *Find:* summing indices $J(i)$, $i = 1, \ldots, p$, that minimize the objective function

$$W = \sum_{i=1}^{p} w_{i,J(i)}$$

under the constraints

$$J(1) = 1, \quad J(p) = q,$$
$$0 \leq J(i) - J(i - 1) \leq 1, i = 2, \ldots, p.$$

The following lemma provides recurrent formulas to solve Problem 2.

Lemma 1. *[11] Let the conditions of Problem 2 hold. Then the optimal value W^* of the objective function of this problem is given by the formula*

$$W^* = W_{p,q}, \tag{12}$$

and the values of $W_{p,q}$ are calculated with the recurrent formulas

$$W_{s,t} = \min\{W_{s-1,t}, W_{s-1,t-1}\} + w_{s,t}, \quad s = 1, \ldots, p, \ t = 1, \ldots, q, \tag{13}$$

with the following initial and boundary conditions

$$W_{s,t} = \begin{cases} 0, & s = 0, \ t = 0, \\ +\infty, & s = 0, \ t = 1, \ldots, q, \\ +\infty, & s = 1, \ldots, p, \ t = 0. \end{cases} \tag{14}$$

The optimal values of the summing indices are determined by the rule

$$J^*(p) = q;$$
$$J^*(i-1) = \begin{cases} J^*(i), & \text{if } W_{i-1,J^*(i)} \leq W_{i-1,J^*(i)-1}, \\ J^*(i) - 1, & \text{if } W_{i-1,J^*(i)} > W_{i-1,J^*(i)-1}, \end{cases} \tag{15}$$
$$i = p, p-1, \dots, 2.$$

The second auxiliary problem is

Problem 3. *Given*: the positive integers $T_{\max}$, ℓ and the collection $\{q_1, \dots, q_K\}$ of positive integers such that $\ell \leq T_{\max}$, $q_k \leq \ell$, $k = 1, \dots, K$, the collection $\{g_k(n,p), \ k = 1, \dots, K, \ p = q_k, \dots, \ell, \ n = 1, \dots, N - p + 1\}$ of numerical sequences. *Find*: the collection $\{(n_1, p_1, k_1), \dots, (n_M, p_M, k_M)\}$ of positive integer triples, where $k_m \in \{1, \dots, K\}$, $n_m \in \{1, \dots, N\}$, $p_m \in \{q_{k_m}, \dots, \ell\}$, $m = 1, \dots, M$, and the size M of this collection that minimize the objective function

$$G((n_1, p_1, k_1), \dots, (n_M, p_M, k_M)) = \sum_{m=1}^{M} g_{k_m}(n_m, p_m),$$

under the restrictions (6) and (7).

The following lemma provides recurrent formulas to solve Problem 3.

Lemma 2. *Let the conditions of Problem 3 hold. Then the optimal value G^* of the objective function of the problem is given by the formula*

$$G^* = \min_{n \in \omega} \min_{p \in \delta(n)} \min_{k \in \mathcal{K}(p)} G_k(n, p), \tag{16}$$

and the value of $G_k(n, p)$ are calculated with the recurrent formula

$$G_k(n, p) = \begin{cases} g_k(n, p), & n \in \omega \setminus \omega^+, \ p \in \delta(n), \ k \in \mathcal{K}(p), \\ \min\left\{0, \ \min_{j \in \gamma(n)} \min_{i \in \theta(n,j)} \min_{t \in \mathcal{K}(i)} G_t(j, i)\right\} + g_k(n, p), \\ \qquad\qquad n \in \omega^+, \ p \in \delta(n), \ k \in \mathcal{K}(p), \end{cases} \tag{17}$$

where

$$\omega = \{1, \dots, N - q_{\min} + 1\}$$

is the set of admissible values of n_m, $m = 1, \dots, M$, with

$$q_{\min} = \min_{k=1,\dots,K} q_k;$$

$$\omega^+ = \{q_{\min} + 1, \dots, N - q_{\min} + 1\}$$

is the set of admissible values of n_m, $m = 2, \dots, M$;

$$\delta(n) = \{q_{\min}, \dots, \min\{\ell, \ N - n + 1\}\}, \quad n \in \omega,$$

is the set of admissible values of p_m, $m = 1, \ldots M$, if $n_m = n$;

$$\gamma(n) = \{\max\{n - T_{\max}, 1\}, \ldots, n - q_{\min}\}, \ n \in \omega^+,$$

is the set of admissible values of n_{m-1}, $m = 2, \ldots, M$, if $n_m = n$;

$$\theta(n, j) = \{q_{\min}, \ldots, \min\{\ell, n - j\}\}, \ n \in \omega^+, \ j \in \gamma(n),$$

is the set of admissible values of p_{m-1}, $m = 2, \ldots, M$, if $n_m = n$ and $n_{m-1} = j$;

$$\mathcal{K}(q) = \left\{ k \ \middle| \ k \in \{1, \ldots, K\}, q_k \leq q \right\}, \ q = 1, \ldots, N - q_{\min} + 1,$$

is the set of indices of elements in $\{q_1, \ldots, q_K\}$, which are not greater than q. The optimal solution to Problem 3 is the collection

$$\{(\nu_M, \pi_M, \kappa_M), \ldots, (\nu_1, \pi_1, \kappa_1)\}$$

of triples, where

$$\nu_1 = \arg \min_{n \in \omega} \left\{ \min_{p \in \delta(n)} \min_{k \in \mathcal{K}(p)} G_k(n, p) \right\}, \tag{18}$$

$$\pi_1 = \arg \min_{p \in \delta(\nu_1)} \min_{k \in \mathcal{K}(p)} G_k(\nu_1, p), \tag{19}$$

$$\kappa_1 = \arg \min_{k \in \mathcal{K}(\pi_1)} G_k(\nu_1, \pi_1), \tag{20}$$

$$\nu_m = I(\nu_{m-1}), \ \pi_m = \pi(\nu_{m-1}), \kappa_m = \arg \min_{k \in \mathcal{K}(\pi_m)} G_k(\nu_m, \pi_m), \ m = 2, \ldots, M,$$
$$\tag{21}$$

$$M = \min\{m \mid m \in \{1, \ldots, N\}, \ I(\nu_m) = 0\},$$

$$I(n) = \begin{cases} 0, & \text{if } n \in \omega \setminus \omega^+, \\ 0, & \text{if } \min_{j \in \gamma(n)} \min_{i \in \theta(n,j)} \min_{t \in \mathcal{K}(i)} G_t(j, i) \geq 0, \ n \in \omega^+, \\ \arg \min_{j \in \gamma(n)} \left\{ \min_{i \in \theta(n,j)} \min_{t \in \mathcal{K}(i)} G_t(j, i) \right\}, & \\ & \text{if } \min_{j \in \gamma(n)} \min_{i \in \theta(n,j)} \min_{t \in \mathcal{K}(i)} G_t(j, i) < 0, \ n \in \omega^+, \end{cases} \tag{22}$$

$$\pi(n) = \begin{cases} 0, & \text{if } I(n) = 0, \\ \arg \min_{i \in \theta(n, I(n))} \left\{ \min_{t \in \mathcal{K}(i)} G_t(I(n), i) \right\}, & \text{if } I(n) > 0, \end{cases} \ n \in \omega. \tag{23}$$

The proof of Lemma 1 is based on the direct derivation of recurrent formulas.

5 Algorithm

Based on the considered auxiliary problems and formulas for their solution, we write down the following algorithm. Unformally, the idea of the algorithm consists in dividing the optimization process into two stages to be carried out sequentially. In the first stage, we solve a family of Problems 2 to find the array

$$\min_{J(i)} \sum_{i=1}^{p} \{u_{J(i)}^2 - 2y_{n+i-1}u_{J(i)}\},$$

$$U \in V, \ n = 1, \ldots, N - q + 1, \ p = q, \ldots, \min\{\ell, N - n + 1\},$$

of minimums and eliminate minimization by $\mathcal{J}$ from the further calculations. In the second stage, we use this array as a part of the input of Problem 3. The solution obtained gives us the optimal value of the objective function of Problem 1 and optimal collections $\mathcal{U}$, $\mathcal{M}$, and $\mathcal{P}$. Then, returning to data obtained at the first stage we recover the optimal collection $\mathcal{J}$.

Algorithm $\mathcal{A}$.

INPUT: positive integers $T_{\max}$, ℓ, a collection V of numerical sequences and a numerical sequence Y.

Forward pass.

STEP 1 (solving the family of Problems 2). For every $U = (u_1, \ldots, u_q) \in V$, every $n = 1, \ldots, N - q + 1$, and every $p = q, \ldots, \min\{\ell, N - n + 1\}$ do:

(1) Put
$$w_{i,j} = u_j^2 - 2y_{n+i-1}u_j, i = 1, \ldots, p, \ j = 1, \ldots, q.$$

(2) Calculate $W_{s,t}$, $s = 1, \ldots, p$, $t = 1, \ldots, q$, using formulas (12), (13), and (14). Find the sequence $J^*(1), \ldots, J^*(p)$ of summing indices by the rule (15).

(3) Put $W(n, p, U) = W_{p,q}$, $J(n, p, U) = \{J^*(i), \ i = 1, \ldots, p\}$.

STEP 2 (solving Problem 3; forward pass).

(1) Renumber the elements of the set V arbitrarily: $V = \{U_1, \ldots, U_K\}$.

(2) Put $q_k = \dim(U_k)$, $k = 1, \ldots, K$, $g_k(n, p) = W(n, p, U_k)$, $p = q_k, \ldots, \ell$, $n = 1, \ldots, N - p + 1$.

(3) Calculate $G_k(n, p)$, $p = q_k, \ldots, \ell$, $n = 1, \ldots, N - p + 1$, using (17) and G^* using (16). Put $F_A = G^*$.

STEP 3 (solving Problem 3; backward pass).

(1) Calculate $I(n)$ and $\pi(n)$, $n = 1, \ldots, N - q_{\min} + 1$, by (22) and (23).

(2) Find the components of the auxiliary collections $(\nu_1, \nu_2, \ldots)$, $(\pi_1, \pi_2, \ldots)$, and $(\kappa_1, \kappa_2, \ldots)$, and their size M by (18), (19), (20), and (21).

STEP 4. Put $M_A = M$, $\mathcal{M}_A = \{\nu_{M_A}, \ldots, \nu_1\}$, $\mathcal{P}_A = \{\pi_{M_A}, \ldots, \pi_1\}$, $J^{(m)} = J(\nu_{M_A-m+1}, \pi_{M_A-m+1})$, $m = 1, \ldots, M_A$, and $\mathcal{J}_A = \{J^{(1)}, \ldots, J^{(M_A)}\}$, $\mathcal{U}_A = \{U_{\kappa_{M_A}}, \ldots, U_{\kappa_1}\}$.

OUTPUT: The positive integer M_A, the collections $\mathcal{M}_A$, $\mathcal{P}_A$, $\mathcal{J}_A$, $\mathcal{U}_A$, and the value F_A.

The main result of the research is the following theorem.

Theorem 1. *Algorithm* $\mathcal{A}$ *finds an exact solution to Problem 1 in time* $\mathcal{O}(T_{\max}^3 K^2 N)$.

The proof of the theorem is based on the following chain of equalities:

$$F^* \overset{(1)}{=} \min_{\mathcal{U},\mathcal{M},\mathcal{P},\mathcal{J}} \sum_{m=1}^{M} \sum_{i=1}^{p_m} \{(u_{J^{(m)}(i)}^{(m)})^2 - 2y_{n_m+i-1} u_{J^{(m)}(i)}^{(m)}\}$$

$$\overset{(2)}{=} \min_{\mathcal{U},\mathcal{M},\mathcal{P}} \min_{J^{(1)},\dots,J^{(M)}} \sum_{m=1}^{M} \sum_{i=1}^{p_m} \{(u_{J^{(m)}(i)}^{(m)})^2 - 2y_{n_m+i-1} u_{J^{(m)}(i)}^{(m)}\}$$

$$\overset{(3)}{=} \min_{\mathcal{U},\mathcal{M},\mathcal{P}} \sum_{m=1}^{M} \left\{ \min_{J^{(m)}} \sum_{i=1}^{p_m} \{(u_{J^{(m)}(i)}^{(m)})^2 - 2y_{n_m+i-1} u_{J^{(m)}(i)}^{(m)}\} \right\}$$

$$\overset{(4)}{=} \min_{\mathcal{U},\mathcal{M},\mathcal{P}} \sum_{m=1}^{M} \left\{ \min_{J} \sum_{i=1}^{p_m} \{(u_{J(i)}^{(m)})^2 - 2y_{n_m+i-1} u_{J(i)}^{(m)}\} \right\}$$

$$\overset{(5)}{=} \min_{\mathcal{U},\mathcal{M},\mathcal{P}} \sum_{m=1}^{M} W(n_m, p_m, U^{(m)}) \overset{(6)}{=} \min_{(k_1,\dots,k_M),\mathcal{M},\mathcal{P}} \sum_{m=1}^{M} W(n_m, p_m, U_{k_m})$$

$$\overset{(7)}{=} \min_{\{(n_1,p_1,k_1),\dots,(n_M,p_M,k_M)\}} \sum_{m=1}^{M} g_{k_m}(n_m, p_m) \overset{(8)}{=} F_A.$$

In this chain, equality (1) is the definition of an optimal solution to Problem 1. Equality (2) is obvious. Equality (3) holds because $J^{(m)}$, $m = 1,\dots,M$, are independent and at fixed $\mathcal{U}$, $\mathcal{M}$, and $\mathcal{P}$, the double sum on the left-hand side of equality (3) is a separable function of $(J^{(1)},\dots,J^{(M)})$. In other words, we formally have $\min_{J^{(1)},\dots,J^{(M)}} \sum_{m=1}^{M}(\bullet) = \sum_{m=1}^{M} \min_{J^{(m)}}(\bullet)$. Equality (4) is just the change of variables. Equality (5) holds in line with the description of Step 1, the definition of the optimal solution to Problem 2, and Lemma 1. Equality (6) is just replacing minimization by alphabet elements with minimization by their ordinal numbers, according to enumeration introduced at Step 1. Equality (7) is valid according to the description of Step 2 and using the regrouping of variables:

$$\min_{(k_1,\dots,k_M),\mathcal{M},\mathcal{P}} \bullet = \min_{(k_1,\dots,k_M),(n_1,\dots,n_M),(p_1,\dots,p_M)} \bullet = \min_{\{(n_1,p_1,k_1),\dots,(n_M,p_M,k_M)\}} \bullet.$$

Finally, equality (8) follows from the description of Step 2 and Step 4, the definition of the optimal solution to Problem 3, and Lemma 2.

6   Numerical Simulation

Theorem 1 proves the optimality of the solution to Problem 1 obtained by Algorithm $\mathcal{A}$, so the numerical simulation results presented below are illustrative. In fact, from the mathematical point of view, it doesn't matter which sequences are included in the alphabet. To illustrate the efficiency of the algorithm, we present two examples of processing modeled quasiperiodic sequences. In each of the examples, the alphabet includes several sampled biomedical pulses: ECG in the first example and PPG in the second. The main reason for choosing namely ECG and PPG pulses to form the alphabet is our desire to illustrate the potential applicability of the algorithm presented to bio-medical applications. We suppose that the algorithm solving Problem 1 can serve as one of the steps to building noise-resistant algorithms for processing real biomedical data.

Figure 4 shows the input data of Algorithm A: an example of the alphabet V of reference sequences consisting of 3 ECG pulses (at the top) and the sequence Y to be processed—at the bottom.

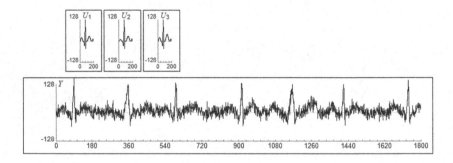

Fig. 4. Example 1. Input data

There are two auxiliary modeled unobservable sequences in Fig. 5. One of them (the top row) is the sequence $\mathcal{U}$ of randomly selected alphabet elements. Another one (the bottom row) is an example of modeled (generated) unobservable pulse sequence $X \in \mathcal{X}$. This sequence contains random extensions of sequences (pulses) from $\mathcal{U}$ with random fluctuations in the time intervals between the pulses. Recall that these sequences are not the part of the problem input. The sequence Y, depicted in the bottom row of Fig. 4, is the element-wise sum of

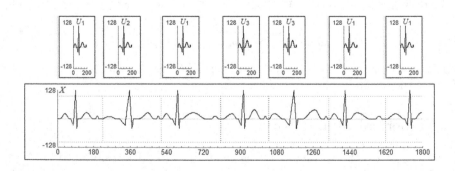

Fig. 5. Example 1. Unobservable data

the modeled sequence X and a sequence of independent identically distributed Gaussian random variables (white noise).

Figure 6 shows the result of Algorithm A operation. In the top row, the collection $\mathcal{U}_A$ of alphabet elements is situated (discrimination result). In the bottom row, we can see the sequence X_A recovered using the rule (8) on the base of four collections obtained as algorithm output. This example is computed for $K = 3$, $V = \{U_1, U_2, U_3\}$, $q_1 = 203$, $q_2 = 203$, $q_3 = 203$, $T_{\max} = 300$, $\ell = 300$, $N = 1800$, maximal amplitude pulse value is 128, and the noise level $\sigma = 15$.

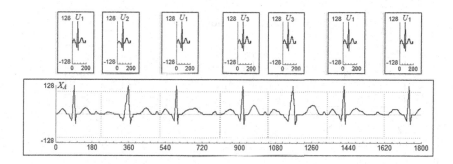

Fig. 6. Example 1. Output data

Similarly, Fig. 7, 8, and 9 illustrate algorithm operation when we use several sampled PPG pulses as alphabet elements. This example is computed for $K = 3$, $V = \{U_1, U_2, U_3\}$, $q_1 = 120$, $q_2 = 120$, $q_3 = 120$, $T_{\max} = 280$, $\ell = 280$, $N = 1000$, maximum amplitude value is 128, the noise level $\sigma = 15$.

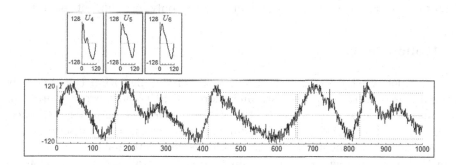

Fig. 7. Example 2. Input data

The numerical simulation results clearly confirm the optimality of the solution found by the algorithm. In such a way, the algorithm suggested is a suitable tool for noise-resistant processing of data in the form of a quasiperiodic sequence with the above-described structure. First of all, in both examples, the modeled

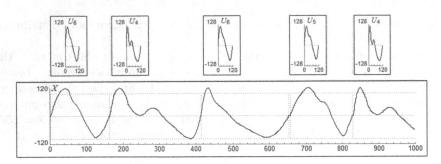

Fig. 8. Example 2. Unobservable data

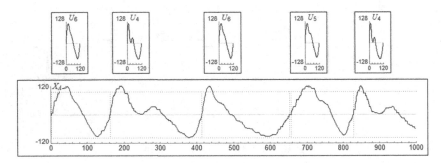

Fig. 9. Example 2. Output data

sequence $\mathcal{U}$ inducing the unobservable sequence X and the sequence $\mathcal{U}_A$ being the part of the output coincide. It means that discrimination is carried out correctly. Secondly, the visual comparison of the two graphs (of the unobservable sequence X and the recovered sequence X_A) demonstrates only insignificant differences.

7 Conclusion

In this paper, we have constructed and justified the algorithm that finds the optimal solution to one previously unexplored discrete optimization problem. This optimization problem arises (in the framework of the a posteriori approach) when solving the applied problem of simultaneous detection and discrimination of subsequences-fragments in a quasiperiodic sequence. We have proved that the algorithm is polynomial-time. As an illustration of the algorithm operation, we have used the examples of processing the sequences in which sampled ECG and PPG pulses were used as elements of the alphabet.

A modification of Problem 1, in which the quantity of subsequences-fragments is a part of the problem input, remains to be studied. Our nearest plans are connected with exploring this modification.

References

1. Rajni, R., Kaur, I.: Electrocardiogram signal analysis – an overview. Int. J. Comput. Appl. **84**(7), 22–25 (2013)
2. Al-Ani, M.S.: ECG waveform classification based on P-QRS-T wave recognition. UHD J. Sci. Technol. **2**(2), 7–14 (2018)
3. Shelley, K., Shelley, S.: Pulse oximeter waveform: photoelectric plethysmography. In: Lake, C., Hines, R., Blitt, C. (eds.) Clinical Monitoring, pp. 420–428. W.B. Saunders Company, Philadelphia (2001)
4. Elgendy, M.: On the analysis of fingertrip photoplethysmogram signals. Curr. Cardiol. Rev. **8**(1), 14–25 (2012)
5. Anderson, B.D., Moore, J.D.: Optimal Filtering. Prentice Hall, Englewood Cliffs (1995)
6. Sparks, T., Chase, G.: Filters and Filtration Handbook. Butterworth-Heinemann, Oxford (2015)
7. Polunchenko, A., Tartakovsky, A.: State-of-the-art in sequential change-point detection. Methodol. Comput. Appl. Probab. **14**(3), 649–684 (2012). https://doi.org/10.1007/s11009-011-9256-5
8. Poor, H.V., Hadjiliadis, O.: Quickest Detection. Cambridge University Press, Cambridge (2008)
9. Fukunaga, K.: Introduction to Statistical Pattern Recognition, 2nd edn. Academic, New York (1990)
10. Duda, R., Hart, P., Stork, D.: Pattern Classification, 2nd edn. Wiley-Interscience, New York (2000)
11. Kel'manov, A., Khamidullin, S., Mikhailova, L., Ruzankin, P.: Polynomial-time solvability of one optimization problem induced by processing and analyzing quasiperiodic ECG and PPG signals. In: Jaćimović, M., Khachay, M., Malkova, V., Posypkin, M. (eds.) OPTIMA 2019. CCIS, vol. 1145, pp. 88–101. Springer, Cham (2020). https://doi.org/10.1007/978-3-030-38603-0_7
12. Kel'manov, A.V., Okol'nishnikova, L.V.: A posteriori simultaneous detection and discrimination of subsequences in a quasiperiodic sequence. Pattern Recogn. Image Anal. **11**(3), 505–520 (2001)
13. Kel'manov, A.V., Khamidullin, S.A.: A posteriori joint detection and discrimination of a given number of subsequences in a quasiperiodic sequence. Pattern Recogn. Image Anal. **10**(3), 379–388 (2000)
14. Kel'manov, A.V., Khamidullin, S.A.: A posteriori concurrent detection and identification of quasiperiodic fragments in a sequence from their pieces. Pattern Recogn. Image Anal. **16**(4), 599–613 (2006)
15. Kel'manov, A.V., Khamidullin, S.A.: Simultaneous a posteriori detection and identification of a predetermined number of quasiperiodic fragments in a sequence based on their segments. Pattern Recogn. Image Anal. **16**(3), 344–357 (2006)

Lurie Systems Stability Approach for Attraction Domain Estimation in the Wheeled Robot Control Problem

Lev Rapoport[1,2,3]([envelope]) [iD] and Alexey Generalov[1,2,3] [iD]

[1] Institute of Control Sciences, 117997 Moscow, Russia
LBRapoport@gmail.com, generalov.alexey@gmail.com
[2] Skolkovo Institute of Science and Technology, 121205 Moscow, Russia
[3] Topcon Positioning Systems, 115114 Moscow, Russia

Abstract. Considered is the problem of the attraction domain estimation in the space "distance to the trajectory – orientation" for the problem of the planar motion control of a wheeled robot. This problem has received much attention in connection with precision farming applications [11]. The mathematical model of the robot takes into account kinematic relationships between velocity of a given target point, orientation of the platform, and control. It is supposed that the four wheels platform moves without slipping. The rear wheels are assumed to be driving while the front wheels are responsible for the rotation of the platform. The control goal is to drive the target point to the desired trajectory and to stabilize its motion. The case of the straight line trajectory is considered in the paper. The control was obtained using the feedback linearization approach [5] and is subject to the two-sided constraints. The system is then rewritten in the so called Lurie form [1,12] and embedded in the class of systems with nonlinearities constrained by the sector condition. Based on this, the method of attraction domain estimation in the state space of the system is proposed. The negativity condition for the derivative of the Lyapunov function with respect to the system's dynamics under sector conditions is formulated in terms of solvability of the linear matrix inequality (LMI) [2]. To take into account quadratic constraints the S-procedure [12] is applied. The Lyapunov function is supposed to be a quadratic form with addition of an integral over nonlinearity (so called Lurie-Postnikov function). Earlier, other classes of Lyapunov functions were used for this problem, see for example [8]. The optimization problem was formulated as a semidefinite programming (SDP) problem with LMI constraints. Numerical results are presented. The estimates achieved show less conservativeness in comparison with ellipsoidal ones.

Keywords: Wheeled robot · Absolute stability · Lurie systems ·
Attraction domain

This work was financially supported by the Russian Foundation for Basic Research, project 18-08-00531.

N. Olenev et al. (Eds.): OPTIMA 2020, LNCS 12422, pp. 224–238, 2020.
https://doi.org/10.1007/978-3-030-62867-3_17

1   Introduction

One of the fundamental problems of nonlinear control systems theory is a description of the attraction domain of the equilibrium state. Usually the equilibrium refers to nominal system's operation, which should be stabilized. Complete solution of this problem in general form is extremely hard. Therefore, an internal (by inclusion) estimate of the attraction domain is important for applications. Usually engineers are looking for an estimate of a domain that combines two properties: it must provide asymptotic attraction and it must be invariant. By invariance we mean such a property of the domain, that the trajectory of the system, once getting inside, will no longer leave it. Standard approach for construction such estimates consists in using Lyapunov functions from certain parametric classes. Let $x \in R^n$ be the system state. Given the Lyapunov function $V(x)$, the estimate of the attraction domain is constructed as a level set $\{x : V(x) \leq c\}$, provided the time derivative with respect to the system dynamics is negative: $\dot{V} < 0$. Candidates for use as a Lyapunov function are selected from some parametric class. Thus, the more general is the parametric class of Lyapunov functions, the more is freedom of choice, and the less conservative is the resulting estimate of the attraction domain. Desire to maximize the volume of the attraction domain leads to the problem of optimal parameters choice.

In this paper we consider the problem of stabilization of the planar motion of a wheeled robot. It is supposed that the four wheels platform moves without cross-track slipping, subject to non-holonomic constraints. The rear wheels are assumed to be driving while the front wheels are responsible for the rotation of the platform. The control goal is to drive the target point to the desired trajectory and to stabilize its motion. In this paper, we restrict ourselves to the case of motion along a straight path for simplicity. The control law is obtained by the feedback linearization approach and is subject to the two-sided constraints as described in the earlier papers [8]. To estimate attraction domain of the closed loop system the quadratic Lyapunov function was earlier used. Parameters to be chosen were entries of the positive definite square matrix. In this paper we consider the more general class of Lyapunov functions consisting of a quadratic form with addition of an integral over a nonlinearity multiplied by an unknown scalar which is also a parameter to be chosen. These functions are called Lurie-Postnikov functions.

The use of attraction domains in automatic control systems of wheeled robots is introduced in papers [7,8,11] as assistance to operators. At the same time, nowadays there has been a movement towards autonomous robots that exclude the presence of people at all. Therefore, the behavior of robots should be even more predictable. There are a lot of potential areas for autonomous robotic operation, e.g. small wheeled robots for precise agriculture, golf course lawn mowers, and so on. The common challenge for these areas is to make the robot behavior reliable and fully predictable, including safe and secure operation [3].

As for the motion planning, it is supposed that the optimal field coverage problem was already solved for the particular working site and desired trajectory was obtained as a result. One can assume, that the robot is delivered first to the neighborhood of the working site and left in arbitrary position near the beginning of the pre-planned path. The autonomous control goal is to bring the robot to the desired start of the planned path and stabilize the motion of a target point along it [8,11]. Also, it should be ensured that the control algorithm will not cause system stability loss. It means, that the robot should get inside the attraction domain of the dynamic system closed by a synthesized control law, thus guaranteeing the asymptotic stability. In addition, even starting sufficiently close to the target trajectory, the robot may perform large oscillations going far enough even if the closed loop system is stable. Thus, the estimate of attraction domain subjected to reasonable geometric constraints is desired to be invariant: once getting inside, the system trajectories should not leave it.

The goal of this study is to propose a method of attraction domain estimation in the state space of the system, that is less conservative in comparison with ellipsoidal estimates, obtained earlier. Description of the attraction domain estimation inscribed into the band of certain width around any state space variables of the system and guaranteeing prescribed exponential convergence rate is also considered. The optimal parameters choice problem is formulated. Results of the paper are illustrated by examples.

Note also that the absolute stability approach to stability analysis of systems with constrained control is not the only possible approach, see for example [9, 10] for other approaches.

2   Problem Statement

Consider the kinematic model of a wheeled robot, described for example in [8]. Let (ξ, η) be the position of the target point and ψ be the angle describing the orientation of the platform with respect to the fixed reference system $O\xi\eta$. For convenience, the middle point of the rear axle is taken as the target point, ψ is the angle between the direction of the velocity vector of the target point and the ξ-axis. The kinematic equations are:

$$\begin{aligned}
\dot{\xi} &= v\cos\psi, \\
\dot{\eta} &= v\sin\psi, \\
\dot{\psi} &= v\,u,
\end{aligned} \tag{1}$$

where the dot denotes the time derivative, $v \equiv v(t)$ is a scalar linear velocity of the target point. The control variable u is the instantaneous curvature of the trajectory circumscribed by the target point, which is related to the angle of rotation of the front wheels. The control goal is to stabilize the motion of the target point along the ξ-axis, when the lateral deviation η and the angular deviation ψ are equal to zero. The constraining condition of the control resources can be written as the two-sided constraints:

$$-\bar{u} \le u \le \bar{u}, \tag{2}$$

where $\bar{u} = 1/R_{\min}$ is the maximum possible curvature of the actual trajectory circumscribed by the target point and $R_{\min}$ represents the minimum possible turning radius of the platform.

Differential Eq. (1) can be generalized for the case of any oblique straight line. It has been shown that, by changing the state variables and the independent variable, the straight path following problem can be written in the canonical form:

$$z_1' = z_2,$$
$$z_2' = u(1 + z_2^2)^{\frac{3}{2}}. \tag{3}$$

In (3), z_1 is the lateral deviation of the robot from the target path and $z_2 = \tan \psi$, where ψ is the angle between the direction of the velocity vector and the target path. The prime denotes differentiation with respect to the new independent variable ξ.

Ignoring two-sided constraints on control and using feedback linearization technique [5] leads to the choice of control u in the form

$$u = -\frac{\sigma}{(1 + z_2^2)^{\frac{3}{2}}} \tag{4}$$

for some $\lambda > 0$ and

$$\sigma = \lambda^2 z_1 + 2\lambda z_2, \tag{5}$$

where λ is the desired rate of exponential decrease. Then the closed loop system (3) takes the form

$$z_1' = z_2,$$
$$z_2' = -\sigma, \tag{6}$$

which is equivalent to the equation $z_1'' + 2\lambda z_1' + \lambda^2 z_1 = 0$. It implies the exponential convergence of the vector $z = (z_1, z_2)^T$ to zero with the rate $-\lambda$. The particular form of function (5) determines poles of the closed-loop system (6). For given σ the system has one pole $-\lambda$ of multiplicity 2. Another forms of σ implies another poles. We suppose that λ is somehow chosen and don't discuss the pole placement problem. For the optimal poles placement discussion see for example [6].

However, in general, control (4) does not satisfy the two-sided constraints (2). Taking control in the form

$$u = -s_{\bar{u}}\left(\frac{\sigma}{(1 + z_2^2)^{\frac{3}{2}}}\right), \tag{7}$$

where

$$s_{\bar{u}}(u) = \begin{cases} -\bar{u} & \text{for } u \leq -\bar{u}, \\ u & \text{for } |u| < \bar{u}, \\ \bar{u} & \text{for } u \geq \bar{u}, \end{cases}$$

may not guarantee that z_1 and z_2 decrease exponentially with given rate of exponential stability. Moreover, undesirable overshoot in variations of the variables is possible. Especially dangerous is overshoot in the variable z_1. This means that even starting sufficiently close to the target trajectory, the robot may perform large oscillations going far enough even if the closed loop system is stable [8]. Thus, it is of interest to describe such a domain satisfying certain geometric constraints starting from which the motion of the system will not leave this domain and will asymptotically converge to zero. In this paper we construct the attraction domain inscribed into the band of certain width around axes z_1, z_2 and guaranteeing prescribed exponential convergence rate.

Rewrite the last Eq. in (3) taking the control u as (7)

$$z_2' = -s_{\bar{u}}\left(\frac{\sigma}{(1+z_2^2)^{\frac{3}{2}}}\right)(1+z_2^2)^{\frac{3}{2}} \equiv -\Phi(z_2, \sigma).$$

Then

$$\Phi(z_2, \sigma) = \begin{cases} \bar{u}(1+z_2^2)^{\frac{3}{2}} & \text{for } \sigma \geq \bar{u}(1+z_2^2)^{\frac{3}{2}}, \\ \sigma & \text{for } |\sigma| < \bar{u}(1+z_2^2)^{\frac{3}{2}}, \\ -\bar{u}(1+z_2^2)^{\frac{3}{2}} & \text{for } \sigma \leq -\bar{u}(1+z_2^2)^{\frac{3}{2}}. \end{cases} \quad (8)$$

The system (3) takes the form of closed-loop system with nonlinear feedback

$$\begin{aligned} z' &= Az + b\Phi(z_2, \sigma), \\ \sigma &= c^T z, \end{aligned} \quad (9)$$

where

$$A = \begin{bmatrix} 0 & 1 \\ 0 & 0 \end{bmatrix}, b = \begin{bmatrix} 0 \\ -1 \end{bmatrix}, c = \begin{bmatrix} \lambda^2 \\ 2\lambda \end{bmatrix}, z = \begin{bmatrix} z_1 \\ z_2 \end{bmatrix}.$$

The system (9) is presented in the form of so called *Lurie* system (see [12]) consisting of the linear part and closed by the nonlinear feedback function $\Phi(z_2, \sigma)$ of the form (8). Thus, the closed loop system (9), (8) is nonlinear with nonlinearity defined by the formula (8). Following the absolute stability approach, along with this certain nonlinear system we consider the whole class of systems consisting of the same linear part, but uncertain nonlinear function. Given a positive value σ_0, assume that the variable σ takes values from the interval

$$-\sigma_0 \leq \sigma \leq \sigma_0. \quad (10)$$

Then the function (8) satisfies the sector constraints

$$\frac{\bar{u}}{\sigma_0} \leq \frac{\Phi(z_2, \sigma)}{\sigma} \leq 1. \quad (11)$$

The absolute stability means the asymptotic stability of the system (9) whatever nonlinearity can be if only it satisfies the sector condition (11). In the absolute stability theory traditionally two classes of nonlinearities are considered:
(a) stationary (or time invariant) nonlinearity of the form $\Phi(\sigma)$,
(b) non-stationary (or time-varying) nonlinearity of the form $\Phi(\xi, \sigma)$.
As shown in [1,12], the systems (a) are usually analyzed using Lyapunov functions from the Lurie-Postnikov class. The systems (b) are treated using quadratic Lyapunov functions. Lurie-Postnikov functions allow for less conservative absolute stability conditions comparing with those obtained by using quadratic functions. The problem with using Lurie-Postnikov functions for analysis of systems (b) is necessity to take derivative over explicitly presented variable ξ. In this

paper we are analyzing the absolute stability of the system (9), with nonlinearity (8), depending on variables z_2 (implicitly depending on ξ) and σ. Thus, we extend using of the Lurie-Postnikov function for the class of nonlinearities (8). Using this approach, we obtain estimates of the attraction domain wider than estimates previously obtained using quadratic Lyapunov functions.

3   Attraction Domain Estimation. Main Result

To construct the attraction domain for the system (9) with nonlinear function $\Phi(z_2, \sigma)$ defined in (8) we introduce the Lyapunov function $V(z)$ of the Lurie-Postnikov form

$$V(z) = z^T P z + \theta \int_0^{c^T z} \Phi(z_2, \sigma) d\sigma. \tag{12}$$

Unknown parameters to be found are: the positive definite matrix $P \succ 0$ and the scalar θ. Everywhere in the paper symbols $\prec$ and $\succ$ ($\preceq$ and $\succeq$) stand for definiteness (semidefiniteness) of matrices. Obviously, if $\theta \geq 0$ then taking into account (11) we conclude that $V(z) > 0$ if only $z \neq 0$. Actually, as it is shown in the book [1], the condition $V(z) > 0$ holds also for negative values of θ if only

$$V'(z) < 0$$

for $z \neq 0$, where $V'(z(\xi)) \equiv \frac{d}{d\xi} V(z(\xi))$ and derivative is taken with respect to the system (9). Aiming to define not just an attraction domain guaranteeing asymptotic stability, but exponential asymptotic stability with a given rate $-\mu$, we are looking for function (12) satisfying condition

$$V'(z) + 2\mu V(z) \leq 0. \tag{13}$$

The exponential stability domain will be estimated as

$$\Omega_0(\nu) = \{z : V(z) \leq \nu^2\} \tag{14}$$

for appropriate choice of the constant ν. Obviously, we have to guarantee that (13) holds inside $\Omega_0(\nu)$. For a given value $\alpha > 0$ consider the band-like region

$$\Pi(\alpha) = \{z : |z_2| \leq \alpha\} \tag{15}$$

and denote

$$\gamma_0 = 1 + \alpha^2. \tag{16}$$

Having desire to present the inequality (13) in the form of LMI we first obtain upper quadratic estimate of $V'(z)$ and $V(z)$. The condition $z \in \Pi(\alpha)$ will simplify this estimation.

Let $Y \in R^3$ be the vector of the form

$$Y = (z^T, \Phi(z_2, \sigma))^T.$$

Let also define the matrix

$$R(\theta, P) = \begin{bmatrix} PA + A^T P & Pb + \frac{\theta}{2} A^T c \\ b^T P + \frac{\theta}{2} c^T A & \theta c^T b \end{bmatrix}. \tag{17}$$

The following statement holds.

Lemma 1. *Assume that* $z \in \Pi(\alpha)$. *If* $\theta \geq 0$ *then*

$$V'(z) \leq \begin{cases} Y^T R Y - 3\theta \bar{u}^2 z_2 (c^T z - \Phi) & for \ z_2(c^T z - \Phi) \geq 0, \\ Y^T R Y - 3\theta \bar{u}^2 \gamma_0^2 z_2 (c^T z - \Phi) & for \ z_2(c^T z - \Phi) < 0. \end{cases} \tag{18}$$

If $\theta \leq 0$ *then*

$$V'(z) \leq \begin{cases} Y^T R Y - 3\theta \bar{u}^2 \gamma_0^2 z_2 (c^T z - \Phi) & for \ z_2(c^T z - \Phi) \geq 0, \\ Y^T R Y - 3\theta \bar{u}^2 z_2 (c^T z - \Phi) & for \ z_2(c^T z - \Phi) < 0. \end{cases} \tag{19}$$

In (18), (19) for brevity R *is used for* $R(\theta, P)$ *and* Φ *is used for* $\Phi(z_2, \sigma)$.

Proof. Derivative of the function $V(z)$ (12) with $\Phi(z_2, \sigma)$ defined in (8) with respect to the system (9) has the following form

$$V'(z) = 2z^T P z' + \theta \Phi(z_2, \sigma) c^T z' + \theta \int_0^{c^T z} \frac{\partial \Phi(z_2, \sigma)}{\partial z_2}(-\Phi(z_2, \sigma)) d\sigma. \tag{20}$$

Consider two cases:

1. Let $|c^T z| < \bar{u}(1 + z_2^2)^{\frac{3}{2}}$. Then we have $\Phi(z_2, \sigma) = c^T z$ and $\frac{\partial \Phi(z_2, \sigma)}{\partial z_2} \equiv 0$ and

$$V'(z) = 2z^T P A z + 2z^T (Pb + \frac{\theta}{2} A^T c) \Phi(z_2, \sigma) + \theta c^T b \Phi^2(z_2, \sigma) = Y^T R(\theta, P) Y.$$

2. Let $|c^T z| \geq \bar{u}(1 + z_2^2)^{\frac{3}{2}}$. Then we have $\Phi(z_2, \sigma) = \pm \bar{u}(1 + z_2^2)^{\frac{3}{2}}$ and

$$V'(z) = Y^T R(\theta, P) Y - 3\theta \bar{u}^2 (1 + z_2^2)^2 z_2 (c^T z - \Phi(z_2, \sigma)).$$

It follows from (15) that $(1 + z_2^2)^2 \leq \gamma_0^2$, which gives estimates (18), (19). Proof of Lemma 1 is completed.

The following statement gives upper and lower estimates of the function (12) in terms of quadratic forms.

Lemma 2. *The following inequalities hold*

$$V(z) \leq Y^T Q(\theta, P) Y \equiv$$
$$\equiv \begin{cases} z^T P z + \theta \Phi(c^T z - \frac{1}{2}\Phi) \leq z^T P z + \frac{\theta}{2}(c^T z)^2 & for \ \theta \geq 0, \\ z^T P z & for \ \theta < 0, \end{cases} \tag{21}$$

$$V(z) \geq z^T M(\theta, P) z \equiv \begin{cases} z^T P z & for \ \theta \geq 0, \\ z^T P z + \frac{\theta}{2}(c^T z)^2 & for \ \theta < 0. \end{cases} \tag{22}$$

Matrices $Q(\theta, P)$ *and* $M(\theta, P)$ *are defined by quadratic forms (21) and (22).*

Proof. Consider following three cases:

1. Let $|c^T z| < \bar{u}(1 + z_2^2)^{\frac{3}{2}}$. Then

$$V = z^T P z + \theta \frac{(c^T z)^2}{2}. \tag{23}$$

2. Let $c^T z \geq \bar{u}(1 + z_2^2)^{\frac{3}{2}}$. Then

$$V = z^T P z + \theta \int_0^{\bar{u}(1+z_2^2)^{\frac{3}{2}}} \sigma d\sigma + \theta \int_{\bar{u}(1+z_2^2)^{\frac{3}{2}}}^{c^T z} \bar{u}(1 + z_2^2)^{\frac{3}{2}} d\sigma$$

$$= z^T P z + \theta \bar{u}(1 + z_2^2)^{\frac{3}{2}} \left(c^T z - \frac{\bar{u}(1+z_2^2)^{\frac{3}{2}}}{2} \right). \tag{24}$$

3. Let $c^T z \leq -\bar{u}(1 + z_2^2)^{\frac{3}{2}}$. Then

$$V = z^T P z + \theta \int_0^{-\bar{u}(1+z_2^2)^{\frac{3}{2}}} \sigma d\sigma + \theta \int_{-\bar{u}(1+z_2^2)^{\frac{3}{2}}}^{c^T z} \left(-\bar{u}(1 + z_2^2)^{\frac{3}{2}} \right) d\sigma$$

$$= z^T P z + \theta \bar{u}(1 + z_2^2)^{\frac{3}{2}} \left(-c^T z - \frac{\bar{u}(1+z_2^2)^{\frac{3}{2}}}{2} \right). \tag{25}$$

Combination of (8), (23), (24), (25) gives (21) and (22). Proof of Lemma 2 is completed.

We will be looking for the attraction domain of the system (3) inscribed into the set

$$\bar{\Pi}(\alpha, \sigma_0) = \Pi(\alpha) \cap \{z : |\sigma| \leq \sigma_0\}. \tag{26}$$

Applying the absolute stability theory argumentation, along with the fixed function $\Phi(z_2, \sigma)$ consider the class of functions $\phi(z_2, \sigma)$ satisfying the sector conditions (11) which are rewritten in the form of quadratic inequality

$$\left(\frac{\bar{u}}{\sigma_0} c^T z - \phi(z_2, \sigma) \right) (\phi(z_2, \sigma) - c^T z) \geq 0. \tag{27}$$

Continuing this argumentation, along with the systems (9) with the function $\Phi(z_2, \sigma)$ of the form (8) we are considering the whole class of systems (9) with an arbitrary nonlinearity $\phi(z_2, \sigma)$ satisfying the sector condition (27). If the inequality (13) holds for all functions $\phi(z_2, \sigma)$ satisfying conditions (27), then property (13) also holds along the trajectories of the system (9) with the nonlinearity $\Phi(z_2, \sigma)$. Thus, absolute stability of the zero point of the system (9) with arbitrary nonlinearity of the case (27) implies exponential stability of the zero point of the system (9) with the nonlinearity $\Phi(z_2, \sigma)$ with exponential rate $-\mu$.

To check the condition (13) under constraint (27) when $z \neq 0$ we will use *the S-procedure* [12]. Let $f_i(z) = z^T F_i z$, $i = 0, \ldots, m$, $z \in R^n$ be quadratic forms. The following question often arises in the control theory: under what conditions on matrices F_i inequalities $f_j(z) \geq 0$, $i = 1, \ldots, m$, and $z \neq 0$ imply

$f_0(z) < 0$. This property is often called the "conditional sign-definiteness". The S-procedure consists in checking if the following condition holds:

$$\exists \tau_i \geq 0, \ i = 1, \cdots, m : \ F_0 + \sum_{i=1}^{m} \tau_i F_i \prec 0. \tag{28}$$

Thus, the S-procedure reduces the conditional sign definiteness problem to solving the conventional LMI (28) with respect to variables $\tau_i \geq 0$ or establishing its infeasibility, see [2]. If $m = 1$ then the S-procedure establishes necessary and sufficient conditions for conditional sign definiteness, see for example [12]. Generally, if $m > 1$ then the S-procedure gives only sufficient conditions for conditional sign definiteness. This property is referred to as *looseness of the S-procedure with multiple constraints*. Nevertheless, even taking into account the looseness of the S-procedure, its application is attractive, since checking the condition (28) can be reduced to a convex programming problem, for which there are efficient polynomial algorithms. In our case there are two quadratic constraints. One constraint follows from (27). Other conditions follow from formulation of the Lemma 1, conditions (18), (19).

Given z_2 and σ the function $\phi(z_2, \sigma)$ can take arbitrary values satisfying the quadratic inequality (27). Thus, ϕ can be considered as an independent variable constrained only by the inequality (27) and inequalities (18), (19) with $\Phi(z_2, \sigma)$ changed by ϕ. Denote

$$\beta_0 = \frac{\bar{u}}{\sigma_0}, \ e = (0, 1)^T. \tag{29}$$

Following the S-procedure approach, let us construct the following quadratic forms that must be negative definite

$$\begin{aligned} S_1(Y) &= V' + 2\mu V + \tau_1(\beta_0 c^T z - \phi)(\phi - c^T z) + \tau_2 e^T z(c^T z - \phi) < 0, \\ S_2(Y) &= V' + 2\mu V + \tau_1(\beta_0 c^T z - \phi)(\phi - c^T z) - \tau_3 e^T z(c^T z - \phi) < 0, \end{aligned} \tag{30}$$

They both are linearly dependent on parameters τ_1, τ_2, τ_3 and θ to be found along with the matrix P. The vector Y is composed of independent variables (z_1, z_2, ϕ). If there exist such values $\tau_i \geq 0 \ (i = 1, 2, 3), \theta$ and matrix $P \succ 0$ that inequalities (30) are satisfied, then the condition (13) holds under constraint (27) when $Y \neq 0$. Denote

$$T = \begin{bmatrix} -\beta_0 c c^T & \frac{1}{2}(\beta_0 + 1)c \\ \frac{1}{2}(\beta_0 + 1)c^T & -1 \end{bmatrix}, \ E = \frac{1}{2} \begin{bmatrix} c e^T + e c^T & -e \\ -e^T & 0 \end{bmatrix},$$

and

$$\text{sgn } \theta = \begin{cases} 1 & \text{for } \theta \geq 0, \\ -1 & \text{for } \theta < 0. \end{cases}$$

The following Theorem establishes the main result.

Theorem 1. *Assume that for given numbers* $\bar{u} > 0, \sigma_0 > \bar{u}, \lambda > 0, 0 \leq \mu < \lambda$ *and* $\alpha > 0$ *the following linear matrix inequalities in variables* P, θ, *and* $\tau_i \geq 0, i = 1, 2, 3$, *are feasible:*

$$R(\theta, P) + 2\mu Q(\theta, P) + \tau_1 T - (3\theta \bar{u}^2 - \tau_2 \text{ sgn } \theta) E \preceq 0, \tag{31}$$

$$R(\theta, P) + 2\mu Q(\theta, P) + \tau_1 T - \left(3\theta \bar{u}^2 \gamma_0^2 + \tau_3 \operatorname{sgn} \theta\right) E \preceq 0, \tag{32}$$

$$\begin{bmatrix} M(\theta, P) & c \\ c^T & \sigma_0^2 \end{bmatrix} \succeq 0, \tag{33}$$

then the domain $\Omega_0(1) \subseteq \bar{\Pi}(\alpha, \sigma_0)$ *is an invariant estimate of attraction domain of the system (3) under control (7). Moreover, for any starting point* $z(0) \in \Omega_0(1)$, *the whole trajectory* $z(\xi)$ *remains in* $\Omega_0(1)$ *for* $\xi \geq 0$ *and decreases exponentially at the rate* $-\mu$.

Proof. The stability of the zero solution of the system (9) for all possible functions $\phi(z_2, \sigma)$ satisfying (27) implies the stability of the zero solution of the system (9) with $\Phi(z_2, \sigma)$ and hence, the stability of the zero solution of system (3) under control (7) and initial conditions satisfying $z(0) \in \Omega_0(1)$. Moreover, condition (13) ensures the following estimate

$$V(z(\xi)) \leq V(z(0)) \exp\left(-2\mu\xi\right), \tag{34}$$

which holds along trajectories of the system (9) for all possible functions $\phi(z_2, \sigma)$ satisfying (27). As it follows from the Lemma 2

$$z^T M z \leq V(z) \leq z^T Q z, \tag{35}$$

where

$$Q = \begin{cases} P + \frac{\theta}{2} c c^T & \text{for } \theta \geq 0, \\ P & \text{for } \theta < 0, \end{cases} \quad M = \begin{cases} P & \text{for } \theta \geq 0, \\ P + \frac{\theta}{2} c c^T & \text{for } \theta < 0, \end{cases} \quad Q \succ 0, M \succ 0.$$

The inequality (35) results in

$$\lambda_0 \|z\|^2 \leq V(z) \leq \lambda_1 \|z\|^2, \tag{36}$$

where λ_1 is the maximum eqigenvalue of the matrix Q and λ_0 is the minimum eigenvalue of the matrix M, $\|z\|$ stands for Euclidean norm. Combination of (36) with (34) gives

$$\|z(\xi)\| \leq \sqrt{\frac{\lambda_1}{\lambda_0}} \|z(0)\| \exp\left(-\mu\xi\right),$$

which ensures the exponential decay of z with the exponential rate $-\mu$.

To satisfy the condition (13) for all functions $\phi(z_2, \sigma)$ satisfying (27) it is sufficient to satisfy conditions (30), either of two depending on the sign of θ. Last condition is guaranteed by LMI's (31), (32). Finally, according to (22) the LMI (33) guarantees that $\Omega_0(1) \subseteq \bar{\Pi}(\alpha, \sigma_0)$. This completes the proof of Theorem 1.

4   Optimal Choice of Parameters and Numerical Examples

The formulation of the Theorem 1 implies not unique choice of parameters. Therefore, possibility of the optimal choice arises. Consider the following problems.

Problem 1. Given $\bar{u} > 0$, $\sigma_0 > \bar{u}$, $\alpha > 0$, $\lambda > 0$, $0 < \mu \leq \lambda$ find such a matrix P and a scalar θ satisfying the conditions of Theorem 1 for which the volume of the ellipsoid $\Omega_0(1)$ takes the maximum value. The last demand can be formulated in the following form:

$$\min_{P,\theta} \operatorname{tr}(M(\theta, P)). \tag{37}$$

But according to (22) maximization of the volume of the ellipsoid defined by the matrix $M(\theta, P)$ does not mean exactly maximization of the volume of the estimate $\Omega_0(1)$, because the last is inscribed into the ellipsoid. Thus, given solution of the Problem 1, we can proceed to maximize the volume of the estimate considering the following problem:

Problem 2. Let P and θ be the solution of Problem 1 satisfying the conditions of Theorem 1 corresponding to the $\nu = 1$. Find the value

$$\nu^* = \sup\{\nu > 1 : \Omega_0(\nu) \subseteq \bar{\Pi}(\alpha, \sigma_0)\}. \tag{38}$$

Note now that formulation of the Theorem 1 contains certain value of σ_0. To explicitly express dependence of solution of the Problem 2 on σ_0 denote it by $\Omega_0(\nu^*, \sigma_0)$. To maximize the volume of $\Omega_0(\nu^*, \sigma_0)$, the optimal choice of σ_0 is needed. Thus, consider the following problem:

Problem 3. Find the value

$$\sigma_0^* = \sup\{\sigma_0 : \Omega_0(\nu^*, \sigma_0) \neq \emptyset\}. \tag{39}$$

For practical issues it is convenient to limit the lateral deviation of the robot from the target trajectory, which means overshooting in the variable z_1. This means that the lateral deviation of the target point from the desired path must be limited as the robot approaches the target path. Let us inscribe the estimate $\Omega_0(\nu)$ into the set $\bar{\bar{\Pi}}(\alpha, \sigma_0, \kappa) = \bar{\Pi}(\alpha, \sigma_0) \cap \{z : |z_1| \leq \kappa\}$. Thus, one more LMI must be added to the formulation of the Theorem 1:

Problem 4. Find the solution of the Problem 1 satisfying the conditions of Theorem 1 and additional linear matrix inequality

$$M(\theta, P) \geq \begin{bmatrix} \frac{1}{\kappa^2} & 0 \\ 0 & 0 \end{bmatrix}. \tag{40}$$

To illustrate the proposed method several numerical examples were considered. For every example Problems 1-4 were solved. We used CVX, a Matlab package for solving convex optimization problems [4]. To illustrate behavior of the system and to show its trajectories with respect to estimates of the attraction domain, in all figures we combine both the phase portrait and the attraction domain.

Thus, Fig. 1, 2, 3 and 4 show the attraction domains estimates together with the phase portrait of the system (3) closed by the control (7). In all cases we set $\bar{u} = 0.5$. All other parameters differ, their values are given in the figure captions.

In all figures the red contour line denotes the attraction domain estimate obtained with the Lurie-Postnikov Lyapunov function based on the formulation of the Theorem 1. It is compared with ellipsoidal estimates shown in blue and obtained earlier in [8] based on quadratic Lyapunov functions. Green lines show the trajectories of the system for the initial conditions from the attraction domain estimates. Magenta lines show the saturation regions of the system. Black dotted lines limit the regions $\bar{\Pi}(\alpha, \sigma_0)$ and $\bar{\bar{\Pi}}(\alpha, \sigma_0, \kappa)$.

Two numerical examples illustrating solution of the Problem 2 corresponding to the certain value of $\sigma_0 = 2$ and optimal values ν^* are shown in Fig. 1 and Fig. 2 for the $\lambda = 1$, $\mu = 0.1\lambda$ and $\lambda = 1.5$, $\mu = 0.2\lambda$ respectively.

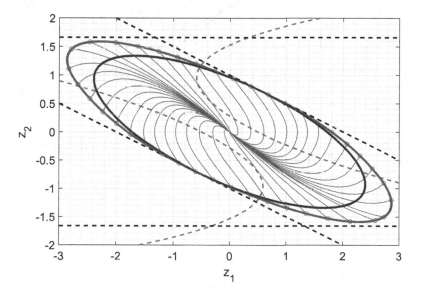

Fig. 1. Attraction domain estimate for given $\bar{u} = 0.5$, $\sigma_0 = 2$, $\lambda = 1$, $\mu = 0.1$, $\alpha = 1.66$ and optimal $\nu^* = 1.16$. (Color figure online)

As easily seen, the obtained attraction domain estimates (red) are invariant for the system and are less conservative in comparison with ellipsoidal estimates (blue).

The numerical example of solving the Problem 3 is shown in Fig. 3. We note that, with the optimal choice of σ_0, the proposed method allows us to considerably expand the volume of $\Omega_0(\nu^*, \sigma_0)$ in comparison with ellispsoidal estimate. The resulting optimal parameters are $\sigma_{0,Q}^* = 2.8$ and $\sigma_{0,LP}^* = 3.6$ for the cases of quadratic Lyapunov functions and Lurie-Postnikov functions respectively.

The proposed method allows for inscription of the attraction domain estimation into the band of a certain width around any state space variables of the system. The numerical example illustrating solution of the Problem 4 for the given $\kappa = 0.2$ is shown in Fig. 4.

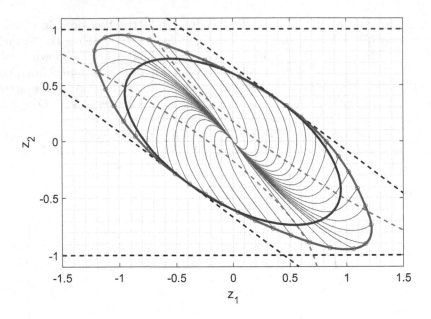

Fig. 2. Attraction domain estimate for given $\bar{u} = 0.5$, $\sigma_0 = 2$, $\lambda = 1.5$, $\mu = 0.3$, $\alpha = 1$ and optimal $\nu^* = 1.65$. (Color figure online)

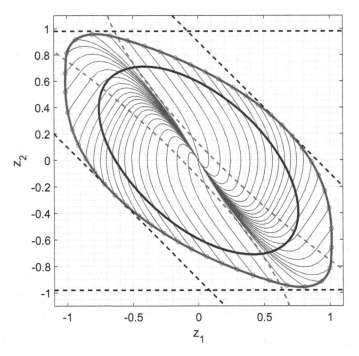

Fig. 3. Attraction domain estimate for given $\bar{u} = 0.5$, $\lambda = 2$, $\mu = 0.2$, $\alpha = 0.98$ and optimal $\nu^* = 1.55$, $\sigma_{0,Q}^* = 2.8$, $\sigma_{0,LP}^* = 3.6$. (Color figure online)

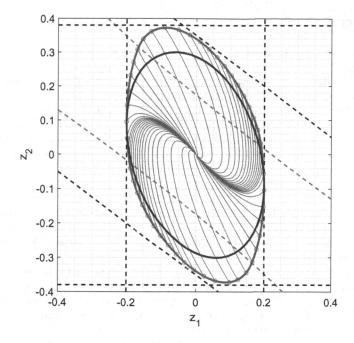

Fig. 4. Attraction domain estimate for given $\bar{u} = 0.5$, $\lambda = 1.5$, $\mu = 0.15$, $\kappa = 0.2$, fixed $\nu = 1$, $\alpha = 0.38$ and optimal $\sigma_0^* = 1.05$. (Color figure online)

The examples considered in this section demonstrate the advantages of using the Lurie-Postnikov Lyapunov functions over the quadratic Lyapunov functions. Using the proposed approach makes it possible to choose Lyapunov functions from a wider parametric class in comparison with quadratic functions. As a result, we get less conservative estimates of regions of attraction.

5   Conclusion

To stabilize the wheeled robot motion along a straight line path, the control law based on the feedback linearization scheme was used. Using the saturation function for modeling a bounded control can destroy the exponential stability. To guarantee the exponential stability, a method of attraction domain estimation based on absolute stability approach was proposed. The invariant sets of initial conditions were estimated, for which the prescribed rate of exponential stability of the system is guaranteed. Numerical examples illustrate advantage of the proposed approach. The estimates achieved demonstrate less conservativeness in comparison with ellipsoidal ones, used earlier.

References

1. Aizerman, M., Gantmacher, F.: Absolute Stability of Regulation Systems. Holden Day (1964)
2. Boyd, S., Ghaoui, L., Feron, E., Balakrishnan, V.: Linear Matrix Inequalities in System and Control Theory. SIAM, Philadelphia (1994)
3. Cui, J., Sabaliauskaite, G.: On the alignment of safety and security for autonomous vehicles. In: Proceedings of IARIA CYBER, pp. 59–64 (2017)
4. Grant, M., Boyd, S.: Graph implementations for nonsmooth convex programs. In: Blondel, V., Boyd, S., and Kimura, H. (eds.) Recent Advances in Learning and Control (a tribute to M. Vidyasagar), Lecture Notes in Control and Information Sciences, pp. 95–110 (2008)
5. Muñoz-Arias, Mauricio., Scherpen, Jacquelien M.A., Dirksz, Daniel A.: Position control via force feedback in the port-hamiltonian framework. In: van de van de Wouw, Nathan, Lefeber, Erjen, Lopez Arteaga, Ines (eds.) Nonlinear Systems. LNCIS, vol. 470, pp. 181–207. Springer, Cham (2017). https://doi.org/10.1007/978-3-319-30357-4_9
6. Pesterev, Alexander: On optimal selection of coefficients of path following controller for a wheeled robot with constrained control. In: Evtushenko, Yury, Jaćimović, Milojica, Khachay, Michael, Kochetov, Yury, Malkova, Vlasta, Posypkin, Mikhail (eds.) OPTIMA 2018. CCIS, vol. 974, pp. 336–350. Springer, Cham (2019). https://doi.org/10.1007/978-3-030-10934-9_24
7. Pesterev, A., Rapoport, L., Morozov, Y.: Control of a wheeled robot following a curvilinear path. In: Proceedings of the 6th EUROMECH Nonlinear Dynamics Conference (ENOC 2008), pp. 1–7 (2008)
8. Rapoport, L.: Estimation of attraction domains in wheeled robot control. Autom. Remote Control **67**(9), 1416–1435 (2006). https://doi.org/10.1134/S0005117906090062
9. Tarbouriech, S., Garcia, G., Gomes da Silva, J.J., Queinnec, I.: Stability and Stabilization of Linear Systems with Saturating Actuators. Springer, Dordrecht (2011). https://doi.org/10.1007/978-0-85729-941-3
10. Tarbouriech, S., Turner, M.: Anti-windup design: an overview of some recent advances and open problems. IET Control Theor. Appl. **3**(1), 1–19 (2009). https://doi.org/10.1049/IET-CTA:20070435
11. Thuilot, B., Cariou, C., Martinet, P., Berducat, M.: Automatic guidance of a farm tractor relying on a single CP-DGPS. Auton. Robots **13**(1), 53–71 (2002). https://doi.org/10.1023/A:1015678121948
12. Yakubovich, V.A., Leonov, G.A., Gelig, A.K.: Stability of Stationary Sets in Control Systems with Discontinuous Nonlinearities. World Scientific, Series on Stability, Vibration and Control of Systems (2004)

Penalty-Based Method for Decentralized Optimization over Time-Varying Graphs

Alexander Rogozin$^{(\boxtimes)}$ and Alexander Gasnikov

Moscow Institute of Physics and Technology, Moscow, Russia
aleksandr.rogozin@phystech.edu, gasnikov@yandex.ru

Abstract. Decentralized distributed optimization over time-varying graphs (networks) is nowadays a very popular branch of research in optimization theory and consensus theory. Applications of this field include drone or satellite networks, as well as distributed machine learning. However, the first theoretical results in this branch appeared only a few years ago. In this paper, we propose a simple method which alternates making gradient steps and special communication procedures. Our approach is based on reformulation of the distributed optimization problem as a problem with linear constraints and then replacing linear constraints with a penalty term.

1 Introduction

The theory of decentralized distributed optimization goes back to [2]. In the last few years this branch of research has aroused great interest in optimization community. A set of papers proposing optimal algorithms for convex optimization problems of sum-type has appeared. See for example [1, 6, 10, 21] and references therein. In all these papers, authors consider sum-type convex target functions and aim at proposing algorithms that find the solution with required accuracy and make the best possible number of communications steps and number of oracle calls (gradient calculations of terms in the sum). In [6], it is mentioned that the theory of optimal decentralized distributed algorithms looks very close to the analogous theory for ordinary convex optimization [4, 16, 17]. Roughly speaking, in a first approximation, decentralized distributed optimization comes down to the theory of optimal methods and this theory is significantly based on the theory of non-distributed optimal methods.

In decentralized distributed optimization over time-varying graphs, another situation takes place. The communication network topology changes from time to time, which can be caused by technical malfunctions such as loss of connection between the agents. Due to the many applications, the interest to these class of problems has grown significantly during the last few years. There appears

The research of A. Rogozin was partially supported by RFBR, project number 19-31-51001. The research of A. Gasnikov was partially supported by RFBR 18-29-03071 mk. and by the Ministry of Science and Higher Education of the Russian Federation (Goszadaniye) No. 075-00337-20-03, project No. 0714-2020-0005.

N. Olenev et al. (Eds.): OPTIMA 2020, LNCS 12422, pp. 239–256, 2020.
https://doi.org/10.1007/978-3-030-62867-3_18

a number of papers with theoretical analysis of rate of convergence for convex type problems: [11, 12, 14, 15, 22]. But there is still a big gap between the theory for decentralized optimization on fixed graphs and the theory over time-varying graphs. The attempt to close this gap (specifically, to develop optimal methods) for the moment required very restricted additional conditions [20].

In this paper, we study a sum-type minimization problem

$$f(x) = \sum_{i=1}^{n} f_i(x) \longrightarrow \min_{x \in \mathbb{R}^d}$$

where each f_i is stored at a separate computational entity and is assumed to be convex. We make a step in the direction of development of optimal methods over time-varying graphs: we propose non-accelerated gradient descent for smooth strongly convex target functions of sum-type. Our analysis is based on reformulation of initial optimization problem as convex optimization problem under affine constraints. These constraints change from time to time but still determine the same hyperplane. Then we use projected gradient descent. In order to solve auxiliary problem (to find a projection on hyperplane) we use non accelerated consensus type algorithms (see [9] and references therein for comparison) that can also be interpreted as gradient descent for special penalized optimization problem [6,7]. Note that proposed analysis of external non-accelerated gradient descent method can be generalized for the case of accelerated gradient method. We plan to do it in subsequent works.

This paper is organized as follows. In Sect. 2, we recall some basic definitions and show how to reformulate a problem with affine constraints by introducing a penalty. In Sect. 3, we analyze the performance of gradient descent on a time-varying function. We introduce decentralized projected gradient method in Sect. 4 and analyze its convergence using results of Sect. 3. Finally, we provide numerical experiments and comparison to other methods in Sect. 5.

2 Preliminaries

2.1 Strong Convexity and Smoothness

Strongly convex and smooth functions are the focus of this paper.

Definition 1. *Let $\mathbb{X}$ be either $\mathbb{R}^d$ with 2-norm or $\mathbb{R}^{d \times n}$ with Frobenius norm. A differentiable function $f : \mathbb{X} \to \mathbb{R}$ is called*

– **convex,** *if for any $x, y \in \mathbb{X}$*

$$f(y) \geqslant f(x) + \langle \nabla f(x), y - x \rangle;$$

– μ-**strongly convex,** *if for any $x, y \in \mathbb{X}$*

$$f(y) \geqslant f(x) + \langle \nabla f(x), y - x \rangle + \frac{\mu}{2} \|y - x\|^2;$$

– **L-smooth**, *if its gradient* $\nabla f(x)$ *is L-Lipschitz, i.e. for any* $x, y \in \mathbb{X}$

$$\|\nabla f(y) - \nabla f(x)\| \leqslant L\|y - x\|,$$

or, equivalently, for all $x, y \in \mathbb{X}$

$$f(y) \leqslant f(x) + \langle \nabla f(x), y - x \rangle + \frac{L}{2}\|y - x\|^2.$$

2.2   Graph Laplacian

In this paper, a communication network is represented by a connected undirected graph $\mathcal{G} = (V, E)$.

Definition 2. *For an undirected graph* $\mathcal{G} = (V, E)$ *with* $|V| = n$ *nodes, its* **Laplacian** *is a matrix* $W \in \mathbb{R}^{n \times n}$ *such that*

$$[W]_{ij} = \begin{cases} \deg i, & i = j \\ -1, & (i, j) \in E \\ 0, & else \end{cases} \tag{1}$$

In the statement below, we list the basic Laplacian properties, which can be obtained using Perron-Frobenius theorem [18].

Proposition 1. – W *is positive semi-definite;*
 – *If graph* $\mathcal{G}$ *is connected, then* $Wx = 0 \Leftrightarrow x_1 = \ldots = x_n$, *i.e.* $\operatorname{Ker} W = Span(\mathbf{1})$. *Moreover,* $\operatorname{Ker} \sqrt{W} = \operatorname{Ker} W = Span(\mathbf{1})$.

2.3   Strongly Convex Problem with Affine Constraints

As will be shown later in the paper, decentralized optimization problems may be reformulated as problems with affine constraints. Consider optimization problem

$$f(x) \longrightarrow \min_{Ax=0} \tag{2}$$

for some μ-strongly convex function $f : \mathbb{R}^d \to \mathbb{R}$ and symmetric positive semi-definite matrix $A \in \mathbb{R}^{d \times d}$. Let y^* be the solution of Lagrange dual to (2) with minimal norm and $R_y = \|y^*\|$. Introduce a penalized problem

$$f_A(x) = f(x) + \frac{R_y^2}{\varepsilon}\|Ax\|^2 \longrightarrow \min_{x \in \mathbb{R}^d}.$$

The quantity R_y can be bounded as [10] $R_y^2 \leqslant \frac{\|\nabla f(x^*)\|^2}{\lambda_{\min}^+(A^2)}$, where x^* is the solution of (2) and $\lambda_{\min}^+(\cdot)$ denotes the minimum positive eigenvalue of the corresponding matrix.

Proposition 2. *Let* $x \in \mathbb{R}^d$ *and* $f_A(x_N) - \min_x f_A(x) < \varepsilon$. *Then*

$$\begin{cases} f(x_N) - \min_{Ax=0} f(x) < \varepsilon \\ \|Ax_N\| < 2\varepsilon/R_y \end{cases}$$

Proof. See [6,7], Theorem 1 in [8]. □

Proposition 3. *Denote* $x^* = \arg\min_{Ax=0} f(x)$ *and* $x_A^* = \arg\min_x f_A(x)$. *Then* $\|x_A^* - x^*\| \leqslant \sqrt{a\varepsilon}$, *where* a *is defined as*

$$a = \frac{4\|\nabla f(x^*)\|}{\mu R_y \lambda_{\min}^+(A)} \tag{3}$$

Proof. Using Proposition 2, since $f(x_A^*) - f(x_A^*) = 0 < \varepsilon$, it holds $\|Ax_A^*\| < \varepsilon/R_y$. Denote $\Delta x = x_A^* - x^* = \Delta x_\perp + \Delta x_{||}$, where $\Delta x_{||} = \Pi_{\mathrm{Ker}\,A}(\Delta x)$.

1. First, we estimate $\|x_\perp\|$.

$$\lambda_{\min}^+(A)\|\Delta x_\perp\| \leqslant \|A\Delta x_\perp\| = \|Ax_A^*\| \leqslant 2\varepsilon/R_y$$
$$\|\Delta x_\perp\| \leqslant \frac{2\varepsilon}{R_y \lambda_{\min}^+(A)}$$

2. Second, let us estimate $\|x_{||}\|$. Since $x^* = \arg\min_{Ax=0}$, we have $\langle \nabla f(x^*), \Delta x_{||} \rangle = 0$ and by strong convexity of f it holds

$$f(x_A^*) \geqslant f(x^*) + \langle \nabla f(x^*), \Delta x \rangle + \frac{\mu}{2}\|\Delta x\|^2$$
$$= f(x^*) + \langle \nabla f(x^*), \Delta x_\perp \rangle + \frac{\mu}{2}\|\Delta x\|^2$$
$$\geqslant f(x^*) - \|\nabla f(x^*)\|_2 \cdot \|\Delta x_\perp\| + \frac{\mu}{2}\|\Delta x\|^2$$

$$f(x^*) = f_A(x^*) \geqslant f_A(x_A^*) = f(x_A^*) + \frac{R_y^2}{\varepsilon}\|Ax_A^*\|^2$$

$$\geqslant f(x^*) - \|\nabla f(x^*)\| \cdot \|\Delta x_\perp\| + \frac{\mu}{2}\|\Delta x\|^2 + \frac{R_y^2}{\varepsilon}\|A\Delta x_\perp\|^2$$

$$\|\nabla f(x^*)\| \cdot \|\Delta x_\perp\|_2 - \frac{R_y^2}{\varepsilon}\lambda_{\min}^+(A^2)\|\Delta x_\perp\|^2 \geqslant \frac{\mu}{2}\|\Delta x\|^2$$

$$\|\Delta x\| \leqslant \sqrt{\frac{2}{\mu}\|\nabla F(x^*)\| \cdot \|\Delta x_\perp\|^2} \leqslant \sqrt{\frac{4\|\nabla F(x^*)\|}{\mu R_y \lambda_{\min}^+(A)}\varepsilon} = \sqrt{a\varepsilon}$$

3   Gradient Descent on a Time-Varying Function

Definition 3. *We call a series of functions $g = \{g_k\}_{k=1}^{\infty}$, $g_k : \mathbb{R}^n \to \mathbb{R}$, a **time-varying function**. If each of g_k is convex/μ-strongly convex/L-smooth, we call g a convex/μ-strongly convex/L-smooth time-varying function.*

Define gradient descent on a time-varying function as

$$x_{k+1} = x_k - \gamma \nabla g_k(x_k) \tag{4}$$

We are interested in convergence of the above method. In order to establish the rate, we need additional assumptions.

Assumption 1

1. *Time-varying function g is μ-strongly convex and L-smooth;*
2. *There exists $x^* \in \mathbb{R}^n$, $\varepsilon > 0$ such that for $k = 1, 2, \ldots$ it holds*
   $$\| \operatorname*{arg\,min}_{x \in \mathbb{R}^n} g_k(x) - x^* \|^2 \leqslant \varepsilon.$$

Assumption 1 becomes realistic when time-varying function g represents a functional with changing penalty. It is discussed later in the paper. First, we formulate preliminary facts that will be needed in analysis.

Proposition 4. *For μ-strongly convex L-smooth function f, it holds*

$$\langle \nabla f(x) - \nabla f(y), x - y \rangle \geqslant \frac{\mu L}{\mu + L} \|x - y\|^2 + \frac{1}{\mu + L} \|\nabla f(x) - \nabla f(y)\|^2$$

Proof. See Theorem 2.1.11 in [17].

Proposition 5. *Let u, v be vectors of $\mathbb{R}^n$ of matrices of $\mathbb{R}^{d \times n}$ and p be a positive scalar constant. Then*

$$\langle u, v \rangle \leqslant \frac{\|u\|^2}{2p} + \frac{p\|v\|^2}{2} \tag{5}$$

Additionally, if $p < 1$ then

$$\|v\|^2 \geqslant p\|u\|^2 - \frac{p}{1-p}\|v - u\|^2 \tag{6}$$

Here, if $u, v \in \mathbb{R}^n$, $\| \cdot \|$ denotes the 2-norm in $\mathbb{R}^n$, and if $u, v \in \mathbb{R}^{d \times n}$, $\| \cdot \|$ denotes Frobenius norm.

Proof. See Appendix A.

Denote $r_k = \|x_k - x^*\|$, $x_k^* = \arg\min g_k(x)$, $\Delta x_k = x_k^* - x^*$.

$$r_{k+1}^2 = \|x_k - x^* - \gamma \nabla g_k(x_k)\|^2 = r_k^2 + \gamma^2 \|\nabla g_k(x_k)\|^2 - 2\gamma\langle \nabla g_k(x_k), x_k - x^*\rangle$$

$$= r_k^2 + \gamma^2 \|\nabla g_k(x_k)\|^2$$

$$- 2\gamma\langle \underbrace{\nabla g_k(x_k) - \nabla g_k(x_k^*)}_{=0}, x_k - x_k^*\rangle - 2\gamma\langle \nabla g_k(x_k), x_k^* - x^*\rangle$$

$$\leqslant r_k^2 + \gamma^2 \|\nabla g_k(x_k)\|^2 - 2\gamma \frac{\mu L}{\mu + L} \underbrace{\|x_k - x_k^*\|^2}_{\text{estimate by (6)}}$$

$$- 2\gamma \frac{1}{\mu + L} \|\nabla g_k(x_k)\|^2 - 2\gamma \underbrace{\langle \nabla g_k(x_k), x_k^* - x^*\rangle}_{\text{estimate by (5)}} \quad (7)$$

Now let us employ Proposition 5 for the under-braced terms.

1. Using (5) with $p = \mu + L$:

$$-2\gamma\langle \nabla g_k(x_k), x_k^* - x^*\rangle \leqslant 2\gamma \cdot \frac{\|\nabla g_k(x_k)\|^2}{2(\mu + L)} + 2\gamma \cdot \frac{\mu + L}{2}\|\Delta x_k\|^2$$

$$= \gamma \frac{\|\nabla g_k(x_k)\|^2}{\mu + L} + \gamma(\mu + L)\|\Delta x_k\|^2$$

2. Using (6) with $p = 1/2, u = x_k - x^*, v = x_k - x_k^*$:

$$\|x_k - x_k^*\|^2 \geqslant \frac{1}{2}\|x_k - x^*\|^2 - \|x^* - x_k^*\|^2$$

Returning to (7) we obtain

$$r_{k+1}^2 \leqslant r_k^2 \left(1 - 2\gamma \frac{\mu L}{\mu + L} \cdot \frac{1}{2}\right) + \|\nabla g_k(x_k)\|^2 \cdot \left(\gamma^2 - \frac{2\gamma}{\mu + L} + \frac{\gamma}{\mu + L}\right)$$

$$+ \|\Delta x_k\|^2 \left(\frac{2\gamma\mu L}{\mu + L} + \gamma(\mu + L)\right)$$

$$= r_k^2 \left(1 - \gamma \frac{\mu L}{\mu + L}\right) + \|\nabla g_k(x_k)\|^2 \cdot \gamma\left(\gamma - \frac{1}{\mu + L}\right)$$

$$+ \|\Delta x_k\|^2 \cdot \gamma\left(\frac{2\mu L}{\mu + L} + \mu + L\right)$$

Recall that here γ is stepsize from (4). Setting $\gamma \in \left(0, \frac{1}{\mu + L}\right]$ leads to

$$r_{k+1}^2 \leqslant r_k^2\left(1 - \gamma\frac{\mu L}{\mu + L}\right) + \|\Delta x_k\|^2 \cdot \gamma\left(\frac{2\mu L}{\mu + L} + \mu + L\right)$$

$$\leqslant r_k^2\left(1 - \gamma\underbrace{\frac{\mu L}{\mu + L}}_{\geqslant \mu/2}\right) + \gamma\left(\underbrace{\frac{2\mu L}{\mu + L} + \mu + L}_{\leqslant L}\right) \cdot \varepsilon$$

$$\leqslant r_k^2\left(1 - \gamma\frac{\mu}{2}\right) + \gamma \cdot 3L\varepsilon. \quad (8)$$

In order to obtain linear convergence, let us formulate the following

Lemma 1. *Let the following inequality hold:*

$$r_k^2 \geqslant b\varepsilon \tag{9}$$

where

$$b = 12\frac{L}{\mu} \tag{10}$$

Then $r_{k+1}^2 \leqslant r_k^2\left(1 - \gamma\frac{\mu}{4}\right)$.

Proof. Consider (8) and rewrite it the following way:

$$r_{k+1}^2 \leqslant r_k^2\left(1 - \gamma\frac{\mu}{4}\right) + \left[3\gamma L\varepsilon - \gamma\frac{\mu}{4}r_k^2\right]$$

The latter term is non-positive due to (9), (10), and the desired result follows.

Finally, we are ready to state a convergence result in terms of number of iterations.

Theorem 1. *Under Assumption 1, after N steps of gradient descent with step-size $\gamma = \frac{1}{\mu+L}$, where $N = O\left(\frac{L}{\mu}\log\frac{r_0^2}{b\varepsilon}\right)$ and b is defined in (10), the following inequality holds:*

$$r_N^2 \leqslant b\varepsilon$$

Proof. If $r_k \geqslant b\varepsilon$, Lemma 1 works. If $r_k < b\varepsilon$, then

$$r_{k+1}^2 \leqslant r_k^2\left(1 - \frac{\gamma\mu}{2}\right) + 3\gamma L\varepsilon$$
$$\leqslant 12\frac{L}{\mu}\left(1 - \frac{\gamma\mu}{2}\right)\varepsilon + 3\gamma L\varepsilon \leqslant 12\frac{L}{\mu}\varepsilon = b\varepsilon. \tag{11}$$

This means that once method achieves $b\varepsilon$ accuracy after some $\tilde{N}$ steps, its trajectory remains in $b\varepsilon$-region of x^*, i.e. $r_k^2 \leqslant b\varepsilon$ for all $k \geqslant \tilde{N}$. By Lemma 1, it is sufficient to make $N = O\left(\frac{L}{\mu}\log\frac{r_0^2}{b\varepsilon}\right)$ in order to obtain $r_N^2 \leqslant b\varepsilon$.

4   Decentralized Projected Gradient Method

4.1   Problem Reformulation and Assumptions

Consider minimization of sum of convex functions:

$$f(x) = \sum_{i=1}^{n} f_i(x) \longrightarrow \min_{x \in \mathbb{R}^d} \tag{12}$$

We assume that every f_i is μ_i-strongly convex and L_i-smooth. We seek to solve problem (12) in a decentralized setup, so that every node locally holds f_i and may exchange data with its neighbors. Moreover, we are interested in the time-varying case. This means the communication network changes with time and is represented by a sequence of graphs $\{\mathcal{G}_k\}_{k=1}^{\infty}$. Our analysis is restricted to the following

Assumption 2 *Each of graphs $\{\mathcal{G}_k\}_{k=1}^{\infty}$ is connected.*

Moreover, we introduce bounds on the graph Laplacian spectrum.

Definition 4. *For each graph $\mathcal{G}_k$, let W_k be its Laplacian. Denote*

$$\theta_{\max} = \max_k \lambda_{\max}(W_k) \tag{13a}$$

$$\theta_{\min} = \min_k \lambda_{\min}^+(W_k) \tag{13b}$$

$$\chi = \frac{\theta_{\max}}{\theta_{\min}} \tag{13c}$$

Since all communication graphs $\{\mathcal{G}_k\}_{k=1}^{\infty}$ have a common set of vertices, there is a finite number of these graphs and corresponding matrices W_k. Therefore, quantities $\theta_{\max}$ and $\theta_{\min}$ in (13) are well-defined. Moreover, χ characterizes network connectivity and typically corresponds to the diameter of the graph [19].

Let us reformulate problem (12) in a following way.

$$F(X) = \sum_{i=1}^{n} f_i(x_i) \longrightarrow \min_{x_1 = \ldots = x_n} . \tag{14}$$

Here $X \in \mathbb{R}^{d \times n}$ is a matrix consisting of columns $x_1, \ldots, x_n$. The above representation means local copies x_i of parameter vector x are distributed over the agents in the network. Now, if every node computes $\nabla f_i(x_i)$, then the gradient $\nabla F(X) = [\nabla f_1(x_1), \ldots, \nabla f_n(x_n)]$ will be distributed all over the network. We will use notation $\nabla F(X)$ in the analysis, although $\nabla F(X)$ is not stored at one computational entity.

We call K a linear subspace in $\mathbb{R}^{d \times n}$ determined by the constraint $x_1 = \ldots = x_n$. Note that F defined in (14) is $\mu_{\min}$-strongly convex and $L_{\max}$-smooth on $\mathbb{R}^{d \times n}$, but (μ_f/n)-strongly convex and (L_f/n)-smooth on K, where

$$\mu_{\min} = \min_i \mu_i, \; L_{\max} = \max_i L_i, \tag{15}$$

and μ_f and L_f are strong convexity and smoothness constants of f. Indeed, note that for any $X, Y \in K$ it holds $X = (x, \ldots, x), Y = (y, \ldots, y)$ and therefore

$$F(X) = \sum_{i=1}^{n} f_i(x) = f(x), \; F(Y) = f(y),$$

$$\|Y - X\|^2 = n\|y - x\|^2$$

$$\langle \nabla F(X), Y - X \rangle = \sum_{i=1}^{n} \langle \nabla f_i(x), y - x \rangle = \langle \nabla f(x), y - x \rangle$$

$$F(Y) \geqslant F(X) + \langle \nabla F(X), Y - X \rangle + \frac{\mu_f}{2n}\|Y - X\|^2$$

$$F(Y) \leqslant F(X) + \langle \nabla F(X), Y - X \rangle + \frac{L_f}{2n}\|Y - X\|^2$$

4.2 Gradient Descent with Exact Projections

Let us consider a projected gradient method applied to problem (14).

$$\Pi_{k+1} = \Pi_k - \gamma \operatorname{Proj}_K(\nabla F(\Pi_k)) \tag{16}$$

Since K is a linear subspace, operator $\operatorname{Proj}_K(\cdot)$ is linear. Thus, procedure (16) equivalent to

$$\Pi_{k+1} = \operatorname{Proj}_K(\Pi_k - \gamma \operatorname{Proj}_K(\nabla F(\Pi_k)))$$

and therefore the method trajectory stays in K. Therefore, the algorithm may be interpreted as a simple gradient descent on K. Since function F defined in (14) is (μ_f/n)-strongly convex and (L_f/n)-smooth on K, the algorithm (16) requires $O(L_f/\mu_f \log(1/\varepsilon))$ iterations to achieve ε-solution of problem (14).

However, exact projected method cannot be run in a decentralized manner. In the next section, we introduce an inexact version of this algorithm and analyze its convergence.

4.3 Inexact Projected Gradient Descent

Algorithm 1. Decentralized Projected GD

Require: Each node holds $f_i(\cdot)$ and iteration number N.
1: Initialize $X_0 = [x_0, \ldots, x_0]$, choose $c > 0$.
2: **for** $k = 0, 1, 2, \cdots, N-1$ **do**
3:   $Y_{k+1} = X_k - \gamma \nabla F(X_k)$
4:   $X_{k+1} \approx \operatorname{Proj}_K(Y_{k+1})$ with accuracy ε_1, i.e. $\|X_{k+1} - \operatorname{Proj}_K(Y_{k+1})\|^2 \leqslant \varepsilon_1$ and
     $X_{k+1} - \operatorname{Proj}_K(Y_{k+1}) \in K^\perp$
5: **end for**

Performing step 4 in a decentralized way on a time-varying graph is done by non-accelerated gradient descent and is discussed in later sections. Here we present a convergence result for Algorithm 1.

Theorem 2. *After* $N = O\left(\frac{L_f}{\mu_f} \log\left(\frac{r_0^2}{\varepsilon}\right)\right)$ *iterations, Algorithm 1 with* $\varepsilon_1 = \frac{\mu_f^2}{13 n^2 L_{\max}^2} \varepsilon$ *yields* X_N *such that*

$$\|X_N - X^*\|^2 \leqslant \varepsilon$$

The proof of Theorem 2 is performed in Appendix B.

4.4 Finding Inexact Projection

In this section, we provide an algorithm for finding approximate value of $\text{Proj}_K(Y)$. We formulate projection as an optimization problem

$$\frac{1}{2}\|X - Y\|^2 \longrightarrow \min_{X \in K}$$

Suppose we are given a static connected graph $\mathcal{G}$. Using the fact that $\text{Ker}\, W = K$, (see Proposition 1), the problem above can be rewritten as

$$\frac{1}{2}\|X - Y\|^2 \longrightarrow \min_{X\sqrt{W}=0} \ .$$

Moreover, we can penalize the constraint $X\sqrt{W} = 0$ (see Proposition 2):

$$\frac{1}{2}\|X - Y\|^2 + \frac{R^2}{\varepsilon_2}\|X\sqrt{W}\|^2 \longrightarrow \min_{X \in \mathbb{R}^{d \times n}} \tag{17}$$

with some $R^2 \geqslant \frac{\|X^* - Y\|^2}{\lambda_{\min}^+(W)}$. However, communication graph changes with time and hence graph Laplacian W changes as well, so we are working with a sequence of Laplacians $\{W_k\}_{k=1}^{\infty}$. That leads to a time-varying function $H(X) = \{H_k(X)\}_{k=1}^{\infty}$, where

$$H_k(X) = \frac{1}{2}\|X - Y\|^2 + \frac{R^2}{\varepsilon_2}\|X\sqrt{W_k}\|^2 \tag{18}$$

We are going to employ a decentralized minimization procedure. In order to do this, the gradient $\nabla H_k(X)$ should be computed in a decentralized setup.

$$\nabla H_k(X) = X - Y + \frac{2R^2}{\varepsilon_2}XW$$

Now recall the structure of X, Y and W. Each of these quantities is a matrix of $\mathbb{R}^{d \times n}$ with the i-th column stored at the i-th computational node. Consider $[\nabla H_k(X)]_i$ (i-th column of gradient).

$$[\nabla H_k(X)]_i = [X]_i - [Y]_i + \frac{2R^2}{\varepsilon_2}[XW]_i$$

Note that $[X]_i$ and $[Y]_i$ are held at node i, and $[XW]_i$ is computed as

$$[XW]_i = \deg i \cdot [X]_i - \sum_{j \neq i, (i,j) \in E_k} [X]_j, \tag{19}$$

where E_k denotes the edge set of communication graph $\mathcal{G}_k$. Equation (19) means that $[XW]_i$ can be computed by agent i via communication with its neighbours. Therefore, $[\nabla H_k(X)]_i$ can be computed locally on node i, which makes $\nabla H_k(X)$ available for decentralized computation.

We employ non-accelerated gradient descent on a time-varying function (18). The analysis of this procedure is performed in Sect. 3. First, we prove auxiliary lemmas.

Lemma 2.

1. Let X^* be the solution of (17). Then $X^* \in Y + K^\perp$.
2. Gradient descent applied to time-varying problem (18) with starting point Y stays in $Y + K^\perp$.

Proof. 1. Consider $\Delta X \in K$. Since X^* is the solution of (17),

$$\frac{R^2}{\varepsilon_2} \|(X^* + \Delta X)\sqrt{W}\|^2 + \frac{1}{2}\|(X^* + \Delta X) - Y\|^2$$

$$\geqslant \frac{R^2}{\varepsilon_2} \|X^*\sqrt{W}\|^2 + \frac{1}{2}\|X^* - Y\|^2$$

$$\|X^* + \Delta X - Y\|^2 \geqslant \|X^* - Y\|^2$$

$$\|X^* - Y\|^2 + \|\Delta X\|^2 + 2\langle X^* - Y, \Delta X\rangle$$

$$\geqslant \|X^* - Y\|^2$$

$$\|\Delta X\|^2 \geqslant 2\langle Y - X^*, \Delta X\rangle$$

The latter inequality holds for any $\Delta X \in K$. If we take ΔX small enough, it is necessary that $\langle Y - X^*, \Delta X\rangle = 0$. Therefore, $X^* - Y \in K^\perp$.

2. It is sufficient to show that for any $X \in Y + K^\perp$, the gradient of function in (18) lies in $K^\perp$. Consider some $\Delta X \in K$:

$$\left\langle \nabla \left(\frac{R^2}{\varepsilon}\|X\sqrt{W}_k\|^2 + \frac{1}{2}\|X - Y\|^2\right), \Delta X \right\rangle$$

$$= \left\langle \frac{2R^2}{\varepsilon} XW_k + (X - Y), \Delta X \right\rangle$$

$$= \frac{2R^2}{\varepsilon}\langle X, \Delta X W_k\rangle + \langle X - Y, \Delta X\rangle = 0$$

Lemma 3. Denote $X^* = \mathrm{Proj}_K(Y)$ and let X^*_W be the solution of (17). Then $\|X^*_W - X^*\|^2 \leqslant 4\varepsilon_2$.

Proof. By Proposition 3, it holds

$$\|X^*_W - X^*\|^2 \leqslant \frac{4\left\|\nabla\left(\frac{1}{2}\|X - Y\|^2\right)|_{X=X^*}\right\|}{R\sqrt{\lambda^+_{\min}(W)}}\varepsilon_2$$

$$= \frac{4\|X^* - Y\|}{R\sqrt{\lambda^+_{\min}(W)}}\varepsilon_2 \leqslant \frac{4\|X^* - Y\| \cdot \sqrt{\lambda^+_{\min}(W)}}{\|X^* - Y\|\sqrt{\lambda^+_{\min}(W)}}\varepsilon_2 = 4\varepsilon_2$$

Finally, using Lemmas 2 and 3 we establish the number of iterations for finding projection.

Theorem 3. After $N = O\left(\chi \log\left(\frac{\|X^* - Y\|^2}{\varepsilon_1}\right)\right)$ iterations (see (13) for definition of χ), gradient descent on problem (18) with $\varepsilon_2 = \frac{\varepsilon_1}{52\chi}$ yields X_N such that

$$\|X_N - X^*\|^2 \leqslant \varepsilon_1.$$

Proof of Theorem 3 is provided in Appendix C.

4.5   Overall Complexity

Summarizing the results of Theorems 2 and 3, we get the final iteration complexity result.

Theorem 4. *Algorithm 1 with $\varepsilon_1 = \frac{\mu_f^2}{13n^2 L_{\max}^2}\varepsilon$ requires*

$$
N = O\left(\frac{L_f}{\mu_f} \chi \log \left(\frac{\|\nabla F(X_0)\|^2 n^2 L_{\max}^2}{\varepsilon \mu_f^2} \right) \log \left(\frac{r_0^2}{\varepsilon} \right) \right)
$$

communication steps, including sub-problem solution, to yield X_N such that

$$
\|X_N - X^*\|^2 \leqslant \varepsilon.
$$

Remark 1. The convergence rate depends on L_f and μ_f instead of $\mu_{\mathrm{sum}} = \sum_{i=1}^n \mu_i$ and $L_{\mathrm{sum}} = \sum_{i=1}^n L_i$. First, note that $\mu_f \geqslant \overline{\mu}$ and $L_f \leqslant \overline{L}$. Second, and most importantly, the ratio L_{sum}/L_f may be of magnitude n, and the ratio μ_f/μ_{sum} may be arbitrary large. We illustrate this observation with the following example.

$$
f(x) = \frac{1}{2}(1+\alpha)\|x\|^2, \alpha > 0;
$$
$$
f_i(x) = \frac{1}{2}x_i^2 + \frac{\alpha}{2n}\|x\|^2.
$$

In this particular case, each $f_i(x)$ has $\mu_i = \alpha/n$ and $L_i = 1 + \alpha/n$, and therefore $L_{\mathrm{sum}} = n + \alpha, \mu_{\mathrm{sum}} = \alpha$. On the other hand, $\mu_f = L_f = 1 + \alpha$. Hence,

$$
\frac{L_{\mathrm{sum}}}{L_f} = \frac{n+\alpha}{1+\alpha} \xrightarrow{\alpha \to +0} n,
$$
$$
\frac{\mu_f}{\mu_{\mathrm{sum}}} = \frac{1+\alpha}{\alpha} \xrightarrow{\alpha \to +0} \infty.
$$

The bound obtained in Theorem 4 is based on L_f/μ_f. The example above shows that using this ratio in the bound may be significantly better than using $L_{\mathrm{sum}}/\mu_{\mathrm{sum}}$.

4.6   Extension to Accelerated Gradient Descent

Algorithm 1 includes gradient descent in the outer loop. It is possible to employ an accelerated scheme instead of a non-accelerated method, which leads to the following algorithm. Here L and κ denote the smoothness constant and condition number of F, respectively. The theoretical analysis of this method, including the choice of ε_1, is left for future work. However, our numerical tests in Sect. 5 show that Algorithm 2 outperforms both Algorithm 1 and DIGing [13].

Algorithm 2. Decentralized Accelerated Projected GD

Require: Each node holds $f_i(\cdot)$ and iteration number N.
1: **for** $k = 0, 1, \ldots, N - 1$ **do**
2:   $Y_{k+1} = X_k - \frac{1}{L}\nabla F(X_k)$
3:   $\tilde{Y}_{k+1} \approx \text{Proj}_K(Y_{k+1})$ with accuracy ε_1
4:   $X_{k+1} = \tilde{Y}_{k+1} + \frac{\sqrt{\kappa}-1}{\sqrt{\kappa}+1}(\tilde{Y}_{k+1} - \tilde{Y}_k)$
5: **end for**

---

5 Numerical Experiments

In this section, we provide numerical simulations of Algorithm 1 on *logistic regression* problem on LibSVM datasets [5]. The objective function is defined as

$$f(x) = \frac{1}{m}\sum_{i=1}^{m}\log\left[1 + \exp(-c_i(\langle a_i, x\rangle + b_i))\right],$$

where $a_i \in \mathbb{R}^d$ are training samples and $c_i \in \{0, 1\}$ are class labels. In decentralized scenario, the training dataset is distributed between the agents in the network.

One of the tuned parameters of Algorithm 1 is the number of inner iterations. On Figs. 1(a) and 1(b), we illustrate different choices of this parameter, and Proj-GD-k denotes projected gradient method with k iterations on each sub-problem. Moreover, we compare our algorithm to DIGing [13].

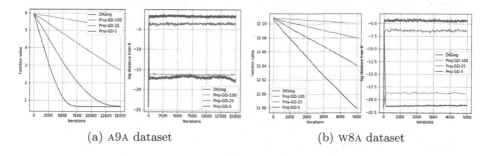

(a) A9A dataset   (b) W8A dataset

Fig. 1. Random graph with 100 nodes

Figures 1(a) and 1(b) suggest that performance of Algorithm 1 is significantly dependent on the number of iterations made on step 4. A large number of steps results in more precise projection procedure, but also requires takes more communication steps. In other words, there is a trade-off between the number of communications and projection accuracy. In practice, one can tune number of iterations for sub-problem and find an optimal value for a specific practical case.

Moreover, we experiment with accelerated gradient descent and compare it to DIGing method. We find that accelerated method performs significantly better.

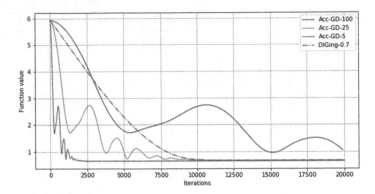

Fig. 2. Random graph with 100 nodes, IJCNN1 dataset.

6   Conclusions and Future Work

Our main result is based on a simple idea – running projected gradient method with inexact projections. This idea is applied to decentralized optimization on time-varying graphs. The proposed method incorporates two different algorithms: projected gradient descent and obtaining mean of values held by agents over the network. The whole procedure is shown to be robust to network changes since non-accelerated schemes are used both for outer and inner loops.

However, the question whether it is possible to employ an accelerated method either for finding projection or for running the outer loop remains open. Moreover, projection may be performed by a variety of algorithms, including randomized and asynchronous gossip algorithms [3]. Investigation of new techniques for finding projection is left for future work.

Supplementary Material

A   Proof of Proposition 5

1. Multiplying both sides by $2p$ yields

$$\|u - pv\|^2 \geqslant 0.$$

2. Analogously, multiplying both sides by $1 - p$ leads to

$$(1 - p)\|v\|^2 \geqslant (p - p^2)\|u\|^2 - p(\|v\|^2 + \|u\|^2 - 2\langle u, v \rangle)$$
$$\|v\|^2 \geqslant -p^2\|u\|^2 + 2p\langle u, v \rangle$$
$$\|v - pu\|^2 \geqslant 0$$

B   Proof of Theorem 2

Denote $\Pi_k = \mathrm{Proj}_K(X_k), X^* = \Pi^* = \arg\min\limits_{K} F(X), r_k = \|\Pi_k - \Pi^*\| = \|\Pi_k - X^*\|$.

$$
\begin{aligned}
r_{k+1}^2 &= \|\Pi_k - \Pi^* - \gamma \,\mathrm{Proj}_K(\nabla F(X_k))\|^2 = r_k^2 + \gamma^2 \|\mathrm{Proj}_K(\nabla F(X_k))\|^2 \\
&\quad - 2\gamma \langle \Pi_k - \Pi^*, \mathrm{Proj}_K(\nabla F(X_k)) \rangle \\
&= r_k^2 + \gamma^2 \|\mathrm{Proj}_K(\nabla F(X_k))\|^2 \\
&\quad - 2\gamma \underbrace{\langle \Pi_k - \Pi^*, \mathrm{Proj}_K(\nabla F(\Pi_k)) - \mathrm{Proj}_K(\nabla F(X^*)) \rangle}_{①} \\
&\quad - 2\gamma \underbrace{\langle \Pi_k - \Pi^*, \mathrm{Proj}_K(\nabla F(X_k)) - \mathrm{Proj}_K(\nabla F(\Pi_k)) \rangle}_{②}
\end{aligned}
\tag{20}
$$

1. First, let us estimate ①. Note that for all $\Pi \in K$, it holds

$$
\Pi = [\pi, \dots, \pi]
$$
$$
\nabla F(\Pi) = [\nabla f_1(\pi), \dots, \nabla f_n(\pi)]
$$

Moreover, for all $X \in \mathbb{R}^{d \times n}$ it holds

$$
\mathrm{Proj}_K(X) = \arg\min\limits_{Z \in K} \|Z - X\|^2 = [\overline{x}, \dots, \overline{x}],
$$

where $\overline{x} = \frac{1}{n} \sum\limits_{i=1}^{n} x_i$. In particular,

$$
\mathrm{Proj}_K(\nabla F(\Pi)) = [\nabla f(\pi)/n, \dots, \nabla f(\pi)/n]
$$

Now we can estimate ① by Proposition 4. For brevity we introduce $\widehat{\mu}_f = \mu_f/n, \widehat{L}_f = L_f/n$.

$$
\begin{aligned}
&\langle \Pi_k - \Pi^*, \mathrm{Proj}_K(\nabla F(\Pi_k)) - \mathrm{Proj}_K(\nabla F(X^*)) \rangle \\
&= n \cdot \langle \pi_k - \pi^*, \nabla f(\pi)/n - \nabla f(x^*)/n \rangle \\
&= \langle \pi_k - \pi^*, \nabla f(\pi) - \nabla f(\pi^*) \rangle \\
&\geqslant \frac{(n\widehat{\mu}_f)(n\widehat{L}_f)}{n\widehat{\mu}_f + n\widehat{L}_f} \|\pi_k - \pi^*\|^2 + \frac{1}{n\widehat{\mu}_f + n\widehat{L}_f} \|\nabla f(\pi_k) - \nabla f(\pi^*)\|^2 \\
&= \frac{\widehat{\mu}_f \widehat{L}_f}{\widehat{\mu}_f + \widehat{L}_f} \|\Pi_k - \Pi^*\|^2 + \frac{1}{\widehat{\mu}_f + \widehat{L}_f} \|\mathrm{Proj}_K(\nabla F(\Pi_k))\|^2
\end{aligned}
$$

2. Let us employ (5) with $p = \frac{\widehat{\mu}_f + \widehat{L}_f}{\widehat{\mu}_f \widehat{L}_f}$ to estimate ②.

$$
\begin{aligned}
&- 2\gamma \langle \Pi_k - \Pi^*, \mathrm{Proj}_K(\nabla F(X_k)) - \mathrm{Proj}_K(\nabla F(\Pi_k)) \rangle \\
&\leqslant \gamma \frac{\widehat{\mu}_f \widehat{L}_f}{\widehat{\mu}_f + \widehat{L}_f} \|\Pi_k - \Pi^*\|^2 + \gamma \frac{\widehat{\mu}_f + \widehat{L}_f}{\widehat{\mu}_f \widehat{L}_f} \|\nabla F(X_k) - \nabla F(\Pi_k)\|^2 \\
&\leqslant \gamma \frac{\widehat{\mu}_f \widehat{L}_f}{\widehat{\mu}_f + \widehat{L}_f} r_k^2 + \gamma \frac{\widehat{\mu}_f + \widehat{L}_f}{\widehat{\mu}_f \widehat{L}_f} L_{\max}^2 \varepsilon_1
\end{aligned}
$$

Now we return to (20).

$$r_{k+1}^2 \leqslant r_k^2 + \underline{\gamma^2 \| \mathrm{Proj}_K(\nabla F(X_k)) \|^2} - 2\gamma \frac{\widehat{\mu}_f \widehat{L}_f}{\widehat{\mu}_f + \widehat{L}_f} r_k^2$$

$$- \underline{\frac{2\gamma}{\widehat{\mu}_f + \widehat{L}_f} \| \nabla \mathrm{Proj}_K(\nabla F(\Pi_k)) \|^2} + \gamma \frac{\widehat{\mu}_f \widehat{L}_f}{\widehat{\mu}_f + \widehat{L}_f} r_k^2 + \gamma \frac{\widehat{\mu}_f + \widehat{L}_f}{\widehat{\mu}_f \widehat{L}_f} L_{\max}^2 \varepsilon_1$$

$$\tag{21}$$

The sum of underlined terms may be estimated by setting $\gamma \in \left(0, \frac{2}{\widehat{\mu}_f + \widehat{L}_f} \right]$ and using (6) with $p = \gamma \frac{\widehat{\mu}_f + \widehat{L}_f}{2} \in (0, 1)$.

$$\gamma^2 \| \mathrm{Proj}_K(\nabla F(X_k)) \|^2 - \frac{2\gamma}{\widehat{\mu}_f + \widehat{L}_f} \| \nabla \mathrm{Proj}_K(\nabla F(\Pi_k)) \|^2$$

$$= \frac{2\gamma}{\widehat{\mu}_f + \widehat{L}_f} \left(\frac{\gamma(\widehat{\mu}_f + \widehat{L}_f)}{2} \| \mathrm{Proj}_K(\nabla F(X_k)) \|^2 - \| \mathrm{Proj}_K(\nabla F(\Pi_k)) \|^2 \right)$$

$$\leqslant \frac{2\gamma}{\widehat{\mu}_f + \widehat{L}_f} \cdot \frac{\gamma(\widehat{\mu}_f + \widehat{L}_f)}{2} \cdot \left(1 - \frac{\gamma(\widehat{\mu}_f + \widehat{L}_f)}{2} \right)^{-1} \cdot$$

$$\cdot \| \mathrm{Proj}_K(\nabla F(X_k)) - \mathrm{Proj}_K(\nabla F(\Pi_k)) \|^2$$

$$\leqslant \frac{2\gamma^2}{2 - \gamma(\widehat{\mu}_f + \widehat{L}_f)} \cdot L_{\max}^2 \varepsilon_1$$

Finally, we return to (21), set $\gamma = \frac{1}{\widehat{\mu}_f + \widehat{L}_f}$ and estimate r_{k+1}.

$$r_{k+1}^2 \leqslant r_k^2 \left(1 - \gamma \frac{\widehat{\mu}_f \widehat{L}_f}{\widehat{\mu}_f + \widehat{L}_f} \right) + \left(\gamma \frac{\widehat{\mu}_f + \widehat{L}_f}{\widehat{\mu}_f \widehat{L}_f} + \frac{2\gamma^2}{2 - \gamma(\widehat{\mu}_f + \widehat{L}_f)} \right) \cdot L_{\max}^2 \varepsilon_1$$

$$= r_k^2 \left(1 - \frac{\widehat{\mu}_f \widehat{L}_f}{(\widehat{\mu}_f + \widehat{L}_f)^2} \right) + \left(\frac{1}{\widehat{\mu}_f \widehat{L}_f} + \frac{2}{(\widehat{\mu}_f + \widehat{L}_f)^2} \right) \cdot L_{\max}^2 \varepsilon_1$$

$$\leqslant r_k^2 \left(1 - \frac{\widehat{\mu}_f}{4\widehat{L}_f} \right) + \frac{3}{2\widehat{\mu}_f \widehat{L}_f} \cdot L_{\max}^2 \varepsilon_1$$

Analogously to Sect. 3, for $r_k^2 \geqslant \frac{12}{\widehat{\mu}_f^2} L_{\max}^2 \varepsilon_1$ it holds

$$r_{k+1}^2 \leqslant r_k^2 \left(1 - \frac{\widehat{\mu}_f}{8\widehat{L}_f} \right).$$

After N iterations, we get

$$\| \Pi_N - X^* \| = \| \Pi_N - \Pi^* \|^2 = r_N^2 \leqslant \frac{12}{\widehat{\mu}^2} L_{\max}^2 \varepsilon_1.$$

Since $\|X_N - \Pi_N\|^2 \leqslant \varepsilon_1$,

$$\|X_N - X^*\|^2 = \|X_N - \Pi_N\|^2 + \|\Pi_N - X^*\|^2$$
$$\leqslant \frac{12}{\bar{\mu}^2}L_{\max}^2\varepsilon_1 + \varepsilon_1 \leqslant \frac{13}{\bar{\mu}^2}L_{\max}^2\varepsilon_1 = \varepsilon,$$

which concludes the proof.

C   Proof of Theorem 3

Gradient descent is run on a time-varying function $H(X) = \{H_k(X)\}_{k=1}^\infty$ with H_k is defined in (18). By Lemma 3, for each k it holds $\|X^* - \arg\min H_k(X)\|^2 \leqslant 4\varepsilon_2$. Moreover, each $H_k(X)$ μ_k-strongly convex on $K^\perp$ and L_k-smooth on $\mathbb{R}^{d \times n}$, where

$$\mu_k = 1 + \frac{2R^2}{\varepsilon_2}\lambda_{\min}^+(W_k) \geqslant 1 + \frac{2R^2}{\varepsilon_2}\theta_{\min}$$
$$L_k = 1 + \frac{2R^2}{\varepsilon_2}\lambda_{\max}(W_k) \leqslant 1 + \frac{2R^2}{\varepsilon_2}\theta_{\max}$$

Therefore, time-varying function $H(X)$ satisfies Assumption 1. Then, by Theorem 1, it follows that

$$\|X_N - X^*\|^2 \leqslant 13\frac{1 + (2R^2/\varepsilon_2)\theta_{\max}}{1 + (2R^2/\varepsilon_2)\theta_{\min}} \cdot 4\varepsilon_2$$
$$\leqslant 52\frac{(2R^2/\varepsilon_2)\theta_{\max}}{(2R^2/\varepsilon_2)\theta_{\min}} \cdot \varepsilon_2 = 52\chi\varepsilon_2 = \varepsilon_1$$

References

1. Arjevani, Y., Shamir, O.: Communication complexity of distributed convex learning and optimization. In: Advances in Neural Information Processing Systems, pp. 1756–1764 (2015)
2. Bertsekas, D.P., Tsitsiklis, J.N.: Parallel and Distributed Computation: Numerical Methods, vol. 23. Prentice hall, Englewood Cliffs (1989)
3. Boyd, S., Ghosh, A., Prabhakar, B., Shah, D.: Randomized gossip algorithms. IEEE/ACM Trans. Netw. 14(SI), 2508–2530 (2006)
4. Bubeck, S.: Convex optimization: algorithms and complexity. Found. Trends Mach. Learn. 8(3–4), 231–357 (2015)
5. Chang, C.C., Lin, C.J.: Libsvm: a library for support vector machines. ACM Trans. Intell. Syst. Technol. (TIST) 2(3), 27 (2011)
6. Dvinskikh, D., Gasnikov, A.: Decentralized and parallelized primal and dual accelerated methods for stochastic convex programming problems. arXiv preprint arXiv:1904.09015 (2019)
7. Gasnikov, A.: Universal gradient descent. arXiv:1711.00394 (2018). [in Russian]
8. Gorbunov, E., Dvinskikh, D., Gasnikov, A.: Optimal decentralized distributed algorithms for stochasticconvex optimization. arXiv preprint arXiv:1911.07363 (2019)

9. Hendrikx, H., Bach, F., Massoulié, L.: Accelerated decentralized optimization with local updates for smooth and strongly convex objectives. arXiv preprint arXiv:1810.02660 (2018)

10. Lan, G., Lee, S., Zhou, Y.: Communication-efficient algorithms for decentralized and stochastic optimization. Mathematical Programming, pp. 1–48 (2018)

11. Lü, Q., Li, H., Xia, D.: Geometrical convergence rate for distributed optimization with time-varying directed graphs and uncoordinated step-sizes. Inf. Sci. **422**, 516–530 (2018)

12. Maros, M., Jaldén, J.: Panda: a dual linearly converging method for distributed optimization over time-varying undirected graphs. 2018 IEEE Conference on Decision and Control (CDC), pp. 6520–6525 (2018)

13. Nedić, A., Olshevsky, A., Shi, W.: Achieving geometric convergence for distributed optimization over time-varying graphs. SIAM J. Optimization **27**(4), 2597–2633 (2017)

14. Nedić, A., Olshevsky, A.: Distributed optimization over time-varying directed graphs. IEEE Trans. Automatic Control **60**(3), 601–615 (2014)

15. Nedić, A., Olshevsky, A., Shi, W., Uribe, C.A.: Geometrically convergent distributed optimization with uncoordinated step-sizes. In: 2017 American Control Conference (ACC), pp. 3950–3955. IEEE (2017)

16. Nemirovskii, A.: Yudin: Problem Complexity and Method Efficiency in Optimization. Wiley (1983)

17. Nesterov, Y.: Introductory Lectures on Convex Optimization. A Basic Course. Springer, Boston (2013). https://doi.org/10.1007/978-1-4419-8853-9

18. Nikaido, H.: Convex Structures and Economic Theory. Academic Press (1968)

19. Olshevsky, A.: Linear time average consensus and distributed optimization on fixed graphs. SIAM J. Control Optimization **55**(6), 3990–4014 (2017)

20. Rogozin, A., Uribe, C., Gasnikov, A., Malkovskii, N., Nedich, A.: Optimal distributed convex optimization on slowly time-varying graphs. In: IEEE Transactions on Control of Network Systems (2019)

21. Scaman, K., Bach, F., Bubeck, S., Lee, Y.T., Massoulié, L.: Optimal algorithms for smooth and strongly convex distributed optimization in networks. In: Precup, D., Teh, Y.W. (eds.) Proceedings of the 34th International Conference on Machine Learning. Proceedings of Machine Learning Research, vol. 70, pp. 3027–3036. PMLR (2017)

22. Van Scoy, B., Lessard, L.: A distributed optimization algorithm over time-varying graphs with efficient gradient evaluations. https://vanscoy.github.io/docs/papers/vanscoy2019distributed.pdf (2019)

The Adaptation of Interior Point Method for Solving the Quadratic Programming Problems Arising in the Assembly of Deformable Structures

Maria Stefanova[✉][iD] and Sergey Lupuleac[iD]

Peter the Great St.Petersburg Polytechnic University,
St.petersburg 195251, Russia
stefanova.m@list.ru

Abstract. The simulation of the airframe assembly process implies the modelling of contact interaction of compliant parts. As every item in mass production deviates from the nominal, the analysis of the assembly process involves the massive solving of contact problems with varying input data. The contact problem may be formulated in terms of quadratic programming. The bottleneck is the computation time that may be reduced by the use of specially adapted optimization methods. The considered problems have an ill-conditioned Hessian and a sparse matrix of constraints. It is necessary to solve a large number of problems with the same constraint matrix and Hessian. This work considers the primal-dual interior-point method (IPM) and proposes its adaptation to the solving of assembly problems. A method is proposed for choosing a feasible starting point based on a physical interpretation for reducing the number of IPM iterations. The numerical comparison of the approaches to solve a system of linear equations at each iteration of IPM is presented, i.e. an augmented system and a normal equation (its reduction) using various preconditioners. Finally, IPM is compared by computation time with an active-set method, a Newton projection method and Lemke's method on a number of aircraft assembly problems.

Keywords: Primal-dual interior-point method · Quadratic programming · Feasible starting point · Augmented system · Normal equation · Contact problem

1 Introduction

The modeling of the assembly process of large-scale compliant parts [10–12,18] involves solving the contact problem. The necessity to obtain a solution with high accuracy entails the use of finite element models with fine meshes, which in turn significantly increases the problem size. By using the variational formulation

---

Supported by RFBR, project number 20-38-90023\20.

N. Olenev et al. (Eds.): OPTIMA 2020, LNCS 12422, pp. 257–271, 2020.
https://doi.org/10.1007/978-3-030-62867-3_19

and the substructuring the contact problem can be reduced to the quadratic programming problem [12]:

$$\min \ \tfrac{1}{2}x^T K x - f^T x,$$
$$\text{s.t.} \quad A^T x - g \leq 0, \tag{1}$$

where $x \in \mathbb{R}^n$ is the vector of geometrically restricted displacements in the area of possible contact (Fig. 1), $K \in \mathbb{R}^{n \times n}$ is the reduced stiffness matrix computed using finite element analysis, $A \in \mathbb{R}^{n \times m}$ is a linear operator that defines the pairs of nodes that may come in contact, $f \in \mathbb{R}^n$ is the vector of loads from fastening elements, and $g \in \mathbb{R}^m$ is the initial gap vector between assembled parts (i.e. before fasteners installation). Note that although the gap is physically always non-negative, the vector g can have components below zero. In this case, the parts installed on the assembly stand come in contact interaction prior to fastening. The stiffness matrix K is ill-conditioned block-diagonal with fully populated blocks, and the matrix of constraints A is sparse. The simulation of the assembly process requires solving of many similar contact problems based on the same assembly model, i.e. with the same matrices K and A, and differing in the various parameters of assembly, such as initial gap between assembled parts, fastening pattern, loads in fastening elements, etc.

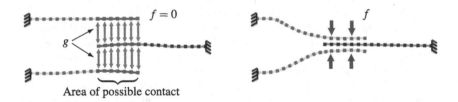

Fig. 1. Schematic representation of assembly cross section. Left is before fasteners installation and right is after fasteners installation.

Interior point methods are successfully used for various types of optimization problems, including contact analysis [13, 14, 22]. The present paper focuses on two aspects of using the interior point method: the choice of the starting point and solving the system of linear equations at each iteration. Two types of interior point methods are distinguished, based on different approaches for choosing the initial point: feasible and infeasible. Feasible methods require the starting point to satisfy all the constraints of the problem. Several approaches to find such an initial point are described in [16, 23]. Finding a feasible starting point can be difficult due to the need to satisfy equality constraints along with the need to obtain a point close to the central path. On the contrary, only the condition of nonnegativity for some of the variables should be guaranteed for an infeasible initial point. This approach gives a perfectly centred starting point. The infeasible algorithms are constructed in such a way that the infeasibility measure is limited by the duality gap and decreases from iteration to iteration.

Nevertheless, the presence of infeasibility in the starting point may affect the computation time. The strategy proposed in [1] allows to control the error in equality constraints for the infeasible starting point. Because of the attractive properties, the infeasible methods are widely used in practice.

The interior point methods are known to converge with a relatively small number of iterations. However, each iteration is computationally expensive. Depending on the scheme of method, at each iteration it is required to solve the linear system(s) of equations, that is the most time consuming operation. Several approaches for solving the linear system of equations in IPM exist [8,24]. The linear system of equations may be reduced to a symmetric augmented system, that does not change the structure of Hessian matrix and the matrix of constraints. Further elimination of variables leads to a normal equation that is symmetric and a positive definite. Both iterative and direct methods are used to solve these two types of linear systems of equations [3,5].

Considering the contact problem (1), the physical meaning of the conditions for the feasible starting point helps to reduce the number of interior point method iterations. Based on the physics of the problem, a heuristic procedure that provides feasible starting point that is closer to the optimum, compared to an infeasible one, is proposed. A comparison of various existing approaches to solve the system of linear equations for typical assembly problems is presented. In addition, the last section of the present paper is devoted to various optimization methods that can be used to solve contact problems. An interior point method, an active set method, a Newton projection method and Lemke's method are compared by the computation time for the contact problems arising in the aircraft assembly simulation.

2   Primal-Dual Interior-Point Method

The primal-dual interior-point method (IPM) is based on the reformulation of the constrained minimization problem (1) into a sequence of unconstrained minimization problems. As a result, the quadratic programming problem is reduced to a sequence of nonlinear systems of equations:

$$S_\mu(x,y,\lambda) = \begin{bmatrix} Kx - f + A\lambda \\ A^T x - g + y \\ \Lambda Y e - \mu e \end{bmatrix} = \begin{bmatrix} 0 \\ 0 \\ 0 \end{bmatrix}, \; y \geq 0, \; \lambda \geq 0. \tag{2}$$

In the above, Λ and Y are diagonal matrices with λ and $y = -A^T x + g$ on their diagonals, $e = 1_m$ is a vector of ones and μ is duality gap. The solution of the system (2) is obtained using Newton's method. At each iteration of the IPM one step of Newton's method is made, and then the duality gap is updated. The iterative process continues until the duality gap μ becomes sufficiently small (Algorithm 1).

Algorithm 1 is based on the Mehrotra predictor-corrector scheme [15]. The scheme involves the search for two directions of the step at each iteration.

The predictor direction is a pure Newton direction. The corrector direction is used for improvement of the centrality and for compensation of the error in the linear approximation. The system of linear equations needs to be solved twice at each iteration in the predictor-corrector scheme. The authors of [6] propose to reduce the complexity of the iteration when using iterative methods for system solving. Specifically, the corrector step may be skipped when the predictor direction ensures that the next iteration is sufficiently close to the central path. Skipping or not the corrector step is determined depending on the step size α.

Algorithm 1: Primal-dual interior-point method

Choose accuracy $\epsilon_\mu > 0$ and boundary closeness parameter $\theta \in (0,1)$, $k = 0$;

Choose initial point $w_0 = (x_0, y_0, \lambda_0)$;

Compute duality gap $\mu_0 = y_0^T \lambda_0 / m$;

while $\mu_k \geq \epsilon_\mu$ **do**

  Compute Newton direction $\Delta w_k^p = (\Delta x_k^p, \Delta y_k^p, \Delta \lambda_k^p)$:

  $$\begin{bmatrix} K & 0 & A \\ A^T & I & 0 \\ 0 & \Lambda_k & Y_k \end{bmatrix} \begin{bmatrix} \Delta x_k^p \\ \Delta y_k^p \\ \Delta \lambda_k^p \end{bmatrix} = \begin{bmatrix} -(Kx_k - f + A\lambda_k) \\ -(A^T x_k - g + y_k) \\ -\Lambda_k Y_k e \end{bmatrix} ;$$

  Compute step length α_k^p which provides $(y_k^p, \lambda_k^p) > 0$;

  Update variables $w_k^p = w_k + \alpha_k^p \Delta w_k^p$;

  Update duality gap $\mu_k^p = (y_k^p)^T \lambda_k^p / m$;

  if $\alpha_k^p < 0.7$ **then**

    Compute centering parameter $\sigma = (\mu_k^p / \mu_k)^3$;

    Compute Newton direction $\Delta w_k^c = (\Delta x_k^c, \Delta y_k^c, \Delta \lambda_k^c)$:

    $$\begin{bmatrix} K & 0 & A \\ A^T & I & 0 \\ 0 & \Lambda_k & Y_k \end{bmatrix} \begin{bmatrix} \Delta x_k^c \\ \Delta y_k^c \\ \Delta \lambda_k^c \end{bmatrix} = \begin{bmatrix} 0 \\ 0 \\ -\Lambda_k^p Y_k^p e + \sigma \mu_k e \end{bmatrix} ;$$

    Compute step length α_k which provides $(y_{k+1}, \lambda_{k+1}) > 0$;

    Update variables $w_{k+1} = w_k + \alpha_k (\Delta w_k^p + \Delta w_k^c)$;

  else

    $w_{k+1} = w_k^p$;

  end

  Update duality gap $\mu_{k+1} = (y_{k+1})^T \lambda_{k+1} / m$;

  $k = k + 1$;

end

---

3   Aircraft Assembly Problems

Consider typical examples of aircraft assembly problems based on the wing-to-fuselage joint [11]. Two types of assembly models are distinguished by the area of possible contact. The first one refers to a simple-type assembly, i.e. two parts are assembled. The second is a complex sandwich-type assembly of at least three assembled layers located one above the other (Fig. 1). The wing-to-fuselage joint consists of upper and lower wing panels (Fig. 2). The upper joint is of simple-type and the lower one is of sandwich-type. Typically, the stiffness matrix K of the lower assembly model is worse conditioned $(cond(K) \approx 10^7 - 10^8)$ than the corresponding matrix of the upper assembly $(cond(K) \approx 10^6 - 10^7)$. The number of unknowns in some particular cases may reach up to 20000 unknowns. However, for simplicity, relatively small problems (≈ 4000 of unknowns) are considered in the sections devoted to IPM adaptation. These problems reflect the behaviour of real assembly problems of larger sizes. In practice, it becomes necessary to calculate the displacements x and the resulting gap y for different initial gaps g and fastener patterns f for the same model, i.e. the same matrices A and K. Therefore, for each test case, we consider a series of contact problems that differ exclusively in the fastener patterns. These problems are ordered by the number of active constraints n_{act} at the optimum point.

Fig. 2. Lower (the sandwich assembly type) and upper (the simple assembly type) wing panel assembly models.

4   Search of Feasible Starting Point

We start by describing the basis for the heuristic procedure used to find a feasible starting point for assembly problems. The feasible starting point must satisfy the following conditions:

$$Kx_0 - f + A\lambda_0 = 0, \tag{3}$$

$$A^T x_0 - g + y_0 = 0, \tag{4}$$

$$y_0 > 0, \lambda_0 > 0. \tag{5}$$

These conditions differ from optimality conditions by the complementarity condition

$$\Lambda Y = 0. \tag{6}$$

Consider the physical meaning of each condition:

- The stationarity condition (3) provides a balance between internal λ_0 and external f forces;
- The geometric restrictions for the vector of displacement x_0 are described by the non-penetration condition (4);
- The positivity condition (5) of the dual variable λ_0 indicates the absence of tensile stresses and $y_0 > 0$ corresponds to the positiveness of the resulting gap between the assembled parts.

The complementarity condition (6) means that if there is some non-zero resulting gap y between the nodes, then the contact forces λ at these nodes must be equal to zero, since the parts do not touch each other. On the contrary, if the gap y at some point is zero, then stresses λ caused by the contact of parts may appear. The slackening of the condition (6) allows 'the contact forces' λ_0 and 'the resulting gap' y_0 to be positive at the same time. This leads to the fact that 'the contact forces' λ_0 are not in agreeance with the external forces f at the starting point $w_0 = (x_0, y_0, \lambda_0)$ and assembled parts can penetrate each other. Therefore, λ_0 can be considered as some forces acting in the opposite to f direction. That is, an increase in the modulus of λ_0 leads to an increase in the gap y_0 between the assembled parts (Fig. 3).

Algorithm 2: STPC1

1 Choose $\lambda > 0$, $\epsilon > 0$, $s > 1$ and it_m;
2 Compute $Q = A^T K^{-1} A$ and $p = A^T K^{-1} f - g$;
3 Initialize $\lambda_0^i = \lambda e$, $it = 0$;
4 **while** $it < it_m$ **do**
5   **for** $i \leftarrow 1$ **to** m **do**
6   Compute $t^i = q_i^T \lambda_0$;
7   **if** $t^i \leq 0$ **then**
8   Compute $\lambda_0^i = -\frac{\sum_{j=1, j \neq i}^{m} q_{ij} \lambda_0^j + \epsilon}{q_{ii}}$;
9   **end**
10   **end**
11   **if** $t^i > 0$ $\forall i$ **then**
12   break;
13   **end**
14      $it = it + 1$;
15 **end**
16 **do**
17   Increase contact force $\lambda_0 = \lambda_0 s$;
18   Compute gap $y_0 = Q\lambda_0 - p$;
19 **while** *Exist* $y_0^i \leq 0$;
20 Compute $x_0 = K^{-1}(f - A\lambda_0)$;
    Result: $w_0 = (x_0, y_0, \lambda_0)$

The method discussed in this paper is based on a modification of the method used for searching a feasible starting point proposed for contact problems in [21]. Let us call it STPC1, that is abbreviation of 'starting point for contact problems'. The practical use of STPC1 has shown its effectiveness for many assembly problems. However, on the part of the assembly models, namely, models having a complex sandwich structure, this approach loses to approaches based on an infeasible starting point. The modification of STPC1 retains its advantages in simple problems with two assembled layers, reducing the number of IPM iterations for more complex problems.

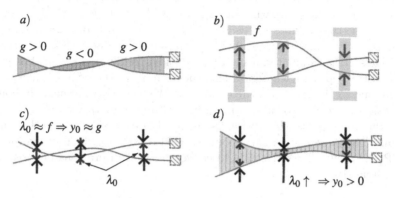

Fig. 3. Schematic representation of STPC2: a) an initial gap g before fasteners installation, b) the loads from fasteners f acting on top for upper panel and on bottom for lower panel, c) the initialization of 'the contact forces' λ_0 in order to compensate external loads f (step 3), d) the correction of λ_0 in order to attain positive gap y_0 (steps 4–15).

The basis for the modified method STPC2 (Algorithm 3, Fig. 3) is the initialization procedure of STPC1 (steps 4–15 in Algorithm 2) which has established itself as an effective and fast way to initialize STPC1. This heuristic procedure provides a choice of λ_0 in such a way that a further increase in 'contact forces' leads to an increase in the gap y_0, i.e. $Q\lambda_0 > 0$. The correction of λ_0 provided by the procedure appears to be possible due to the positiveness of all the diagonal elements of Q and does not require many iterations. Conditions for feasible starting point (3)–(5) allow weakening the condition $Q\lambda_0 > 0$ using instead $Q\lambda_0 - p > 0$. In addition, the contact forces acting in the assembled structure strongly depend on the applied external forces, and often correlate with it. So, based on the specifics of the problem, it can be expected that contact will most likely occur and reaction will not be zero at the installation site of the fasteners. Therefore, it is proposed to initialize λ_0, based on the vector of external forces f (step 3 of Algorithm 3). Note, that if the two parts are assembled and the correction of λ_0 is not required, then y_0 will be equal to the initial gap g. Indeed, in this case $AA^T = 2I$ and $y_0 = A^T K^{-1} A A^T f/2 - A^T K^{-1} f + g = g$.

5   Comparison of Feasible and Infeasible Starting Point Approaches

The ideal starting point is a feasible point from a neighborhood of the central path and the one close to the optimal solution of the problem.

Definition 1. Central path is a set of feasible points $C = \{(x_\mu, y_\mu, \lambda_\mu) : \mu > 0\}$ such that $(x_\mu, y_\mu, \lambda_\mu)$ solves system (2).

The STPC1 and STPC2 give a feasible, but not necessarily close to central path initial point. However, in practice, the solution in the first few iterations goes to the vicinity of the central path (Fig. 4). Infeasible starting point [1] belongs to a central path (that admits infeasible points in this case), but has a significant error in equality constraints. It behaves similarly, giving a quick decrease in the residuals at the first iterations (Fig. 5). The residuals decrease faster than the off-central approximation goes into the vicinity of the central path. This is primarily due to the fact that the condition $\Lambda Y e = \mu e$ is nonlinear, in contrast to the residuals of Eqs. (3) and (4). An important characteristic of the starting point is the closeness to the optimal solution. Figure 4 shows that the value of the duality gap μ at the starting point is several orders of magnitude lower for the STPC2, which allows the IPM to converge several iterations faster than for the infeasible IPM.

Algorithm 3: STPC2

---

1 Choose $\epsilon > 0$, it_m;

2 Compute $Q = A^T K^{-1} A$ and $p = A^T K^{-1} f - g$;

3 Initialize $\lambda_0^i = \begin{cases} a_i^T f / 2 & \text{if } a_i^T f > 0 \\ 1 & \text{otherwise} \end{cases}$, $i = 1, 2.., m$, $it = 0$;

4 **while** $it < it_m$ **do**

5   **for** $i \leftarrow 1$ **to** m **do**

6   Compute $y_0^i = q_i^T \lambda_0 - p^i$;

7   **if** $y_0^i \leq 0$ **then**

8   Compute $\lambda_0^i = -\dfrac{\sum_{j=1, j \neq i}^{m} q_{ij} \lambda_0^j - p^i + \epsilon}{q_{ii}}$;

9   **end**

10   **end**

11   **if** $y_0^i > 0$ $\forall i$ **then**

12   break;

13   **end**

14     $it = it + 1$;

15 **end**

16 Compute $x_0 = K^{-1}(f - A\lambda_0)$;

    Result: $w_0 = (x_0, y_0, \lambda_0)$

Table 1. Number of IPM iterations for different starting point approaches

	Simple assembly type						Sandwich assembly type					
	$n = 4386, m = 2193$						$n = 4268, m = 2716$					
n_{act}	STPC1		STPC2		Infeasible		n_{act}	STPC1		STPC2		Infeasible	
	Iter.	Time,s.	Iter.	Time,s.	Iter.	Time,s.		Iter.	Time,s.	Iter.	Time,s.	Iter.	Time,s.
4	18	27.6	17	23.1	17	24.1	74	44	217.2	31	189.1	31	197.6
374	14	22.8	14	20.7	21	29.2	347	41	434.3	32	480.2	29	400.3
686	14	30.9	15	26.5	23	36.6	1058	37	402.4	28	456.2	30	376.9
1833	14	16.4	14	15.5	19	31.3	1303	38	374.9	27	411.4	29	350.6
2193	13	6.3	13	5.2	18	14.2	2474	36	175	25	119.5	28	124.2

Table 1 compares various starting points by the number of IPM iterations and the computation time in seconds. The time required to search for the initial point is also taken into account. It can be seen that the modified STPC2 algorithm provides an improvement in the number of iterations in almost all test problems. However, considerable time is wasted on the first iterations for the sandwich assembly type of problems when solving a linear system of equations. A large spread of the components at the starting point worsens the condition number of the system, which affects the iterative system solver and the time for solving the problem. For a model with a simple assembly type, the computation time is 7–10% less.

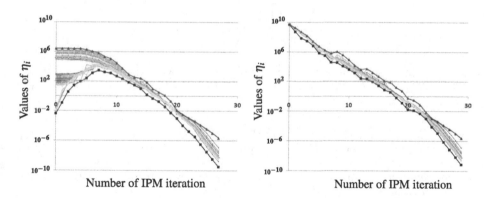

Fig. 4. Change in values $\eta_i = y_i \lambda_i$ with the iteration number of the IPM for different starting points. Feasible STPC2 is left, infeasible STP2 [1] is right. Different lines on the figure correspond to different values of $i = 1, 2, .., m$. The results refer to a lower panel assembly problem with $n = 4268$, $m = 2716$.

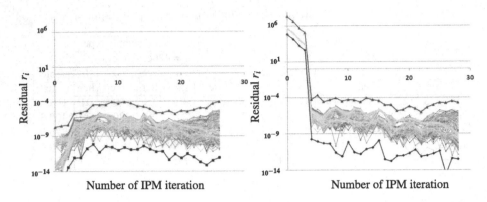

Fig. 5. Change in residual $r_i = \max_i\{|s_1^i|, |s_2^i|\}$, s_1 is residual of (3) and s_2 is residual of (4), with the iteration number of the IPM for different starting points. Feasible STPC2 is left, infeasible STP2 [1] is right. Different lines on the figure correspond to different values of $i = 1, 2, .., m$. The results refer to a lower panel assembly problem with $n = 4268$, $m = 2716$.

6   Solving a Linear System of Equations

The most time consuming IPM step is the solving of the linear system of equations for searching Newton direction:

$$\begin{bmatrix} K & 0 & A \\ A^T & I & 0 \\ 0 & \Lambda & Y \end{bmatrix} \begin{bmatrix} \Delta x \\ \Delta y \\ \Delta \lambda \end{bmatrix} = \begin{bmatrix} s_1 \\ s_2 \\ s_3 \end{bmatrix}, \tag{7}$$

where I is an identity matrix, $s_1 = -(Kx - f + A\lambda)$, $s_2 = -(A^Tx - g + y)$, $s_3 = -(\Lambda Y e - \sigma\mu e)$. The system (7) may be transformed into a symmetric augmented system by eliminating Δy:

$$\begin{bmatrix} K & A \\ A^T & -\Lambda^{-1}Y \end{bmatrix} \begin{bmatrix} \Delta x \\ \Delta \lambda \end{bmatrix} = \begin{bmatrix} s_1 \\ s_2 - \Lambda^{-1}s_3 \end{bmatrix}. \tag{8}$$

Further reduction of augmented system (8) gives two variants of normal equations with respect to $\Delta\lambda$ and Δx:

$$\left[A^T K^{-1} A + \Lambda^{-1}Y \right] [\Delta\lambda] = [p_1], \tag{9}$$

$$\left[K + AY^{-1}\Lambda A^T \right] [\Delta x] = [p_2], \tag{10}$$

where $p_1 = A^T K^{-1} s_1 + \Lambda^{-1}s_3 - s_2$ and $p_2 = s_1 + AY^{-1}\Lambda s_2 - AY^{-1}s_3$. The advantages of solving the augmented system are the symmetry of the system and keeping the original structure of the matrices K and A. The latter is especially valuable when solving large-scale sparse problems. Note, that for an assembly problem matrix A is sparse, however, Hessian K has a block-diagonal structure with fully populated blocks. The number of blocks corresponds to the number of assembled parts, which can vary from 2 to 8. The size of each block is determined by the number of nodes of the computational mesh for each specific part. Oppositely to the augmented system, the normal equation is more likely to have a dense structure. Both (9) and (10) are symmetric positive definite and the size of the system is reduced from $n + m$ to m and n, correspondingly.

The systems of linear Eqs. (8), (9) and (10) are ill-conditioned. On the one hand, this unpleasant feature comes from the ill-conditioned Hessian matrix, specifically $cond(K)$ is of order $10^6 - 10^8$. On the other hand, some of the diagonal terms Y and Λ become close to zero when approaching the optimum, which worsens the conditioning of the system at the last iterations of the IPM. Thus, when using iterative methods to solve a linear system of equations, the preconditioner is required. The system (8) is solved with the preconditioner P_{aug} proposed in [3]:

$$P_{aug} = \begin{bmatrix} D & A \\ A^T & -\Lambda^{-1}Y \end{bmatrix} = $$
$$= \begin{bmatrix} I & 0 \\ A^T D^{-1} & I \end{bmatrix} \begin{bmatrix} D & 0 \\ 0 & -A^T D^{-1}A - \Lambda^{-1}Y \end{bmatrix} \begin{bmatrix} I & D^{-1}A \\ 0 & I \end{bmatrix}, \qquad (11)$$

where $D = diag(K)$. Note, that matrix $-A^T D^{-1}A - \Lambda^{-1}Y$ is diagonal for the problems with two assembled parts. Both Eqs. (9) and (10) are solved with the help of the Jacobi preconditioner $P_J = diag(H)$, where H is a matrix of the corresponding system of Eqs. (9) or (10), and with an incomplete Cholesky preconditioner $P_{ichol} = L_H L_H^T$. The incomplete Cholesky factorisation is applied to the sparsity pattern H_s, that includes only the largest elements of matrix H and all other elements are replaced by zeros. The sparsity pattern H_s contains 50% elements for the Eq. (9), since it is dense, and all non-zeros for the Eq. (10). The regularization parameter is used if the incomplete Cholesky factorization fails [9].

Consider two more approaches for the Eq. (9), that has the smallest size m and differs at neighboring iterations only by diagonal term $\Lambda^{-1}Y$. The first, proposed in [21] for an equation of this type, uses combined preconditioner $P_C = (L_Q + \sqrt{\Lambda^{-1}Y})(L_Q + \sqrt{\Lambda^{-1}Y})^T$, where L_Q is the Cholesky decomposition of matrix $Q = A^T K^{-1}A$. The Cholesky factorisation L_Q is performed once for a group of similar contact problems, i.e. problems with the same matrices A and K. The second, is direct solving of Eq. (9) using Cholesky factorisation.

The computational time required to solve the assembly problem is summarized in Table 2. The table presents the total time ('T') and the average time required for one IPM iteration ('Avr.'). As seen from the table, solving the Eq. (9) is faster than (10) or (8). The iterative approach is generally faster than the direct one for (9). Among the iterative approaches, the conjugate gradient method with preconditioner P_C gives the best computation time.

Table 2. Computation time (T) in seconds of IPM with different approaches for solving a linear system of equations. 'Avr.' is average computation time for one iteration. The results refer to a upper panel assembly problem with $n = 4386$, $m = 2193$ and $cond(K) = 2.63 \cdot 10^6$.

n_{act}	Augmented system (8)		Normal equation (9)								Normal equation (10)			
			Direct		Iterative P_J		Iterative P_{ichol}		Iterative P_C		Iterative P_J		Iterative P_{ichol}	
	T	Avr.	T	Avr.	T	Avr.	T	Avr.	T	Avr.	T	Avr.	T	Avr.
20	1681	112	663	35	55	2.9	217	11	48	2.5	2372	119	1873	94
378	2363	113	736	33	126	5.7	444	20	59	2.7	3761	145	2627	101
695	2678	112	771	34	210	9.1	489	21	68	3	4542	162	3183	110
1830	1927	84	502	34	279	19	597	40	43	2.8	2542	98	2942	113
2193	1809	79	502	33	337	22	532	35	14	0.93	2435	87	3266	117

7 IPM Vs Other Type of Methods for Solving Assembly Problems

In the current section we compare the developed IPM approach with the Newton projection method (NPM) [4], Lemke's method (LM) [17], and an active set method, namely Goldfarb-Idnani algorithm (GIP) [7,19]. All listed algorithms are adapted to the features of the considered type of problems [2,18,20].

The fact that the matrices K and A are usually the same for the series of computations during aircraft assembly simulation, allows to use more attractive formulations of the original problem (1) even though the preparation of data may take additional time. Consider two different reformulations of problem (1), namely the dual problem and the relative problem. The dual formulation of problem (1) is given by

$$\max \quad -\tfrac{1}{2}\lambda^T Q\lambda + p^T\lambda + s,$$
$$\text{s.t.} \quad \lambda \geq 0, \tag{12}$$

where $\lambda \in R^m$ is the vector of Lagrange multipliers corresponding to contact forces in the area of possible contact, $Q = A^T K^{-1}A \in R^{m\times m}$ is a symmetric positive-definite fully populated matrix, $p = A^T K^{-1}f - g \in R^m$ and $s = -\tfrac{1}{2}f^T K^{-1}f \in R^1$. The relative formulation of problem (1) [20] is a variant of primal formulation based on a vector of relative displacements defined as $u = A^T x \in R^m$:

$$\min \quad \tfrac{1}{2}u^T \tilde{K}u - \tilde{f}^T u,$$
$$\text{s.t.} \quad u - g \leq 0, \tag{13}$$

where $\tilde{K} = (A^T K^{-1}A)^{-1} \in \mathbb{R}^{m\times m}$ is a symmetric positive-definite fully populated matrix and $\tilde{f} = \tilde{K}A^T K^{-1}f \in \mathbb{R}^m$. Lemke's method (LM) is used to solve the linear complementarity problem that for given vector $q \in \mathbb{R}^N$ and matrix $M \in \mathbb{R}^{N\times N}$, consists in searching of vectors $w, z \in \mathbb{R}^N$ such that

$$w = q + Mz,$$
$$z^T w = 0, \tag{14}$$
$$w \geq 0, z \geq 0$$

or in concluding that no such pair of vectors w, z exists. The linear complementarity problem, corresponding to the problem (1) is given by

$$
\begin{aligned}
M &= A^T K^{-1} A, \\
q &= -A^T K^{-1} f + g, \\
w &= y, \\
z &= \lambda.
\end{aligned}
\tag{15}
$$

Note, that the primal problem (1), the dual problem (12) and the linear complementarity formulation lead to the same system of linear equations for IPM, i.e. the normal Eq. (9).

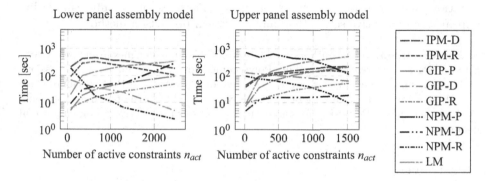

Fig. 6. Comparison of optimization methods for aircraft assembly problems. The lower panel assembly model has $n = 4268, m = 2716, cond(K) = 1.28 \cdot 10^8$ and the upper panel assembly model has $n = 7308, m = 3654, cond(K) = 1.24 \cdot 10^6$. The postfix '-P' stands for the primal problem (1), '-D' stands for the dual problem (12), '-R' stands for the relative problem (13).

Figure 6 presents the computation time for all discussed types of methods applied to different formulations of the contact problem. For the considered examples of the assembly problems, IPM loses by the computation time to Goldfarb-Idnani algorithm and Newton projection method. This happens mainly because of ill-conditioned stiffness matrix K that is typical for assembly problems. However, unlike other methods, IPM is less affected by the size of the problem. Indeed, the computation time is even less for bigger upper panel assembly problem. As seen from Fig. 6 the use of alternative formulations, i.e. dual and relative, significantly reduces the computation time.

8   Conclusion

The adaptation of an interior point method for the assembly of deformable structures, e.g. aircraft assembly, is presented. The adapted algorithm is based on the proposed search method of the initial point and on solving the normal equation with a combined preconditioner, proposed earlier [21]. Using the physical side of the optimization problem helps to construct a starting point that is closer to the optimum and reduces the computational costs. Namely, some operations needed to build the preconditioner and to find the initial point may be done once at the stage of an assembly model preparation, i.e. computation of $A^T K^{-1}$ and Cholesky decomposition of $A^T K^{-1} A$. Necessity to solve a large number of problems, based on the same model with the same matrices K and A, allows the use of the relative formulation that further helps to reduce computation time.

Comparative analysis of different types of methods for solving assembly problems shows that the IPM is not the best one for such an application. The ill-conditioned Hessian of the contact problems worsens the properties of a linear system of equations need to be solved at each iteration, thus affecting the computation time in general. The more appropriate methods for assembly simulation appear to be an active set method and a Newton projection method.

Acknowledgments. The authors would like to thank our colleagues from both Airbus SAS and Peter the Great St.Petersburg Polytechnic University for numerous discussions and helpful suggestions. We also want to express our gratitude to Prof. J. Gondzio for his valuable advices that have helped us improve the quality of the research. We thank O. Tarunina and T. Pogarskay for careful reading of our manuscript and giving useful comments.

References

1. D'Apuzzo, M., De Simone, V., di Serafino, D.: Starting-point strategies for an infeasible potential reduction method. Optim. Lett. **4**(1), 131–146 (2010)
2. Baklanov, S., Stefanova, M., Lupuleac, S.: Newton projection method as applied to assembly simulation. Optim. Methods Softw. 1–28 (2020). https://doi.org/10.1080/10556788.2020.1818079
3. Bergamaschi, L., Gondzio, J., Zilli, G.: Preconditioning indefinite systems in interior point methods for optimization. Comput. Optim. Appl. **28**, 149–171 (2004)
4. Bertsekas, D.P.: Projected newton methods for optimization problems with simple constraints. Siam J. Control Optim. **20**(2), 221–246 (1982)
5. Domahidi, A., Zgraggen, A.U., Zeilinger, M.N., Morari, M., Jones, C.N.: Efficient interior point methods for multistage problems arising in receding horizon control. In: IEEE 51st IEEE Conference on Decision and Control (CDC), 668–674. IEEE, Maui (2012)
6. Fountoulakis, K., Gondzio, J., Zhlobich, P.: Matrix-free interior point method for compressed sensing problems. Math. Program. Comput. **6**(1), 1–31 (2013). https://doi.org/10.1007/s12532-013-0063-6
7. Goldfarb, D., Idnani, A.: A numerically stable dual method for solving strictly quadratic programs. Math. Program. **27**(1), 1–33 (1983)
8. Gondzio, J.: Interior point methods 25 years later. Eur. J. Oper. Res. **218**(3), 587–601 (2012)
9. Lin, C., Saigal, R.: An incomplete cholesky factorization for dense symmetric positive definite matrices. BIT Numer. Math. **40**, 536–558 (2000). https://doi.org/10.1023/A:1022323931043

10. Lupuleac, S., et al.: Software complex for simulation of riveting process: concept and applications. In: SAE Technical Paper, 2016-01-2090 (2016)
11. Lupuleac, S., et al.: Simulation of the wing-to-fuselage assembly process. J. Manuf. Sci. Eng. **141**(6), 061009 (2019)
12. Lupuleac, S., Smirnov, A., Churilova, M., Shinder, J., Bonhomme, E.: Simulation of body force impact on the assembly process of aircraft parts. In: Proceedings of the ASME 2019 International Mechanical Engineering Congress and Exposition. 2B: Advanced Manufacturing, V02BT02A057. ASME, Salt Lake City, Utah, USA (2019)
13. Mangoni, D., Tasora, A., Garziera, R.: A primal-dual predictor-corrector interior point method for non-smooth contact dynamics. Comput. Methods Appl. Mech. Eng. **330**, 351-367 (2018)
14. Mazorche, S.R., Herskovits, J., Canelas, A., Guerra, G.M.: Solution of contact problems in linear elasticity using a feasible interior point algorithm for nonlinear complementarity problems. In: Proceedings of the seventh world congress on structural and multidisciplinary optimization, Seoul, South Korea (2007)
15. Mehrotra, S.: On the implementation of a primal-dual interior point method. SIAM J. Optim. **2**(4), 575-601 (1992)
16. Monteiro, R.D.C., Adler, I.: Interior path following primal-dual algorithms. Part II: convex quadratic programming. Math. Program. **44**(1–3), 43-66 (1989)
17. Murty, K.G.: Linear Complementarity, Linear and Nonlinear Programming. Helderman-Verlag, Berlin (1988)
18. Petukhova, M.V., Lupuleac, S.V., Shinder, Y.K., Smirnov, A.B., Yakunin, S.A., Bretagnol, B.: Numerical approach for airframe assembly simulation. J. Math. Ind. **4**(1), 1–12 (2014). https://doi.org/10.1186/2190-5983-4-8
19. Powell, M.J.D.: On the quadratic programming algorithm of Goldfarb and Idnani. In: Cottle, R.W. (ed.) Mathematical Programming Essays in Honor of George B Dantzig Part II. Mathematical Programming Studies, vol. 25. Springer, Berlin (1985)
20. Stefanova, M., et al.: Convex optimization techniques in compliant assembly simulation. Optim. Eng. **21**(4), 1665-1690 (2020). https://doi.org/10.1007/s11081-020-09493-z
21. Stefanova, M., Yakunin, S., Petukhova, M., Lupuleac, S., Kokkolaras, M.: An interior-point method-based solver for simulation of aircraft parts riveting. Eng. Optim. **50**(5), 781-796 (2018)
22. Tanoh, G., Renard, Y., Noll, D.: Computational experience with an interior point algorithm for large scale contact problems. Optim. Online **10**(2), 1–18 (2004). http://www.optimization-online.org/DB_HTML/2004/12/1012.html
23. Voytov, O., Zorkaltsev, V., Filatov, A.: Oblique path algorithms for solving linear programming problems. Discrete Anal. Oper. Res. **2**(2), 17–26 (2001)
24. Wright, S.J.: Primal-Dual Interior Point Method. SIAM, Philadelphia (1997)

On Optimizing Electricity Markets Performance

Alexander Vasin[iD] and Olesya Grigoryeva[(✉)][iD]

Lomonosov Moscow State University, Moscow 119991, Russia
olesyagrigez@gmail.com

Abstract. Electric energy generation is an important sector of Russian economy. Its optimizing is an important measure for increasing the social welfare and the GDP. This study aims to propose a mathematical model for optimizing a wholesale electricity market's performance by means of modern economic and technical tools: consumption tariffs with multiple rates that aim to shift daily consumption from the pike times, renewable energy generators, energy storages.

Keywords: Electricity market · Optimal tariff rates · Welfare theorem

1 Introduction

Electricity is one of the most important sectors of the Russian economy. Its optimization is an important task in the context of accelerating GDP growth. The solution to this problem is associated with investments in the generation and expansion of networks, as well as with the use of new economic and technical tools to optimize the production and consumption of electricity. These tools include:

- tariff regulation aimed at transferring part of consumption from the peak zone of the schedule to off-peak time of the day,
- use of small generation and renewable energy sources,
- use of electric capacity storage devices.

The density and unevenness of the curves of the daily load of consumers strongly affect the economic performance of energy systems. Aligning the curves reduces the demand for generating capacity, transmission and production costs. This change, however, may cause inconvenience to consumers (and their neighbors).

To replace more expensive energy sources, the capacities of producers using renewable energy sources can be used. But the volume of power they supply is a random variable, depending on weather conditions. In a situation where it is

Supported by RFBR No. 19-01-00533 A.

N. Olenev et al. (Eds.): OPTIMA 2020, LNCS 12422, pp. 272–286, 2020.
https://doi.org/10.1007/978-3-030-62867-3_20

necessary to guarantee the supply of energy to all consumers under the concluded agreements, under adverse conditions it should be replaced by reserve capacities with conventional technologies.

Energy storage devices, in particular lithium-ion batteries, can play a useful role in energy systems. On the part of manufacturers, storing energy can reduce the cost of its production, since energy generation is cheaper for some periods of time. On the part of consumers, storage systems solve the problem of smoothing loads; thanks to them, it becomes possible to transfer some part of the consumption from the peak zones of the schedule to off-peak times of the day.

Well-known models of the electricity market (see [3]) have been developed in a number of scientific papers, taking into account the mentioned new factors.

In [1], inelastic demand from consumers is considered, which includes the hourly components of the required volume, as well as the shiftable load, which can be redistributed during the day, taking into account the cost of transferring from the most favorable time at less convenient. Using the theory of contracts, the authors study the problem of optimizing the operation of the energy system by introducing tariffs that encourage consumers to shift the shiftable load at off-peak times.

The article [6] discusses the problem of creating an optimal generation schedule in terms of minimizing costs and emissions. The optimal planning schedule is based on the use of different electricity prices for consumers in the day-ahead market, as well as energy storage systems to achieve optimal volumes of production and consumption. The results show that shiftable loads should be moved at off-peak periods, in particular at midnight. It is also noted that the presence of a large number of available blocks with low generating power is preferable to the presence of one block with a large output power. While cost minimization and emission minimization are conflicting goals, a solution can be found that optimizes the vector criterion. Paper [13] considers a similar problem of minimizing costs and emission in framework of a stochastic model and employs the probabilistic concept of confidence interval in order to evaluate forecasting uncertainty.

Particular problems of optimizing energy systems of various scales taking these tools into account were also considered in [7–9,12].

In this paper, we develop a mathematical model of the optimal functioning of the wholesale electricity market in terms of increasing public welfare. Section 2 discusses the task of optimizing the production and consumption of electricity for the point market, taking into account shiftable loads and renewable sources. Section 3 discusses economic mechanisms that can ensure its optimal functioning. Section 4 generalizes the results for a power storage system. In conclusion, the main results and some tasks for the future are formulated.

2   Spot Market Model and Formulation of Optimization Problem

Since the demand for electricity varies significantly during the day, we consider the functioning of the system depending on time $t \in \{1, T\}$, where t is a period of time with approximately constant needs, T is the length of the planning interval. In particular, an hour of the day may be denoted by t. In a similar model, one can also take into account the dependence of consumption on the season. In this case, t reflects both the time of day and the season of year.

Consider the main groups of agents (producers and consumers) operating in the market. Let A_1 denote the set of electricity producers with traditional technologies. Each of them is characterized by the cost function $C_a(v)$ for providing the supplied capacity v. The supply function $S_a(p) = \underset{v \geq 0}{Arg\max}(vp - c_a(v))$ determines the optimal volume of production for a given period depending on the price. In a particular case, producer a has several generators characterized by constant marginal costs $c_a^1 < c_a^2 ... < c_a^m$ and capacities $V_a^1, .., V_a^m$. Then the supply function $S_a(p)$ is determined as Fig. 1 shows.

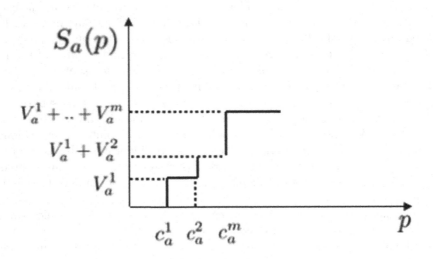

Fig. 1. The supply function $S_a(p)$

Note: when considering a large planning interval, the price of raw materials (gas, fuel oil, etc.) may depend on t, in which case the supply function S_a^t will change accordingly.

Denote A_2 as the set of producers using renewable energy sources. For each of them, the amount of supplied power is a random variable depending on weather conditions and time. It is determined by the function $v_a^t(\psi^t)$, where ψ^t is a random factor value at this time t. In a situation where it is necessary to guarantee the supply of energy to all consumers under the concluded agreements, the capacities of these producers can be used to replace more expensive energy sources, but under adverse conditions they must be replaced by reserve capacities with conventional technologies. The variable costs for solar panels and wind turbines are close to zero and are not considered further.

Consider electricity consumption. Denote B_1 as the set of consumers representing the population, B_2—the set of industrial consumers. Each consumer $b \in B_1$ has needs associated with a specific time t (for example, space heating and lighting), as well as several types of needs $l = 1, .., L$, which can be implemented in different time (cooking, washing, etc.). In the general case, the utility functions of consumers can be described as follows. Consumption for individual $b \in B_1$ is characterized by vector $\overrightarrow{v}_b = (v_{bl}^t, t = 1, ..T, l = 0, .., L)$, where the volume v_{b0}^t is the consumption associated with the needs for a given hour t. The functions $u_{b0}^t(v_{b0}^t, \psi_b^t), t = 1, .., T$, show the utility of such consumption depending on the random factor ψ_b^t characterizing the weather conditions and other random events affecting the need for electricity. The randomness in the volume of consumption is also due to the fact that many consumers have their own sources of generation (for example, solar panels), and the volume of production on them is random. The volume $v_{bl}^t \geq 0$ determines the energy consumption associated with the target l in the period t. For each type of l, the utility of consumption depends on the total amount of allocated to the corresponding target l, taking into account the costs e_{bl}^t for transferring to a less convenient time.

Thus, the total utility of the consumer is as follows:

$$U_b(\overrightarrow{v}_b, \overrightarrow{\psi}_b) = \sum_{t=1}^{T} u_{b0}^t(v_{b0}^t, \psi_b^t) + \sum_{l=1}^{L}(u_{bl}(\sum_{t=1}^{T}(v_{bl}^t)) - \sum_{t=1}^{T} v_{bl}^t e_{bl}^t).$$

In the whole paper below, the cost and the utility functions meet standard assumptions for micro-economic models (see [4]): they are monotonously increasing, cost functions are convex, utility functions are concave.

The demand function $\overrightarrow{D}_b(\overrightarrow{p}, \overrightarrow{\psi}_b)$, where $\overrightarrow{D}_b = (D_{bl}^t, t = 1, .., T, l = 0, .., L)$, shows the optimal volume consumer b purchases at each time period t depending on the price vector $\overrightarrow{p} = (p^t, t = 1, .., T)$ and random factor values. The vector $\overrightarrow{D}_b$ is the solution to the following optimization problem:

$$\overrightarrow{D}_b(\overrightarrow{p}, \overrightarrow{\psi}_b) = Arg \max_{\overrightarrow{v}_b}[U_b(\overrightarrow{v}_b, \overrightarrow{\psi}_b) - \sum_{t=1}^{T} p^t \sum_{l=0}^{L} v_{bl}^t]. \tag{1}$$

For smooth utility functions, first-order conditions for the problem (1) are as follows:

$$\frac{\partial u_{b0}^t}{\partial v_{b0}^t}(v_{b0}^t, \psi_b^t) = p^t, v_{bl}^{\bar{t}} > 0 \Rightarrow \bar{t} \to \min_t(p^t + e_{bl}^t) = u_{bl}'(v_{bl}), l = 1, .., L.$$

Consider the behavior of industrial consumers $b \in B_2$. Company b produces commodity with a price of p_b and is characterized by fixed energy costs D_{b0}^t (associated with management, heating, etc.) and the production function $F_b^t(v_b^t, w_b^t)$, which determines the output in a given period, depending on the consumed amount of energy v_b^t and labor w_b^t. For example, consider an enterprise with a given set of production capacities M_b. Each capacity $m \in M_b$ is characterized by the maximum output v_b^m and constant unit costs of energy ν_b^m and labor ω_b^m per unit of the manufactured product. (For industrial consumers, the parameters are usually independent of t, and for agricultural production the dependence on t is significant.)

The optimal employed capacity during the period t in this case is easily determined by the following algorithm. Available capacities are ordered by increasing total unit costs $\nu_b^m p_e^t + \omega_b^m p_w^t$, where p_e^t, p_w^t are electricity and labor costs per hour t (prices change significantly during the day). Each capacity is employed while the price of the manufactured commodity exceeds the unit costs: $p_b \geq \nu_b^m p_e^t + \omega_b^m p_w^t$.

Thus, the demand function for the company b during period t is as follows:

$$D_b^t(p_e^t, p_w^t) = \sum_{m:\nu_b^m p_e^t + \omega_b^m p_w^t \leq p_b} v_b^m \cdot \nu_b^m + D_{b0}^t. \tag{2}$$

In the general case, the problem of determining the demand function can be non-convex. In particular, for some generators, the fixed costs of their launch play an important role.

The problem of maximizing the social welfare for the market can be formulated as the optimal balanced choice of the generation and the consumption during the planning interval $t \in \{1, .., T\}$. The optimized criterion is a total utility of the consumers minus generation costs. Each feasible strategy is characterized by vector of volumes of generated power $\{v_a^t(\psi^t), a \in A_1 \cup A_2, t = 1, .., T\}$ and vectors of consumption for the population $\{v_b^t(\psi_b^t), b \in B_1, t = 1, .., T\}$ and for industrial consumers $\{v_b^t, b \in B_2, t = 1, .., T\}$, satisfying balance conditions: $\forall t$,

$$\psi^t \in \Psi_t, \ (\psi_b^t \in \Psi_b^t; \ b \in B_1), \ \sum_{a \in A_1 \cup A_2} v_a^t(\psi^t) = \sum_{b \subset B_1}\left(v_{b0}^t(\psi_b^t) + \sum_{l=1}^{L} v_{bl}^t\right) + \sum_{b \subset B_2} v_b^t.$$

The optimized social welfare function is as follows:

$$\sum_{b \in B_1}\left(\sum_{t=1}^{T} u_{b0}^t(v_{b0}^t, \psi_b^t) + \sum_{l=1}^{L}\left(u_{bl}(\sum_{t=1}^{T} v_{bl}^t) - \sum_{t=1}^{T} v_{bl}^t e_{bl}^t\right)\right) +$$
$$+ \sum_{b \in B_2}\sum_{t=1}^{T}(F_b^t(v_b^t, w_b^t)p_b^t - p_w^t w_b^t) - \sum_{a \in A_1}\sum_{t=1}^{T} c_a(v_a^t(\psi^t)), \tag{3}$$

variable costs for renewable sources are not taken into account, since they are close to zero.

When formulating the optimization problem, the following should be taken into account. Firstly, deliveries to consumers are usually planned in advance, before the value of the random factor $\psi(t)$ becomes known. Similarly, production volumes are planned for generators from the set $a \in A_1 \backslash A_0$, except for the backup generators $a \in A_0$. In practice, these values are determined in the day-ahead market. For backup generators, load volumes are determined in the balancing market.

How to optimally distribute generators between the balancing market and the day-ahead market? In order to minimize the expected costs of the market, generators with minimal unit costs are selected for the latter, and generators with high costs are left for the balancing market and employed depending on the random factor value. The technical characteristics of real generators usually permit such a distribution.

Next, consider several particular formulations of the welfare optimization problem in order to study in more detail some important issues of calculating optimal strategies.

From the definition of supply functions S_a, $a \in A$, in the case when their values are unique, the following result is obtained.

Lemma 1. Let the required total generation volume be V_A. Then its optimal distribution

$$\overrightarrow{v}_A^* \to \min_{\overrightarrow{v}} \sum_{a \in A} C_a(v_a)$$

under condition $\sum_a v_a = V_A$ is determined as $v_a^* - S_a(p(V_A))$, $a \in A$, where the value $p(V_A)$ proceeds from the condition: $\sum_a S_a(p(V_A)) = V_A$

Based on Lemma 1, determine the cost function for the set of companies A_1: $C(v) = \sum_{a \in A_1} C_a(S_a(p(v)))$. Similarly, establish the relationship between the aggregate utility function of consumption and demand functions.

Lemma 2. Let total consumption V_B need to be distributed among consumers $b \in B$ with utility functions $u_b(v_b)$ so as to maximize the total utility. Then the optimal distribution

$$\overrightarrow{v}_B^* \to \max_{\overrightarrow{v}} \sum_{b \in B} u_b(v_b)$$

under condition $\sum_b v_b = V_B$ is calculated as $v_b^* - D_b(\overrightarrow{p}(V_B))$, $b \in B$, where the value $p(V_B)$ proceeds from the condition: $\sum_b S_b(p(V_B)) = V_B$, and the demand functions are related to the utility functions by the equations

$$u_b(v_b) = \int_0^{v_b} D_b^{-1}(\overrightarrow{v})d\overrightarrow{v}.$$

Based on Lemma 2, determine the aggregate utility functions of the current population consumption

$$u_{B_1}^t(v, \overrightarrow{\psi}^t) = \sum_{b \in B_1} u_{b_0}^t(D_b^t(\overrightarrow{p}^t(v, \overrightarrow{\psi}^t), \psi_b^t), \psi_b^t),$$

where
$$\forall t, b \int_0^{\bar{v}} (D_b^t)^{-1}(v, \psi_b^t) dv = u_{b_0}^t(v, \psi_b^t),$$

and industrial consumption
$$u_{B_2}^t(v) = \sum_{b \in B_2} F_b^t(D_b^t(p^t(v), p_\omega^t)),$$

where the demand functions D_b^t, $b \in B$, are given by (2). Below we omit the dependence on p_ω^t, assuming these prices be fixed.

Finally, define the general utility function of current consumption
$$u_0^t(v, \overrightarrow{\psi}^t) = u_{B_1}^t(D_{B_1}^t(p^t(v, \overrightarrow{\psi}^t), \overrightarrow{\psi}^t), \overrightarrow{\psi}^t) + u_{B_2}^t(D_{B_2}^t(p^t(v, \overrightarrow{\psi}^t))),$$

where price $p^t(v, \overrightarrow{\psi}^t)$ is determined from the balance condition:
$$D_{B_1}^t(p^t(v, \overrightarrow{\psi}^t), \overrightarrow{\psi}^t) + D_{B_2}^t(p^t(v, \overrightarrow{\psi}^t)) = v,$$

where
$$D_{B_1}^t(p^t(v, \overrightarrow{\psi}^t), \overrightarrow{\psi}^t) = \sum_{b \in B_1} D_b^t(p^t, \psi_b^t), \quad D_{B_2}^t(p^t(v, \overrightarrow{\psi}^t)) = \sum_{b \in B_2} D_b^t(p^t).$$

2.1 Deterministic Model

We start with the simplest setting when supply and demand do not include stochastic components. Let values $v_{A_2}^t$, $t = 1, .., T$ be fixed. Below every couple $\bar{b} = (b, l)$ is considered as a separate consumer with utility function
$$\bar{u}_{\bar{b}}(\overrightarrow{v}_{\bar{b}}) = u_{\bar{b}}(\sum_t v_{\bar{b}}^t) - \sum_t e_{\bar{b}}^t v_{\bar{b}}^t, \quad \bar{b} \in BL.$$

The problem of optimizing the welfare function (3) is determined as follows:
$$\sum_t (u_0^t(v_0^t) + \sum_{\bar{b} \in BL} [u_{\bar{b}}(\sum_t(v_{\bar{b}}^t)) - \sum_t e_{\bar{b}}^t v_{\bar{b}}^t] - \tag{4}$$
$$- \sum_t c(\sum_{\bar{b} \in BL \cup \{0\}} v_{\bar{b}}^t - v_{A_2}^t)) \to \max_{\overrightarrow{v} = (\overrightarrow{v}_0, \overrightarrow{v}_{\bar{b}}, \bar{b} \in BL) \geq 0}.$$

With this optimization problem, we can match a competitive market with many products $\Theta = \{1, .., T\}$, where the commodity t corresponds to the energy consumed in the period t. Each product is produced independently of the others, the cost function is determined as $c^t(v^t) = c(v^t - v_{A_2}^t)$, where $c()$ is the minimum cost function above for the set of companies A_1. The set of consumers is $\bar{B} = BO \cup BL$, where $BO = \bigcup_{t=1}^T BO_t$, the set BO_t is characterized by demand function $D_0^t(p^t) = D_{B_1}^t(p^t) + D_{B_2}^t(p^t)$, corresponding to utility function u_0^t. For these consumers, only the commodity t is useful. Every consumer $\bar{b} \in BL$ corresponds to shiftable load (b, l) and is characterized by demand function $D_{\bar{b}}(\overrightarrow{p})$, corresponding to utility function $\overline{u}_{\bar{b}}(\overrightarrow{v}_{\bar{b}})$. The following relations hold:

$$(v_{\bar{b}}^t > 0) \Rightarrow p^t + e_{\bar{b}}^t = \arg\min_{\tau \in \Theta}(p^\tau + e_{\bar{b}}^\tau) = u_{\bar{b}}'(\sum_{\tau \in \Theta} v_{\bar{b}}^\tau) \tag{5}$$

$$\forall t \in \Theta, \bar{b} \in BL, \quad \overrightarrow{v}_{\bar{b}} \in D_{\bar{b}}(\overrightarrow{p}).$$

That is, every consumer $\bar{b}$ chooses a consumption vector that maximizes his utility, taking into account the cost of purchasing energy.

The competitive equilibrium of a given market is a collection $(\overrightarrow{p}, \overrightarrow{v}_0, \overrightarrow{v}_{\bar{b}}, \bar{b} \in BL)$, including equilibrium price vector $\overrightarrow{p} = (\tilde{p}^t, t = 1, .., T)$, consumption volume vectors $\overrightarrow{v}_0 = (\tilde{b}_0^t, t = 1, .., T)$ for consumers BO and $\overrightarrow{v}_{\bar{b}} = (\tilde{v}_{\bar{b}}^t, t = 1, .., T)$ for each $\bar{b} \in BL$.

In addition to the ratio (5), these vectors must also satisfy the following conditions:

$$\forall t \in \Theta \quad \tilde{v}_0^t = D_0^t(p^t), \text{ or } u_0^{t\,'}(\tilde{v}_0^t) = p^t, \tag{6}$$

that is, the volume of current consumption in the period t is determined from the equity of the marginal utility of consumption and the energy price;

$$\forall t \in \Theta \quad c'(\sum_{\bar{b} \in B} \tilde{v}_{\bar{b}}^t - v_{A_2}^t) = \tilde{p}^t, \tag{7}$$

that is, in each period the marginal cost of production is equal to the price of energy.

Theorem 1. *The problem (4) of the social welfare optimizing is a convex programming problem. The set of its solutions corresponds to the set of competitive equilibria of the above market, given by the relations (5–7).*

Proof. The set of strategies in the problem (4) is convex and closed. Given that $c'(v) \to \infty$ under $v \to \infty$, the area where the objective function is positive is limited. The properties of functions $c(v)$, $u_0^t(v)$, $u_{\bar{b}}(v)$ imply concavity of the objective function. Therefore (4) is a convex programming problem, and its solution is determined from the first-order conditions, which are written in the form (5–7) and establish the fit of the solution to the competitive equilibrium. For a general proof of the Welfare Theorem, see [2].

3   Choosing the Optimal Tariffs

What is the way to ensure the implementation of optimal or near optimal market conditions in the real economy? According to Theorem 1, under conditions of perfect competition, rationally acting producers and consumers will bring the market into a state of competitive equilibrium corresponding to the maximum of social welfare. Two questions arise. First: how to ensure perfect competition?

Second: how to encourage manufacturers to maximize profits, and consumers to maximize their surpluses? Regarding the first question, there is an extensive literature, both general and specifically devoted to the electricity market (see [5]). Along with general measures of antitrust regulation, publicly available information on marginal costs for large generators can be used, based on the rate of fuel consumption and other resources. Each case of decommissioning of a large generator, which entailed a significant jump in price, should be the subject of a special investigation of the antitrust agency. Similar regulatory measures should be also applied to large consumers. All consumers with shiftable loads are to buy electricity at prices equivalent to market prices. Thus, it is necessary to introduce three to four rates for the population, corresponding to the equilibrium prices for different times of the day, and encourage the people to install the appropriate counters. These measures would ensure the efficiency of the uniform price bilateral auction of suppliers and consumers with a single price for each period t.

However, this conclusion is valid only in the framework of the simplified deterministic model considered above. The presence of random factors significantly complicates the situation. Consider the general model (3). Random factors act both on the supply side and on the demand side. The choice of production volumes for many generators and consumption volumes for some consumers occurs in advance. If, however, the marginal costs of such generators are obviously lower than the equilibrium price, and marginal utilities of consumers are obviously higher, then they can be included in the optimal volumes of generation and consumption at the first stage, which corresponds to the day-ahead market. The volumes of production and consumption for the remaining agents are determined after the implementation of random factors, proceeding from the balance of supply and demand. Thus, in theory it is possible to realize the maximum social welfare depending on the values of random factors.

In reality, the main contradiction with this ideal picture is the behavior of mass consumers (population). They cannot plan daily consumption schedules based on current wholesale prices, but determine them, proceeding from tariff rates set for a long time (about a year), taking into account random factors affecting their consumption during the day. Consider the corresponding optimization problem:

$$\overrightarrow{v}_b^* \to \max \Big[\sum_{t=1}^{T} u_{b_0}^t(v_{b_0}^t) + \sum_l (u_{bl}(\sum_t v_{bl}^t) - \sum_{t=1}^{T} v_{bl}^t e_{bl}^t) - $$
$$- \sum_t \overline{p}_b^t max(0, v_{b_0}^t + \sum_l v_{bl}^t - \psi_b^t) \Big]. \tag{8}$$

We bound our study with the case where during all periods the consumption volume of each consumer remains positive, that is, a random factor does not nullify its demand. Then the optimal total consumption (excluding random factors) is determined depending on tariffs $\overrightarrow{p}_b$ in the same way as for the deterministic problem, according to the demand functions $D_b(\overrightarrow{p}_b)$, satisfying relations similar to (5, 6).

Consider the problem of the social welfare optimizing without industrial consumers. The corresponding production volumes are random and are defined as $\sum_b (\vec{D}_b(\vec{p}_b) - \vec{\psi}_b)$. The social welfare is a random variable determined as follows:

$$W(\vec{p}, \psi_b^t) = \sum_{\bar{b} \in B_0 \cup BL} u_{\bar{b}}(D_{\bar{b}}(\vec{p})) - \sum_t c(\sum_{\bar{b}} D_{\bar{b}}^t(\vec{p})) - \sum_b \psi_b^t.$$

Consider the problem of choosing tariffs that optimize the mathematical expectation of the social welfare:

$$\mathbb{E}\{\sum_t u_0^t(D_{B_0}^t(\vec{p})) + \sum_{bl} u_{bl}(\sum_t D_{bl}^t(\vec{p}) - \sum_t D_{bl}^t(\vec{p})e_{bl}^t) -$$
$$- \sum_t c(D_{B_0}^t(\vec{p}) + \sum_{bl} D_{bl}^t(\vec{p}) - \psi^t)\} \to \max_{\vec{p}}. \tag{9}$$

The corresponding problem of optimizing the mathematical expectation of the social welfare by consumption volumes is:

$$\mathbb{E}\{\sum_t u_0^t(v_{B_0}^t) + \sum_{bl} u_{bl}(\sum_t v_{bl}^t - \sum_t v_{bl}^t e_{bl}^t) - \sum_t c(v_{B_0}^t + \sum_{bl} v_{bl}^t - \psi^t)\} \to \max_{\vec{v}}. \tag{10}$$

Theorem 2. *Let $\vec{v^*}$ denote a solution of problem (10). Then the optimal rates for problem (9) are:* $\vec{p^*} = \mathbb{E}\,\vec{c}'(\vec{v^*} - \vec{\psi})$.

Proof. The solution to problem (10) is determined from first-order conditions, which have the form:

$$u_0^{t\,'}(v_0^t) = \mathbb{E}\,c'(v^t - \psi^t),$$
$$u_{bl}^{t\,'}(v_{bl}) - e_{bl}^t = \mathbb{E}\,c'(v^t - \psi^t).$$

The optimal value of problem (9) does not exceed the optimal value in problem (10), since the corresponding volume vector that we get through the demand function is a particular case of arbitrary selection of volume vectors in problem (10).

Consider the optimal value of problem (10), which satisfies the first-order conditions, and select the tariff vector specified in Theorem 2. Then from the first-order conditions (5, 6) it turns out that the corresponding value of the demand function will give us the vector $\vec{v^*}$. Thus, under this choice of tariffs, the value of the objective function (9) will coincide with the value of the objective function (10). So, this is a solution of problem (9).

If the marginal cost function c' is linear, then the optimal tariffs take the form: $\vec{p^*} = \vec{c}'(v^*) - \vec{c}'(\mathbb{E}\,\vec{\psi})$. What is the way to organize a market that implements the optimal situation maximizing the expected social welfare? The regulator does not know the optimal vector v^*, but can observe under known tariff rates $\vec{p}$ the volumes $\vec{v}(\vec{p})$ and the average marginal cost $\mathbb{E}\,\vec{c}'(\vec{v}(\vec{p}) - \vec{\psi})$. In order to increase the social welfare, vector $\vec{p}$ should be shifted towards $\mathbb{E}\,\vec{c}'(\vec{v}(\vec{p}) - \vec{\psi})$. Study of specific options for iterative tariff optimization is an important problem for the future.

4 Deterministic Market Model with Energy Storage

In this section, the model takes into account the possibility of using a large energy storage to optimize the functioning of the system. The characteristics of the storage are the capacity, speed and efficiency coefficients of charging and discharging. Following the works [6,13], describe them as follows. Denote E^{min} and E^{max} respectively, the minimum and maximum allowable charge level of the storage, V_{ch}^{max} and V_{dis}^{max}—maximum speed of charging and discharging, η_{ch} and η_{dis}—charging and discharging efficiency coefficients, respectively. Denote as v_{Bat}^t the amount of energy the battery charges or discharges during period t, a positive value corresponds to charging, v_{Bat}^0 shows the initial battery charge. The storage control strategy is specified by vector $\overrightarrow{v}_{Bat} = (v_{Bat}^t, t = 1, .., T)$. Feasible controls satisfy the following restrictions:

$$0 \le -v_{Bat}^t/\eta_{dis} \le V_{dis}^{max} \quad \text{to discharge;} \tag{11}$$

$$0 \le \eta_{ch} v_{Bat}^t \le V_{ch}^{max} \quad \text{for charging;} \tag{12}$$

$$E^{min} \le \sum_{k=0}^t v_{Bat}^k \le E^{max} \quad \forall t = 0, ..., T. \tag{13}$$

We bound our study with the deterministic model for maximizing the social welfare (see Sect. 2.1 above). In this case, feasible vectors $\overrightarrow{v}_{Bat}$ must also satisfy the condition of the charge balance of the storage for the planning period:

$$\sum_{t=1}^T v_{Bat}^t = 0. \tag{14}$$

The problem setting changes in the following respects: vector $\overrightarrow{v}_{Bat}$, satisfying restrictions (11–14) is included in the strategy, and when calculating the required production volume, the amount of energy supplied or consumed by the storage is taken into account every hour t. As a result, the welfare function takes the following form:

$$\sum_t u_0^t(v_0^t) + \sum_{\overline{b} \in BL} [u_{\overline{b}}(\sum_t (v_{\overline{b}}^t)) - \sum_t e_{\overline{b}}^t v_{\overline{b}}^t] -$$
$$- \sum_t c(\sum_{\overline{b} \in BL \cup \{0\}} v_{\overline{b}}^t + \eta^t v_{Bat}^t - v_{A2}^t) \to \max_{\overrightarrow{v} = (\overrightarrow{v}_0, \overrightarrow{v}_{\overline{b}}, \overrightarrow{v}_{Bat}, \overline{b} \in BL) \ge 0}. \tag{15}$$

Theorem 3. *The problem (15) of the social welfare optimizing is a convex programming problem.*

Proof. The proof is similar to the proof of Theorem 1.

Introduce a number of notations:

$$W(\vec{v}_0, \vec{v}_{\overline{b}}, \vec{v}_{Bat}) = \sum_t u_0^t(v_0^t) + \sum_{\overline{b} \in BL} [u_{\overline{b}}(\sum_t (v_{\overline{b}}^t)) - \sum_t e_{\overline{b}}^t v_{\overline{b}}^t] -$$

$$- \sum_t c(\sum_{\overline{b} \in BL \cup \{0\}} v_{\overline{b}}^t + \eta^t v_{Bat}^t - v_{A_2}^t);$$

$$g_1^t(v_{Bat}^t) = \frac{v_{Bat}^t}{\eta_{dis}} + V_{dis}^{max}, \quad \forall t = 1, ..., T;$$

$$g_2^t(v_{Bat}^t) = V_{ch}^{max} - \eta_{ch} v_{Bat}^t, \quad \forall t = 1, ..., T;$$

$$g_3^t(\vec{v}_{Bat}) = \sum_{k=0}^t v_{Bat}^k - E^{min}, \quad \forall t = 1, ..., T;$$

$$g_4^t(\vec{v}_{Bat}) = E^{max} - \sum_{k=0}^t v_{Bat}^k, \quad \forall t = 1, ..., T;$$

$$g_5(\vec{v}_{Bat}) = \sum_{t=1}^T v_{Bat}^t.$$

Inequalities $g_k^t(...) \leq 0, t = 1, ..., T, k = \overline{1,4}$ correspond to restrictions (11–13), imposed on the storage, and equality $g_5(\vec{v}_{Bat}) = 0$ - to constraint (14). Consider the Lagrange function

$$L(\vec{v}_0, \vec{v}_{\overline{b}}, \vec{v}_{Bat}, \vec{\lambda}) = W(\vec{v}_0, \vec{v}_{\overline{b}}, \vec{v}_{Bat}) + \sum_{t=1}^T \lambda_1^t g_1^t(v_{Bat}^t) + \sum_{t=1}^T \lambda_2^t g_2^t(v_{Bat}^t) +$$

$$+ \lambda_3^t g_3(\vec{v}_{Bat}) + \lambda_4^t g_5(\vec{v}_{Bat}) + \lambda_5 g_5(\vec{v}_{Bat}).$$

Let utility functions $u_{b0}^t(v), u_{bl}(v), b \in B_1, l \in 1, ..., L$ and the cost function $c(v)$ be continuously differentiable. Then the functions $g_k^t(...)$ are also continuously differentiable and their partial derivatives are not all equal to zero at any point. Thus, there exist Lagrange multipliers $\lambda_k^t \geq 0$ and $\lambda_5 \in \mathbb{R}$, such that the optimal point $M^* = (\vec{v}_{b0,0}^t, \vec{v}_{\overline{b},0}^t, \vec{v}_{Bat,0}^t)$ meets the following conditions:

$$\begin{cases} \dfrac{\partial L}{\partial v_0^t}(M^*) = 0 \quad \forall t = 1, ..., T; \\[2mm] \dfrac{\partial L}{\partial v_{\overline{b}}^t}(M^*) = 0 \quad \forall \overline{b} \in BL, \forall t = 1, ..., T; \\[2mm] \dfrac{\partial L}{\partial v_{Bat}^t}(M^*) = 0 \quad \forall t = 0, ..., T. \end{cases}$$

Consider the partial derivatives of the Lagrange function:

$$\begin{cases} u_0^{t'}(v_0^t) - c'(\displaystyle\sum_{\overline{b} \in BL \cup \{0\}} v_{\overline{b}}^t + \eta^t v_{Bat}^t - v_{A_2}^t)|_{(M_0)} = 0 \quad \forall t = 1, ..., T; \\[2mm] u_{\overline{b}}^{t'}(\displaystyle\sum_{t=1}^T v_{\overline{b}}^t) - e_{\overline{b}}^t - c'(\displaystyle\sum_{\overline{b} \in BL \cup \{0\}} v_{\overline{b}}^t + \eta^t v_{Bat}^t - v_{A_2}^t)|_{(M_0)} = 0 \quad \forall \overline{b} \in BL, \forall t = 1, ..., T; \\[2mm] \eta^t c'(\displaystyle\sum_{\overline{b} \in BL \cup \{0\}} v_{\overline{b}}^t + \eta^t v_{Bat}^t - v_{A_2}^t) + (-\lambda_1^t + \lambda_2^t)\eta^t - \displaystyle\sum_{k=0}^t \lambda_3^k + \displaystyle\sum_{k=0}^t \lambda_4^k|_{(M_0)} = 0 \quad \forall t = 1, ..., T. \end{cases}$$

The last system implies that:

$$\eta^t p^t = (\lambda_1^t - \lambda_2^t)\eta^t + (\sum_{k=0}^t \lambda_3^k - \sum_{k=0}^t \lambda_4^k + \lambda_5) \quad \forall t = 1, ..., T. \tag{16}$$

The conditions of complementary slackness to the problem (11, 12, 14, 15) are as follows:

$$\begin{cases} \lambda_1^t(V_{dis}^{max} + \dfrac{v_{Bat}^t}{\eta_{dis}}) = 0 \quad \forall t = 1, ..., T; \\[3mm] \lambda_2^t(V_{ch}^{max} - \eta_{ch} v_{Bat}^t) = 0 \quad \forall t = 1, ..., T. \end{cases}$$

Consider the problem (15) provided that the storage capacity is large enough, that is, constraints (13) are never binding.

The given relations imply that if the battery does not discharge at the maximal rate during period t $(v_{Bat}^t > -V_{dis}^{max})$, then $\lambda_1^t = 0$, since the expression in parentheses becomes strictly positive. Similarly, if the battery does not charge at the maximal rate during t, then $\lambda_2^t = 0$.

That is, equation (16) takes the form $p^t = -\lambda_2^t + \frac{\lambda_3}{\eta_{ch}}$ in periods of maximal charging, and $p^t = \lambda_1^t + \lambda_3 \eta_{dis}$ in periods of maximal discharging, and $\eta^t p^t = \lambda_3$ in other periods. Finally, we obtain the following result.

Theorem 4. *The optimal strategy v^* for the problem (15) satisfies the following conditions: there is a price $p^* = \lambda_3$, such that:*

1. *if the price p^t in period t meets condition $p^t > p^* \eta_{dis}$, then the battery is discharging at the maximal rate: $v_{Bat}^t = -V_{dis}^{max} \eta_{dis}$;*
2. *if $p^t = p^* \eta_{dis}$, then the battery is discharging and the discharging rate meets relations (11, 12);*
3. *if $\frac{p^*}{\eta_{ch}} < p^t < p^* \eta_{dis}$, then the battery does not charge or discharge, $v_{Bat}^t = 0$;*
4. *if $p^t = \frac{p^*}{\eta_{ch}}$, then the battery is charging and the charging rate meets relations (11, 12);*
5. *if $p^t < \frac{p^*}{\eta_{ch}}$, then the battery is charging at the maximal rate: $v_{Bat}^t = \frac{V_{dis}^{max}}{\eta_{ch}}$.*

With respect to the optimal regulation of the wholesale market, the last result means the following. If the energy price in each period t corresponds to the marginal cost of its production, then the optimal control of the energy storage corresponds to the strategy of maximizing its profit from the resale of energy. The threshold value p^* should be determined so that the balance of purchases and sales in the planning interval holds.

5   Conclusion

This paper describes a mathematical model of the wholesale electricity market functioning with account of the new economic and technical tools. The model takes into account special features of industrial consumers and the population consumption. Several problems of the social welfare optimizing are examined for a wholesale market where renewable energy sources, tariff regulation and energy storages may be employed. The convexity of these problems is proved, and first-order conditions for calculating their solutions are obtained. Based on these conditions, we also discussed economic mechanisms that can ensure the optimal functioning of such markets.

It is of interest to study the problems of the electric power industry long-term development with account of the new tools. Models of such sort for traditional markets were considered in [10,11]. Their generalization in the context of the present work is an important task for future research.

References

1. Aizenberg, N., Stashkevich, T., Voropai, N.: Forming rate options for various types of consumers in the retail electricity market by solving the adverse selection problem. Int. J. Public Adm. **2**(5), 99–110 (2019)
2. Arrow, K.J., Debreu, G.: Existence of an equilibrium for a competitive economy. Econometrica **22**, 265–290 (1954)
3. Davidson, M.R., Dogadushkina, Y.V., Kreines, E.M., et al.: Mathematical model of the competitive wholesale electricity market in Russia. Izvestiya RAN. Theor. Control Syst. **3**, 72–83 (2004)
4. Denzau, A.T.: Microeconomic Analysis: Markets and Dynamics. Richard d Irwin, Homewood (1992)
5. Dolmatova, M., Kozlovskiy, D., Khrustaleva, O., Sultanova, T., Vasin, A.: Market parameters dependent indices for competition evaluation in electricity market. Electr. Power Syst. Res. **190**, 106762 (2021). https://doi.org/10.1016/j.epsr.2020.106762
6. Nazari, A., Keypour, R.: Participation of responsive electrical consumers in load smoothing and reserve providing to optimize the schedule of a typical microgrid. Energy Syst. **11**, 1–24 (2019)
7. Gellings, C.W.: The concept of demand-side management for electric utilities. Proc. IEEE **73**(10), 1468–1570 (1985). https://doi.org/10.1109/PROC.1985.13318
8. Conejo, J., Morales, J.M., Baringo, L.: Real time demand response model. IEEE Trans. Smart Grid **1**(3), 236–242 (2010). https://doi.org/10.1109/TSG.2010.2078843

9. Samadi, P., Mohsenian-Rad, H., Schober, R., Wong, V.W.: Advanced demand side management for the future smart grid using mechanism design. IEEE Trans. Smart Grid **3**(3), 1170–1180 (2012). https://doi.org/10.1109/TSG.2012.2203341

10. Vasin, A.A., Grigoryeva, O.M., Tsyganov, N.I.: Optimization of an energy market transportation system. Doklady Math. **96**(1), 411–414 (2017). https://doi.org/10.1134/S1064562417040202

11. Vasin, A.A., Grigoryeva, O.M., Tsyganov, N.I.: Energy markets: optimization of transmission networks. Int. J. Public Adm. **42**(15–16), 1311–1322 (2019). The Management of Large Scale Energy Projects: Opportunities and Challenges. An Introduction to the Special Issue

12. Yaagoubi, N., Mouftan, H.T.: User-aware game theoretic approach for demand management. IEEE Trans. Smart Grid **6**(2), 716–725 (2015). https://doi.org/10.1109/TSG.2014.2363098

13. Motevasel, M., Seifi, A.: Expert energy management of a micro-grid considering wind energy uncertainty. Energy Convers Manag. **83**, 58–72 (2013)

Adaptive Extraproximal Algorithm for the Equilibrium Problem in Hadamard Spaces

Yana Vedel$^{(\boxtimes)}$ (iD) and Vladimir Semenov (iD)

Taras Shevchenko National University of Kyiv, Kyiv, Ukraine
yana.vedel@gmail.com, semenov.volodya@gmail.com

Abstract. In this paper, we consider equilibrium problems in Hadamard metric spaces. For an approximate solution of problems, a new iterative adaptive extra-proximal algorithm is proposed and studied. In contrast to the previously used rules for choosing the step size, the proposed algorithm does not calculate bifunction values at additional points and does not require knowledge of information on of bifunction's Lipschitz constants. For pseudo-monotone bifunctions of Lipschitz type, the theorem on weak convergence of sequences generated by the algorithm is proved. The proof is based on the use of the Fejer property of the algorithm with respect to the set of solutions of problem. It is shown that the proposed algorithm is applicable to pseudo-monotone variational inequalities in Hilbert spaces.

Keywords: Hadamard space · Equilibrium problem ·
Pseudo-monotonicity · Extraproximal algorithm · Adaptivity ·
Convergence

1 Introduction

One of the most popular directions of the first modern application of nonlinear analysis is the study of problems of equilibrium (Ky Fan's inequality, equilibrium programming problems) form [1–12] :

$$find \ x \in C : \ F(x, y) \geq 0 \ \forall y \in C, \tag{1}$$

where C – is a nonempty subset of a Hilbert space H, $F : C \times C \to \mathrm{R}$ – is a function such that $F(x, x) = 0 \ \forall x \in C$ (called a bifunction). In the form (1) one can formulate mathematical programming problems, variational inequalities, and many game problems. We give three typical formulations [1,4].

This work was supported by the Ministry of Education and Science of Ukraine (project "Mathematical Modeling and Optimization of Dynamical Systems for Defense, Medicine and Ecology", 0219U008403).

© Springer Nature Switzerland AG 2020
N. Olenev et al. (Eds.): OPTIMA 2020, LNCS 12422, pp. 287–300, 2020.
https://doi.org/10.1007/978-3-030-62867-3_21

1. If $F(x, y) = \varphi(y) - \varphi(x)$, where $\varphi : C \to R$, then the problem (1) is the constraint minimisation problem $\varphi \to \min_C$.

2. If $F(x, y) = (Ax, y - x)$, where $A : C \to H$, then the problem (1) is the classical variational inequality

$$find \ x \in C : \ (Ax, y - x) \geq 0 \ \forall y \in C.$$

3. Let I be a finite set of indices. For each $i \in I$ a set C_i and a function $\varphi_i : C \to R$ are defined, where $C = \prod_{i \in I} C_i$. For $x = (x_i)_{i \in I} \in C$ we denote $x^i = (x_j)_{j \in I \setminus \{i\}}$. A point $\bar{x} = (\bar{x}_i)_{i \in I} \in C$ is called Nash equilibrium if for all $i \in I$ if inequalities $\varphi_i(\bar{x}) \leq \varphi_i(\bar{x}_i, y_i)$ are hold $\forall y_i \in C_i$. We define the function $F : C \times C \to R$ as follows

$$F(x, y) = \sum_{i \in I} \left(\varphi_i \left(x^i, y_i \right) - \varphi_i (x) \right).$$

A point $\bar{x} \in C$ is Nash equilibrium if and only if it is a solution to problem (1).

The study of algorithms for solving equilibrium and similar problems is actively continuing. A special case of equilibrium problems is variational inequalities [13]. To solve them, G. Korpelevich proposed an extragradient method [14]. For variational inequalities, one of the modern versions of the extragradient method is the proximal mirror method by A. Nemirovsky [15]. This method can be interpreted as a variant of the extragradient method with projection understood in the sense of the Bregman Divergence. In recent papers [16–19] are suggested adaptive modifications of mirror proximal method which do not require knowledge of the Lipschitz constants operators to determine the step size. An interesting one-step proximal extrapolated gradient method with a simple adaptive selection of the step size is studied in [20]. The analogues of the extragradient method for equilibrium problems and related questions are the subject of [5–7, 21–25].

Following A. Antipin, we will call the following analogue of the extragradient method for equilibrium problems extraproximal [2, 5]

$$\begin{cases} y_n = \mathrm{prox}_{\lambda_n F(x_n, \cdot)} x_n, \\ x_{n+1} = \mathrm{prox}_{\lambda_n F(y_n, \cdot)} x_n, \end{cases}$$

where $\lambda_n \in (0, +\infty)$, prox_φ – is the proximal operator for function φ.

In 1980 L. Popov [26] proposed an interesting modification of the Arrow – Hurwitz method for finding saddle points of convex-concave functions defined in a finite-dimensional Euclidean space. In [8], for solving equilibrium problems in a Hilbert space, a two-stage proximal algorithm was proposed

$$\begin{cases} y_n = \mathrm{prox}_{\lambda_n F(y_{n-1}, \cdot)} x_n, \\ x_{n+1} = \mathrm{prox}_{\lambda_n F(y_n, \cdot)} x_n, \end{cases}$$

where $\lambda_n \in (0, +\infty)$, which is an adaptation of L. Popov method for the general problems of equilibrium programming (see also [9, 27, 28]).

Recently, due to the problems of mathematical biology and machine learning interest has arisen to build the theory and algorithms for solving mathematical programming problems in metric Hadamard spaces [29] (also known as $CAT(0)$ spaces). Another strong motivation for studying these problems is the ability to write down some non-convex problems in the form of convex (more precisely, geodesically convex) in a space with a specially selected Riemannian metric [10,29].

Some authors began to study equilibrium problems in Hadamard spaces [10–12]. In [10], existence theorems were obtained for equilibrium problems on Hadamard manifolds, applications to variational inequalities were considered, and the resolvent method for approximating solutions to equilibrium problems and variational inequalities were substantiated. In [11], for more general equilibrium problems with pseudo-monotone bifunctions in Hadamard spaces, existence theorems were obtained and a proximal algorithm was proposed and its convergence was proved. A more constructive approach is considered in [12], where the authors, starting from the results of [5], proposed and substantiated for pseudomonotone equilibrium problems in Hadamard spaces an analog of the extragradient (or extraproximal) method.

In this work, following [17,25], we proposed a new adaptive extraproximal algorithm for the approximate solution of the equilibrium problem in Hadamard spaces. In the proposed algorithm we do not need to compute values of bifunction at additional points and do not require knowledge of Lipschitz constants for bifunction. For pseudo-monotone bifunctions of Lipschitz type the theorem on weak convergence (or Δ-convergence) generated by the algorithm of sequences is proved. The proof is based on the use of the Fejer property of the algorithm with respect to the set of solutions to the problem. It is shown that the proposed algorithm is applicable to pseudo-monotone variational inequalities in Hilbert spaces.

2   Preliminaries

We give several concepts and facts related to Hadamard metric spaces. Details can be found in [29–31].

Let (X, d) be a metric space and $x,\ y \in X$. Geodesic path connecting points x and y is an isometry $\gamma : [0, d(x, y)] \to X$ such that $\gamma(0) = x$, $\gamma(d(x, y)) = y$. Set $\gamma([0, d(x, y)]) \subseteq X$ is denoted $[x, y]$ and called geodesic segment with ends x and y (or geodesic). Metric space (X, d) is called geodesic space if any two points X can be connected by geodesic, and uniquely geodesic space if for any two points X there is exactly one geodesic connecting them.

Geodesic space (X, d) is called $CAT(0)$ space if for any three points y_0, y_1, $y_2 \in X$ such that $d^2(y_1, y_0) = d^2(y_2, y_0) = \frac{1}{2}d^2(y_1, y_2)$ the inequality holds

$$d^2(x, y_0) \leq \frac{1}{2}d^2(x, y_1) + \frac{1}{2}d^2(x, y_2) - \frac{1}{4}d^2(y_1, y_2) \quad \forall x \in X. \tag{2}$$

Inequality (2) is called CN inequality [30] (in Euclidean space (2) turns into identity), and a point y_0 is the middle between y_1 and y_2 (it always exists in geodesic space).

It is known that $CAT(0)$ space is uniquely geodesic [29].

For two points x and y from $CAT(0)$ space (X, d) and $t \in [0, 1]$ we denote $tx \oplus (1 - t) y$ unique point z of the segment $[x, y]$ such that $d(z, x) = (1 - t) d(x, y)$ and $d(z, y) = td(x, y)$. Set $C \subseteq X$ is called convex (geodesically convex) if for all $x, y \in C$ and $t \in [0, 1]$ $tx \oplus (1 - t) y \in C$ is satisfied.

A useful tool for working in $CAT(0)$ space (X, d) is the following inequality

$$d^2 (tx \oplus (1 - t) y, z) \leq td^2 (x, z) + (1 - t) d^2 (y, z) - t (1 - t) d^2 (x, y), \qquad (3)$$

where $\{x, y, z\} \in X, t \in [0, 1]$.

Important examples of $CAT(0)$ spaces are Euclidian R-trees, Hadamard manifolds (complete connected Riemannian manifolds of nonpositive curvature), and Hilbert ball with a hyperbolic metric [29–31].

Complete $CAT(0)$ space is called Hadamard space.

Let (X, d) be a metric space and (x_n) be bounded sequence of elements X. Let $r(x, (x_n)) = \varlimsup_{n \to \infty} d(x, x_n)$. A number $r((x_n)) = \inf_{x \in X} r(x, (x_n))$ is called asymptotic radius (x_n) and set $A((x_n)) = \{x \in X : r(x, (x_n)) = r((x_n))\}$ is asymptotic center (x_n). It is known that in Hadamard spaces $A((x_n))$ consists of one point [29].

Sequence (x_n) of elements from Hadamard space (X, d) converges weakly (or as sometimes called Δ-converges [30]) to element $x \in X$ if $A((x_{n_k})) = \{x\}$ for any subsequence (x_{n_k}). It is known that any sequence of elements from bounded closed and convex set K from Hadamard space has a subsequence which converges weakly to element from K [29, 30].

In the proof of the weak convergence of the sequence of elements from the Hadamard space we use analogue of Opial lemma.

Lemma 1 ([29], **p. 60.**). *Let the sequence (x_n) of elements from the Hadamard space (X, d) converges weakly to the element $x \in X$. Then for all $y \in X \backslash \{x\}$ we have*

$$\lim_{n \to \infty} d(x_n, x) < \lim_{n \to \infty} d(x_n, y).$$

Let (X, d) be a Hadamard space. Function $\varphi : X \to \overline{R} = R \cup \{+\infty\}$ is called convex (geodesically convex) if for all $x, y \in X$ and $t \in [0, 1]$ it holds

$$\varphi(tx \oplus (1 - t) y) \leq t\varphi(x) + (1 - t) \varphi(y).$$

For example in Hadamard space functions $y \mapsto d(y, x)$ are convex. If there exists constant $\mu > 0$ such that for all $x, y \in X$ and $t \in [0, 1]$ it holds

$$\varphi(tx \oplus (1 - t) y) \leq t\varphi(x) + (1 - t) \varphi(y) - \mu t (1 - t) d^2 (x, y),$$

then function φ is called strongly convex. It is known that for convex functions lower semicontinuity and weak lower semicontinuity are equivalent [29], and a strongly convex lower semicontinuous function reaches a minimum at a unique point.

Remark 1. Many constructions important for applications in Hadamard spaces are associated with the minimum points of convex functions [29,31]. For example let a set of points $\{x_i\}_{i=\overline{1,m}}$ of metric space (X,d) and set of positive numbers $\{\alpha_i\}_{i=\overline{1,m}}$ be given. The barycenter (center of mass, Frechet mean) of points $\{x_i\}$ with weights $\{\alpha_i\}$ is the point

$$z \in \arg\min_{y \in X} \sum_{i=1}^{m} \alpha_i d^2(y, x_i).$$

In Hadamard space functions $y \mapsto d^2(y, x_i)$ are strongly convex (in follows from the inequality (3)) therfore the function $y \mapsto \sum_{i=1}^{m} \alpha_i d^2(y, x_i)$ is also strongly convex. It follows that the barycenter exists and it is unique.

For a convex, proper, and lower semicontinuous function $\varphi : X \to \overline{R} = R \cup \{+\infty\}$ is defined as follows [29]

$$\text{prox}_{\varphi} x = \arg\min_{y \in X} \left(\varphi(y) + \frac{1}{2} d^2(y, x) \right).$$

Since functions $\varphi + \frac{1}{2} d^2(\cdot, x)$ are strongly convex the definition of proximal operator is correct, that is, for each $x \in X$ there exists an unique element $\text{prox}_{\varphi} x \in X$.

We proceed to the formulation of equilibrium problem in Hadamard space.

3   The Equilibrium Problem in Hadamard Space

Let (X, d) be a Hadamard space. For non-empty convex closed set $C \subseteq X$ and bifunction $F : C \times C \to R$ we consider the equilibrium problem (or equilibrium programming problem [2,4,8]):

$$find \ x \in C : \ F(x, y) \geq 0 \ \forall y \in C \tag{4}$$

Assume that the following conditions are met:

1. $F(x, x) = 0$ for all $x \in C$;
2. functions $F(x, \cdot) : C \to R$ are convex and lower semicontinuous for all $x \in C$;
3. functions $F(\cdot, y) : C \to R$ are weakly upper semicontinuous for all $y \in C$;
4. bifunction $F : C \times C \to R$ is pseudomonotone, e. q. for all $x, y \in C$ from $F(x, y) \geq 0$ it follows that $F(y, x) \leq 0$.
5. bifunction $F : C \times C \to R$ is Lipschitz type, so that there exist two constants $a > 0$, $b > 0$, such that

$$F(x, y) \leq F(x, z) + F(z, y) + ad^2(x, z) + bd^2(z, y) \ \forall x, y, z \in C. \tag{5}$$

Remark 2. Condition (5) of the Lipschitz type in Euclidean space was introduced by G. Mastroeni [3].

Consider the dual equilibrium problem:

$$find \ x \in C \ : \ F(y, x) \leq 0 \ \forall y \in C. \tag{6}$$

We denote set of solutions for problems (4) and (6) by S and S^*. If conditions 1)–4) are satisfied we have $S = S^*$ [11]. Moreover, set S^* is convex and closed.
Further we will assume that $S \neq \emptyset$.

4   Adaptive Extraproximal Algorithm

For approximate solution of problem (4) we consider extraproximal algorithm with adaptive step.

Algorithm 1
Initialization. Choose element $x_1 \in C$, $\tau \in (0, 1)$, $\lambda_1 \in (0, +\infty)$. Let $n = 1$.
Step 1. Calculate

$$y_n = \text{prox}_{\lambda_n F(x_n, \cdot)} x_n = \arg \min_{y \in C} \left(F(x_n, y) + \frac{1}{2\lambda_n} d^2(y, x_n) \right).$$

If $x_n = y_n$, we stop and $x_n \in S$. Otherwise, go to step 2.
Step 2. Calculate

$$x_{n+1} = \text{prox}_{\lambda_n F(y_n, \cdot)} x_n = \arg \min_{y \in C} \left(F(y_n, y) + \frac{1}{2\lambda_n} d^2(y, x_n) \right).$$

Step 3. Calculate

$$\lambda_{n+1} = \begin{cases} \lambda_n, & \text{if } F(x_n, x_{n+1}) - F(x_n, y_n) - F(y_n, x_{n+1}) \leq 0, \\ \min \left\{ \lambda_n, \frac{\tau}{2} \frac{d^2(x_n, y_n) + d^2(x_{n+1}, y_n)}{(F(x_n, x_{n+1}) - F(x_n, y_n) - F(y_n, x_{n+1}))} \right\}, & \text{otherwise.} \end{cases}$$

Let $n := n + 1$ and go to step 1.
At each step of Algorithm 1 we solve two convex problems with strongly convex functions. Assume they can be solved effectively. In the proposed algorithm parameter λ_{n+1} depends on location of points x_n, y_n, x_{n+1}, values of $F(x_n, x_{n+1})$, $F(x_n, y_n)$ and $F(y_n, x_{n+1})$. No information on constants a and b from inequality (5) is not used. Obviously, the sequence (λ_n) is non-decreasing. It also has a lower limit $\min \left\{ \lambda_1, \frac{\tau}{2 \max\{a, b\}} \right\}$. Indeed, we have

$$F(x_n, x_{n+1}) - F(x_n, y_n) - F(y_n, x_{n+1}) \leq ad^2(x_n, y_n) + bd^2(x_{n+1}, y_n) \leq$$

$$\leq \max\{a, b\} \left(d^2(x_n, y_n) + d^2(x_{n+1}, y_n) \right).$$

For variational inequalities in a Hilbert space, Algorithm 1 takes the following form.

Algorithm 2.
Initialization. Choose element $x_1 \in C$, $\tau \in (0,1)$, $\lambda_1 \in (0,+\infty)$. Let $n = 1$.
Step 1. Calculate

$$y_n = P_C (x_n - \lambda_n A x_n).$$

If $x_n = y_n$, we stop and x_n is the solution. Otherwise, go to step 2.
Step 2. Calculate

$$x_{n+1} = P_C (x_n - \lambda_n A y_n).$$

Step 3. Calculate

$$\lambda_{n+1} = \begin{cases} \lambda_n, & \text{if } (A x_n - A y_n, x_{n+1} - y_n) \leq 0, \\ \min \left\{ \lambda_n, \frac{\tau}{2} \frac{\|x_n - y_n\|^2 + \|x_{n+1} - y_n\|^2}{(A x_n - A y_n, x_{n+1} - y_n)} \right\}, & \text{otherwise.} \end{cases} \tag{7}$$

Let $n := n + 1$ and go to 1.

Remark 3. Algorithm 2 differs from the algorithm studied in [17,25] by parameter selection rule λ_{n+1}. In [17,25] instead of (7) the following rule was considered

$$\lambda_{n+1} = \begin{cases} \min \left\{ \lambda_n, \tau \frac{\|x_n - y_n\|}{\|A x_n - A y_n\|} \right\}, & \text{if } A x_n \neq A y_n, \\ \lambda_n, & \text{otherwise.} \end{cases}$$

Now we go to the convergence of Algorithm 1.

5   Algorithm Convergence

First we prove an important inequality.

Lemma 2. *For $x \in C$ and $x^+ = \text{prox}_{\lambda F(x,\cdot)} x$, where $\lambda > 0$, the following inequality holds*

$$F(x, x^+) - F(x, y) \leq \frac{1}{2\lambda} \left(d^2(y, x) - d^2(x, x^+) - d^2(x^+, y) \right) \forall y \in C. \tag{8}$$

Proof. From the definition of $x^+ = \arg\min_{y \in C} \left(F(x, y) + \frac{1}{2\lambda} d^2(y, x) \right)$ it follows

$$F(x, x^+) + \frac{1}{2\lambda} d^2(x^+, x) \leq F(x, p) + \frac{1}{2\lambda} d^2(p, x) \forall p \in C. \tag{9}$$

Assuming in (9) $p = t x^+ \oplus (1 - t) y$, $y \in C$, $t \in (0,1)$, we obtain

$$F(x, x^+) + \frac{1}{2\lambda} d^2(x^+, x)$$

$$\leq F(x, t x^+ \oplus (1 - t) y) + \frac{1}{2\lambda} d^2(t x^+ \oplus (1 - t) y, x)$$

$$\leq t F(x, x^+) + (1 - t) F(x, y)$$

$$+ \frac{1}{2\lambda} \left(t d^2(x^+, x) + (1 - t) d^2(y, x) - t(1 - t) d^2(x^+, y) \right).$$

Therefore,
$$(1-t) F\left(x, x^{+}\right)-(1-t) F\left(x, y\right)$$

$$\leq \frac{1}{2\lambda}\left(-(1-t) d^{2}\left(x^{+}, x\right)+(1-t) d^{2}\left(y, x\right)-t(1-t) d^{2}\left(x^{+}, y\right)\right). \quad (10)$$

Reducing in (10) $1-t$ and going to the limit value when $t \to 1$ we obtain (8).

From Lemma 2 it follows that for sequences (x_n), (y_n), generated by Algorithm 1, the following inequality holds $(\forall y \in C)$

$$F\left(x_{n}, y_{n}\right)-F\left(x_{n}, y\right) \leq \frac{1}{2\lambda_{n}}\left(d^{2}\left(y, x_{n}\right)-d^{2}\left(x_{n}, y_{n}\right)-d^{2}\left(y_{n}, y\right)\right), \quad (11)$$

$$F\left(y_{n}, x_{n+1}\right)-F\left(y_{n}, y\right) \leq \frac{1}{2\lambda_{n}}\left(d^{2}\left(y, x_{n}\right)-d^{2}\left(x_{n}, x_{n+1}\right)-d^{2}\left(x_{n+1}, y\right)\right). \quad (12)$$

Inequality (11) gives a substantiation for the stopping rule of Algorithm 1. Indeed, for $x_n = y_n$ from (11) it follows

$$-F\left(x_{n}, y\right) \leq 0 \ \forall y \in C,$$

i.e., $x_n \in S$

Remark 4. Actually, the equivalence holds:

$$x \in S \ \Leftrightarrow \ x = \operatorname{prox}_{\lambda F(x, \cdot)} x, \ \lambda > 0.$$

Let us prove an important estimate connecting the distances between the points generated by Algorithm 1 and an arbitrary element of the set of solutions S.

Lemma 3. *For sequences (x_n), (y_n), generated by algorithm 1, the inequality takes place*

$$d^{2}\left(x_{n+1}, z\right) \leq d^{2}\left(x_{n}, z\right)$$

$$-\left(1-\tau \frac{\lambda_{n}}{\lambda_{n+1}}\right) d^{2}\left(x_{n+1}, y_{n}\right)-\left(1-\tau \frac{\lambda_{n}}{\lambda_{n+1}}\right) d^{2}\left(y_{n}, x_{n}\right), \quad (13)$$

where $z \in S$.

Proof. Let $z \in S$. From pseudomonotonicity of bifunction F it follows

$$F\left(y_{n}, z\right) \leq 0. \quad (14)$$

From (14) and (12) it follows

$$2\lambda_{n} F\left(y_{n}, x_{n+1}\right) \leq d^{2}\left(z, x_{n}\right)-d^{2}\left(x_{n}, x_{n+1}\right)-d^{2}\left(x_{n+1}, z\right). \quad (15)$$

From λ_{n+1} it follows

$$F\left(x_{n}, x_{n+1}\right)-F\left(x_{n}, y_{n}\right)-F\left(y_{n}, x_{n+1}\right)$$

$$\leq \frac{\tau}{2\lambda_{n+1}} \left(d^2\left(x_n, y_n\right) + d^2\left(x_{n+1}, y_n\right)\right). \tag{16}$$

Estimating left part of (15) using (16), we obtain

$$2\lambda_n \left(F\left(x_n, x_{n+1}\right) - F\left(x_n, y_n\right)\right) - \tau\frac{\lambda_n}{\lambda_{n+1}} \left(d^2\left(x_n, y_n\right) + d^2\left(x_{n+1}, y_n\right)\right)$$

$$\leq d^2\left(z, x_n\right) - d^2\left(x_n, x_{n+1}\right) - d^2\left(x_{n+1}, z\right). \tag{17}$$

For lower bound $2\lambda_n \left(F\left(x_n, x_{n+1}\right) - F\left(x_n, y_n\right)\right)$ in (17) we use inequality (11). We have

$$d^2\left(x_n, y_n\right) + d^2\left(y_n, x_{n+1}\right) - d^2\left(x_{n+1}, x_n\right)$$

$$-\tau\frac{\lambda_n}{\lambda_{n+1}} \left(d^2\left(x_n, y_n\right) + d^2\left(x_{n+1}, y_n\right)\right) \leq$$

$$\leq d^2\left(z, x_n\right) - d^2\left(x_n, x_{n+1}\right) - d^2\left(x_{n+1}, z\right). \tag{18}$$

Regrouping (18), we obtain (13).

To prove the convergence of Algorithm 1, we need an elementary lemma on numerical sequences.

Lemma 4. *Let (a_n), (b_n) be two sequences of non-negative numbers satisfying the inequality*

$$a_{n+1} \leq a_n - b_n \ \forall n \in \mathrm{N}. \tag{19}$$

Then the limit $\lim\limits_{n\to\infty} a_n$ exists and $(b_n) \in \ell_1$.

Let us formulate the main result of the work.

Theorem 1. *Let (X, d) is Hadamard space $C \subseteq X$ is a nonempty convex closed set, for bifunction $F : C \times C \to \mathrm{R}$ conditions 1)–5) are satisfied and $S \neq \emptyset$. Then sequences generated by Algorithm 1 (x_n), (y_n) converge weakly to the solution $z \in S$ of equilibrium problem (4), wherein*

$$\lim_{n\to\infty} d\left(y_n, x_n\right) = \lim_{n\to\infty} d\left(y_n, x_{n+1}\right) = 0.$$

Proof. Let $z \in S$. Assume

$$a_n = d\left(z, x_n\right),$$

$$b_n = \left(1 - \tau\frac{\lambda_n}{\lambda_{n+1}}\right) d^2\left(x_{n+1}, y_n\right) - \left(1 - \tau\frac{\lambda_n}{\lambda_{n+1}}\right) d^2\left(y_n, x_n\right).$$

Inequality (13) takes the form

$$a_{n+1} \leq a_n - b_n.$$

Since there exists $\lim\limits_{n\to\infty} \lambda_n > 0$, then

$$1 - \tau\frac{\lambda_n}{\lambda_{n+1}} \to 1 - \tau \in (0, 1) \ at \ n \to \infty. \tag{20}$$

From Lemma 4 there exists limit

$$\lim_{n \to \infty} d^2 (z, x_n)$$

and

$$\sum_{n=1}^{\infty} \left(d^2 (x_{n+1}, y_n) + d^2 (y_n, x_n) \right) < +\infty.$$

From that we obtain that (x_n) is bounded and

$$\lim_{n \to \infty} d(y_n, x_n) = \lim_{n \to \infty} d(x_{n+1}, y_n) = \lim_{n \to \infty} d(x_{n+1}, x_n) = 0. \qquad (21)$$

Consider the subsequence (x_{n_k}), which converges weakly to some point $z \in C$. Then from (21) it follows that (y_{n_k}) converges weakly to z. Let us show that $z \in S$. We have

$$F(y_{n_k}, y) \geq F(y_{n_k}, x_{n_k+1})$$

$$-\frac{1}{2\lambda_{n_k}} \left(d^2(y, x_{n_k}) - d^2(x_{n_k}, x_{n_k+1}) - d^2(x_{n_k+1}, y) \right)$$

$$\geq F(x_{n_k}, x_{n_k+1}) - F(x_{n_k}, y_{n_k}) - \frac{\tau}{2\lambda_{n_k+1}} \left(d^2(x_{n_k}, y_{n_k}) + d^2(x_{n_k+1}, y_{n_k}) \right)$$

$$-\frac{1}{2\lambda_{n_k}} \left(d^2(y, x_{n_k}) - d^2(x_{n_k}, x_{n_k+1}) - d^2(x_{n_k+1}, y) \right)$$

$$\geq -\frac{1}{2\lambda_{n_k}} \left(d^2(x_{n_k+1}, x_{n_k}) - d^2(x_{n_k}, y_{n_k}) - d^2(y_{n_k}, x_{n_k+1}) \right)$$

$$-\frac{\tau}{2\lambda_{n_k+1}} \left(d^2(x_{n_k}, y_{n_k}) + d^2(x_{n_k+1}, y_{n_k}) \right)$$

$$-\frac{1}{2\lambda_{n_k}} \left(d^2(y, x_{n_k}) - d^2(x_{n_k}, x_{n_k+1}) - d^2(x_{n_k+1}, y) \right) \ \forall y \in C. \qquad (22)$$

Passing to the limit in (22) taking into account (21) and weak upper semicontinuity of function $F(\cdot, y) : C \to \mathrm{R}$, we obtain

$$F(z, y) \geq \varlimsup_{k \to \infty} F(y_{n_k}, y) \geq 0 \ \forall y \in C,$$

i.e., $z \in S$.

Applying a variant of the Opial lemma for Hadamard spaces (Lemma 1), we obtain the weak convergence of the sequence (x_n) to the point $z \in S$. Indeed, we judge from the contrary. Suppose that there exists a subsequence (x_{m_k}), which converges weakly to some point $\bar{z} \in C$ and $\bar{z} \neq z$. It is clear that $\bar{z} \in S$. Then, we have

$$\lim_{n \to \infty} d(x_n, z) = \lim_{k \to \infty} d(x_{n_k}, z)$$

$$< \lim_{k \to \infty} d(x_{n_k}, \bar{z}) = \lim_{n \to \infty} d(x_n, \bar{z}) = \lim_{k \to \infty} d(x_{m_k}, \bar{z})$$

$$< \lim_{k \to \infty} d(x_{m_k}, z) = \lim_{n \to \infty} d(x_n, z),$$

which is impossible. Consequently, (x_n) converges weakly to $z \in S$. From (21) it follows that sequence (y_n) converges weakly to $z \in S$.

Remark 5. As we can see from proof of Theorem 1 for sequence (x_n) starting from some number N the Fejer condition with respect to the set of solutions S is satisfied.

We now consider the special case of the problem of equilibrium: variational inequality in Hilbert space H:

$$find\ x \in C:\ (Ax, y - x) \geq 0\ \forall y \in C. \tag{23}$$

And from Theorem 1, the following result holds.

Theorem 2. *Let H be a Hilbert space, $C \subseteq X$ nonempty convex closed set, operator $A : C \to H$ pseudomonotone, Lipschitz continuous, and sequentially weakly continuous, and there exist solutions (23). Then sequences generated by Algorithm 2 (x_n), (y_n) converge weakly to the solution of variational inequality (23), and*

$$\lim_{n \to \infty} \|y_n - x_n\| = \lim_{n \to \infty} \|y_n - x_{n+1}\| = 0.$$

6   Conclusions

In this work, continuing guide to become and [17,25], we proposed a new adaptive extraproximal algorithm for the approximate solution of the equilibrium problem in Hadamard spaces. The algorithm has the following structure

$$\begin{cases} y_n = \mathrm{prox}_{\lambda_n F(x_n,\cdot)} x_n = \arg\min_{y \in C} \left(F(x_n, y) + \frac{1}{2\lambda_n} d^2(y, x_n) \right), \\ x_{n+1} = \mathrm{prox}_{\lambda_n F(y_n,\cdot)} x_n = \arg\min_{y \in C} \left(F(y_n, y) + \frac{1}{2\lambda_n} d^2(y, x_n) \right), \end{cases}$$

where $\lambda_n > 0$ is selected adaptively. In the proposed algorithm does not calculate bifunction values at additional points and does not require knowledge of Lipschitz bifunction constants. For pseudo-monotone bifunctions of Lipschitz type, the theorem on weak convergence (Δ-convergence) generated by the sequence algorithm is proved. The proof is based on the use of the Fejer property of the algorithm with respect to the set of solutions to the problem. It is shown that the proposed algorithm is applicable to pseudo-monotone variational inequalities in Hilbert spaces.

In one of the next works we are planning to introduce adaptive modification of two-stage proximal algorithm

$$\begin{cases} y_n = \mathrm{prox}_{\lambda_n F(y_{n-1},\cdot)} x_n = \arg\min_{y \in C} \left(F(y_{n-1}, y) + \frac{1}{2\lambda_n} d^2(y, x_n) \right), \\ x_{n+1} = \mathrm{prox}_{\lambda_n F(y_n,\cdot)} x_n = \arg\min_{y \in C} \left(F(y_n, y) + \frac{1}{2\lambda_n} d^2(y, x_n) \right). \end{cases}$$

This algorithm for problems in a Hilbert space was proposed in [8] (see also [9, 27,28]). Note that recently a variant of this algorithm for variational inequalities has become known among machine learning specialists as "Extrapolation from the Past" [32]. It is also of interest to build randomized adaptive versions of algorithms.

In addition, in [33] (see also [18,19]), an inexact oracle model for variational inequalities with monotone operator was proposed and adaptive versions of the Mirror Prox algorithm for variational inequalities and saddle point problems in finite-dimensional vector spaces were studied. Recent work [34] contains an interesting result about the first near-optimal parameter-free method for the setting of strongly monotone Lipschitz operators and convex-concave min-max optimization. It is interesting to understand how to extend the technique of these works to equilibrium problems in Hadamard spaces.

References

1. Kassay, G., Radulescu, V.D.: Equilibrium Problems and Applications. Academic Press, London (2019). xx+419 p

2. Antipin, A.S.: Equilibrium programming: Proximal methods. Comput. Math. Math. Phys. **37**, 1285–1296 (1997). https://doi.org/10.1134/S0965542507120044

3. Mastroeni, G.: On auxiliary principle for equilibrium problems. In: Daniele, P., et al. (eds.) Equilibrium Problems and Variational Models, pp. 289–298. Kluwer Academic Publishers, Dordrecht (2003). https://doi.org/10.1007/978-1-4613-0239-1

4. Combettes, P.L., Hirstoaga, S.A.: Equilibrium programming in Hilbert spaces. J. Nonlinear Convex Anal. **6**, 117–136 (2005)

5. Quoc, T.D., Muu, L.D., Hien, N.V.: Extragradient algorithms extended to equilibrium problems. Optimization **57**, 749–776 (2008). https://doi.org/10.1080/02331930601122876

6. Lyashko, S.I., Semenov, V.V., Voitova, T.A.: Low-cost modification of Korpelevich's methods for monotone equilibrium problems. Cybernet. Syst. Anal. **47**(4), 631–639 (2011). https://doi.org/10.1007/s10559-011-9343-1

7. Semenov, V.V.: Strongly convergent algorithms for variational inequality problem over the set of solutions the equilibrium problems. In: Zgurovsky, M.Z., Sadovnichiy, V.A. (eds.) Continuous and Distributed Systems. Solid Mechanics and Its Applications, vol. 211, pp. 131–146. Springer, Heidelberg (2014). https://doi.org/10.1007/978-3-319-03146-010

8. Lyashko, S.I., Semenov, V.V.: A new two-step proximal algorithm of solving the problem of equilibrium programming. In: Goldengorin, B. (ed.) Optimization and Its Applications in Control and Data Sciences. Springer Optimization and Its Applications, vol. 115, pp. 315–325. Springer, Cham (2016). https://doi.org/10.1007/978-3-319-42056-110

9. Chabak, L., Semenov, V., Vedel, Y.: A new non-euclidean proximal method for equilibrium problems. In: Chertov, O., Mylovanov, T., Kondratenko, Y., Kacprzyk, J., Kreinovich V., Stefanuk V. (eds.) Recent Developments in Data Science and Intelligent Analysis of Information. ICDSIAI 2018. Advances in Intelligent Systems and Computing, vol. 836, pp. 50–58. Springer, Cham (2019). https://doi.org/10.1007/978-3-319-97885-76

10. Colao, V., Lopez, G., Marino, G., Martin-Marquez, V.: Equilibrium problems in Hadamard manifolds. J. Math. Anal. Appl. **388**, 61–77 (2012). https://doi.org/10.1016/j.jmaa.2011.11.001

11. Khatibzadeh H., Mohebbi V. Monotone and pseudo-monotone equilibrium problems in Hadamard spaces. J. Austr. Math. Soc. 1–23 (2019). https://doi.org/10.1017/S1446788719000041
12. Khatibzadeh, H., Mohebbi, V.: Approximating solutions of equilibrium problems in Hadamard spaces. Miskolc Math. Not. **20**(1), 281–297 (2019). https://doi.org/10.18514/MMN.2019.2361
13. Kinderlehrer, D., Stampacchia, G.: An introduction to variational inequalities and their applications. Academic Press, New York (1980). Russian transl., Moscow: Mir, 1983. 256 p
14. Korpelevich, G.M.: An extragradient method for finding saddle points and for other problems. Matecon. **12**(4), 747–756 (1976)
15. Nemirovski, A.: Prox-method with rate of convergence O(1/T) for variational inequalities with Lipschitz continuous monotone operators and smooth convex-concave saddle point problems. SIAM J. Optim. **15**(1), 229–251 (2004). https://doi.org/10.1137/S1052623403425629
16. Bach, F., Levy, K.Y.: A universal algorithm for variational inequalities adaptive to smoothness and noise. arXiv preprint arXiv:1902.01637 (2019)
17. Denisov, S.V., Semenov, V.V., Stetsyuk, P.I.: Bregman extragradient method with monotone rule of step adjustment*. Cybern. Syst. Anal. **55**(3), 377–383 (2019). https://doi.org/10.1007/s10559-019-00144-5
18. Stonyakin, F.S.: On the adaptive proximal method for a class of variational inequalities and related problems. Trudy Inst. Mat. i Mekh. UrO RAN **25**(2), 185–197 (2019). https://doi.org/10.21538/0134-4889-2019-25-2-185-197
19. Stonyakin, F.S., Vorontsova, E.A., Alkousa, M.S.: New version of mirror prox for variational inequalities with adaptation to inexactness. In: Jaćimović, M., Khachay, M., Malkova, V., Posypkin, M. (eds.) OPTIMA 2019. Communications in Computer and Information Science, vol. 1145, pp. 427–442. Springer, Cham (2020). https://doi.org/10.1007/978-3-030-38603-031
20. Malitsky, Y.: Proximal extrapolated gradient methods for variational inequalities. Optim. Methods Softw. **33**(1), 140–164 (2018). https://doi.org/10.1080/10556788.2017.1300899
21. Nadezhkina, N., Takahashi, W.: Weak convergence theorem by an extragradient method for nonexpansive mappings and monotone mappings. J. Optim. Theory Appl. **128**, 191–201 (2006). https://doi.org/10.1007/s10957-005-7564-z
22. Vuong, P.T., Strodiot, J.J., Nguyen, V.H.: Extragradient methods and linesearch algorithms for solving Ky Fan inequalities and fixed point problems. J. Optim. Theory Appl. **155**, 605–627 (2012). https://doi.org/10.1007/s10957-012-0085-7
23. Verlan, D.A., Semenov, V.V., Chabak, L.M.: A strongly convergent modified extragradient method for variational inequalities with non-lipschitz operators. J. Autom. Inf. Sci. **47**(7), 31–46 (2015). https://doi.org/10.1615/JAutomatInfScien.v47.i7.40
24. Semenov, V.V.: Modified extragradient method with Bregman divergence for variational inequalities. J. Autom. Inf. Sci. **50**(8), 26–37 (2018). https://doi.org/10.1615/JAutomatInfScien.v50.i8.30
25. Denisov, S.V., Nomirovskii, D.A., Rublyov, B.V., Semenov, V.V.: Convergence of extragradient algorithm with monotone step size strategy for variational inequalities and operator equations. J. Autom. Inf. Sci. **51**(6), 12–24 (2019). https://doi.org/10.1615/JAutomatInfScien.v51.i6.20
26. Popov, L.D.: A modification of the Arrow-Hurwicz method for search of saddle points. Math. Not. Acad. Sci. USSR **28**(5), 845–848 (1980). https://doi.org/10.1007/BF01141092

27. Semenov, V.V.: A version of the mirror descent method to solve variational inequalities*. Cybern. Syst. Anal. **53**(2), 234–243 (2017). https://doi.org/10.1007/s10559-017-9923-9

28. Nomirovskii, D.A., Rublyov, B.V., Semenov, V.V.: Convergence of two-stage method with bregman divergence for solving variational inequalities*. Cybern. Syst. Anal. **55**(3), 359–368 (2019). https://doi.org/10.1007/s10559-019-00142-7

29. Bacak, M.: Convex Analysis and Optimization in Hadamard Spaces. De Gruyter, Berlin-Boston (2014). viii+185 p

30. Kirk,W., Shahzad, N: Fixed Point Theory in Distance Spaces. Springer, Cham (2014). https://doi.org/10.1007/978-3-319-10927-5

31. Burago, D., Burago, Yu., Ivanov, S.: A Course in Metric Geometry. Graduate Studies in Mathematics, vol. 33. AMS, Providence (2001). xiv+415 p

32. Gidel, G., Berard, H., Vincent, P., Lacoste-Julien, S.: A Variational Inequality Perspective on Generative Adversarial Networks. arXiv preprint arXiv:1802.10551 (2018)

33. Stonyakin, F., Gasnikov, A., Dvurechensky, P., Alkousa, M., Titov, A.: Generalized Mirror Prox for Monotone Variational Inequalities: Universality and Inexact Oracle. arXiv preprint arXiv:1806.05140 (2019)

34. Diakonikolas, J.: Halpern iteration for near-optimal and parameter-free monotone inclusion and strong solutions to variational inequalities. arXiv preprint arXiv:2002.08872 (2020)

The Dual Simplex-Type Method for Linear Second-Order Cone Programming Problem

Vitaly Zhadan[(✉)] [ID]

Dorodnicyn Computing Centre, FRC "Computer Science and Control" of RAS, 40, Vavilova str., Moscow 119333, Russia
zhadan@ccas.ru

Abstract. The linear second-order cone programming problem is considered. For its solution, the dual simplex-type method is proposed. The method is the generalization of the standard dual simplex method for linear programming for cone programming. At each iteration the primal variable is defined, and pivoting of the dual variable is carried out. The proof of the local convergence is given.

Keywords: Linear second-order cone programming · Dual simplex-type method · Finite and infinite convergence

1 Introduction

Linear second-order cone program (SOCP) is the most important generalizations of linear programming problems. Many optimization problems, including, in particular, combinatorial optimization problems, may be modeled in the form of SOCP [1,2]. The linear SOCP is a problem, where the linear objective function is minimized over intersection of the affine manifold with the Cartesian product of some second-order cones. Many theoretical results and numerical techniques, concerning the linear SOCP, are given in the overview paper [1] and in the handbooks [3,4].

The most popular methods for solving the SOCP are the primal-dual interior point techniques [5,6]. These methods are generalizations of the corresponding primal-dual methods for linear programming. In contrast to interior point methods, the simplex-type methods are less developed for the SOCP. The matter of the essence is that the second-order cone is not of the polyhedron type. As a consequence, there are infinitely many extreme points of the feasible set. Nevertheless, there are some variants of the simplex-type methods for the SOCP. In [7] the simplex method for solving linear optimization problems with the general convex cones is proposed. Also variants of the primal simplex-type methods

This investigation was supported by the Ministry of Science and Higher Education of the Russian Federation, project No. 075-15-2020-799.

for SOCP had been proposed in [8,9]. In the first paper, the SOCP having one second-order cone, but possibly with additional non-negative variables, is considered. Moreover, the objective function was of the special form. In the second paper, the variant of the simplex-type method for solving the general SOCP is worked out. This algorithm is based on the reformulation of SOCP as a linear semi-infinite program and on consequent application of the dual-simplex primal-exchange method from [10] for solving the reformulated problem. The variant of the primal simplex-type method for solving the SOCP was proposed also in [11]. The pivoting procedure in this method is similar to the standard procedure using in the primal simplex method for linear programming.

In the present paper, the dual simplex-type method is suggested. The method is constructed in the way similar to one used in [11]. In the pivoting procedure all dual slack variables belonging to the second order cones are treated as a single variable. The method can be considered also as special approach for solving the system of equations from the optimality conditions. The dual feasibility with the complementarity conditions are preserved during of the iterative process. At each iteration the primal variable is computed and updating of the dual variables is carried out. The local convergence of the method is discussed, and the example of the SOCP where this convergence takes place is given.

The paper is organized as follows. In Sect.2, we formulate the primal and dual SOCP. Also some basic definitions are given in this section. In particular, the notions of regular and irregular extreme points of the feasible set in the dual problem are introduced. The pivoting in the case of the regular extreme point is described in Sect. 3. The dual simplex-type algorithm is presented in Sect. 4. Finally, in Sect. 5 the convergence of the dual simplex-method is proved.

Give some notations which will be used throughout the paper without explanations. The second-order cone (the Lorentz cone) in $\mathbb{R}^n$ is defined by

$$\mathbb{K}^n = \left\{ [x^1; \bar{x}] \in \mathbb{R} \times \mathbb{R}^{n-1} : \ x^1 \geq \|\bar{x}\| \right\},$$

where $\bar{x} = [x^2; \ldots; x^n]$, and $\| \cdot \|$ is the standard Euclidean norm. Here and in what follows we use ";" for adjoining vectors or components of a vector in a column. The cone $\mathbb{K}^n$ is self-dual, i.e. $(\mathbb{K}^n)^* = \mathbb{K}^n$. Moreover, it induces in $\mathbb{R}^n$ the partial order, namely, $x_1 \succeq_{\mathbb{K}^n} x_2$, if $x_1 - x_2 \in \mathbb{K}^n$. The identity matrix of order n is denoted by I_n. By 0_{nm} we will denote the zero $n \times m$ matrix, and by 0_n we will denote the zero vector of dimension n. The angle brackets denote the usual Euclidean scalar product in $\mathbb{R}^n$.

2   Primal and Dual Second-Order Cone Programming Problems

Consider the linear cone programming problem

$$\min \sum_{i=1}^r \langle c_i, x_i \rangle, \qquad \qquad (1)$$
$$\sum_{i=1}^r A_i x_i = b, \quad x_1 \succeq_{\mathbb{K}^{n_1}} 0_{n_1}, \ \ldots, \ x_r \succeq_{\mathbb{K}^{n_r}} 0_{n_r}.$$

Here: $b \in \mathbb{R}^m$ and $c_i \in \mathbb{R}^{n_i}$, $1 \le i \le r$. The matrices A_i have dimensions $m \times n_i$, $1 \le i \le r$. Moreover, we suppose that $r > 1$ and $n_i > 1$ for all $1 \le i \le r$.

The following problem is dual to (1)

$$\max \langle b, u \rangle, \tag{2}$$
$$A_i^T u + y_i = c_i, \ 1 \le i \le r; \quad y_1 \succeq_{\mathbb{K}^{n_1}} 0_{n_1}, \ \ldots, \ y_r \succeq_{\mathbb{K}^{n_r}} 0_{n_r},$$

where $u \in \mathbb{R}^m$.

Let $n = n_1 + \cdots + n_r$. Denote

$$c = [c_1; \ldots; c_r] \in \mathbb{R}^n, \quad x = [x_1; \ldots; x_r] \in \mathbb{R}^n; \quad y = [y_1; \ldots; y_r] \in \mathbb{R}^n.$$

Denote also $\mathcal{A} = [A_1, \ldots A_r]$, $\mathcal{K} = \mathbb{K}^{n_1} \times \cdots \times \mathbb{K}^{n_r}$. Then it is possible to rewrite problems (1) and (2) in the following form:

$$\min \langle c, x \rangle, \quad \mathcal{A}x = b, \quad x \succeq_{\mathcal{K}} 0_n, \tag{3}$$

$$\max \langle b, u \rangle, \quad \mathcal{A}^T u + y = c, \quad y \succeq_{\mathcal{K}} 0_n. \tag{4}$$

Assume that solutions of problems (3), (4) exist. Assume also that $m < n$ and that rows of the matrix $\mathcal{A}$ are linear independent. By $\mathcal{F}_P$ and $\mathcal{F}_D$ we denote the feasible sets in problems (3) and (4), respectively:

$$\mathcal{F}_P = \{x \in \mathcal{K} : \ \mathcal{A}x = b\}, \quad \mathcal{F}_D = \{[u, y] \in \mathbb{R}^m \times \mathcal{K} : \ y = c - \mathcal{A}^T u\}.$$

Furthermore, by $\mathcal{F}_{D,u}$ denote the projection of the set $\mathcal{F}_D$ onto the space $\mathbb{R}^m$, that is the set

$$\mathcal{F}_{D,u} = \{u \in \mathbb{R}^m : \ c - \mathcal{A}^T u \succeq_{\mathcal{K}} 0_n\}.$$

If $x \in \mathcal{F}_P$ and $[u, y] \in \mathcal{F}_D$ are solutions of problems (3) and (4), then they satisfy to the following system of equalities

$$\langle x, y \rangle = 0, \quad \mathcal{A}x = b, \quad \mathcal{A}^T u + y = c \tag{5}$$

and to inclusions: $x \in \mathcal{K}$, $y \in \mathcal{K}$.

Our aim is to develop the dual simplex-type method for solving problems (3) and (4). At each iteration of the method the passage from one extreme point of the set $\mathcal{F}_{D,u}$ to another one is carried out with increasing the value if the objective function in the dual problem (4). The point $u \in \mathcal{F}_{D,u}$ is extreme if and only if the point $[u, y] \in \mathcal{F}_D$ is also extreme, where $y = y(u) = c - \mathcal{A}^T u$.

Let us give the characterization of an extreme point $[u, y] \in \mathcal{F}_D$, following to [1]. The vector $y \in \mathcal{K}$ we decompose onto three blocks of components: y_F, y_I and y_O. Without loss of generality, we assume that

$$y = [y_F; y_I; y_O]. \tag{6}$$

The first block y_F consists of r_F components $y_F = [y_1; \ldots; y_{r_F}]$. The second block y_I consists of r_I components $y_I = [y_{r_F+1}; \ldots; y_{r_F+r_I}]$. Lastly, the third block y_O consists of r_O remaining components $y_O = [y_{r_F+r_I+1}; \ldots; y_r]$. Each component y_i from the block y_F is nonzero and belongs to the boundary $\partial \mathbb{K}^{n_i}$ of the cone $\mathbb{K}^{n_i}$. Each component y_i incoming in the block y_I is a zero vector, i.e. $y_i = 0_{n_i}$. All components y_i from the block y_O belong to interiors $\text{int}\mathbb{K}^{n_i}$ of the cones $\mathbb{K}^{n_i}$. Below, by $J_F^r = J_F^r(y)$, $J_I^r = J_I^r(y)$ and $J_O^r = J_O^r(y)$ we denote the corresponding index sets:

$$J_F^r = [1 \; : \; r_F], \quad J_I^r = [r_F + 1 \; : \; r_F + r_I], \quad J_O^r = [r_F + r_I + 1 \; : \; r].$$

Any nonzero component $y_i \in \mathbb{R}^{n_i}$ permits the spectral decomposition (see [1])

$$y_i = \theta_{i,1} \mathbf{h}_{i,1} + \theta_{i,n_i} \mathbf{h}_{i,n_i}, \tag{7}$$

where $\mathbf{h}_{i,1}$ and $\mathbf{h}_{i,n_i}$ are the pair of vectors

$$\mathbf{h}_{i,1} = \frac{1}{\sqrt{2}} \left[1; \; \frac{\bar{y}_i}{\|\bar{y}_i\|} \right], \qquad \mathbf{h}_{i,n_i} = \frac{1}{\sqrt{2}} \left[1; \; -\frac{\bar{y}_i}{\|\bar{y}_i\|} \right],$$

calling by the Jordan frame. The coefficients $\theta_{i,1}$ and θ_{i,n_i} in (7) are following:

$$\theta_{i,1} = \frac{1}{\sqrt{2}} \left(y_i^1 + \|\bar{y}_i\| \right), \qquad \theta_{i,n_i} = \frac{1}{\sqrt{2}} \left(y_i^1 - \|\bar{y}_i\| \right).$$

The component y_i belongs to the cone $\mathbb{K}^{n_i}$ if and only if both coefficients $\theta_{i,1}$ and θ_{i,n_i} are nonnegative.

Associated with each nonzero y_i there is the symmetric arrow-shaped matrix

$$\text{Arr}\,(y_i) = \begin{bmatrix} y_i^1 & \bar{y}_i^T \\ \bar{y}_i & y_i^1 I_{n-1} \end{bmatrix}$$

of order n. This matrix plays a very important role in the theory of the SOCP (see [1]). It can be shown that $y_i \in \mathbb{K}^{n_i}$ if and only if the matrix $\text{Arr}\,(y_i)$ is positive semi-definite.

In what follows, we assume without loss of generality that $y_i \in \partial \mathbb{K}^{n_i}$ and $y_i = \theta_{i,1} \mathbf{h}_{i,1}$ for all $i \in J_F^r$. Denote by $\mathbf{H}_i$ the orthogonal matrix

$$\mathbf{H}_i = \left[\mathbf{h}_{i,1}, \; \hat{\mathbf{H}}_i, \; \mathbf{h}_{i,n_i} \right],$$

composed from the eigenvectors of $\text{Arr}\,(y_i)$. The vectors $\mathbf{h}_{i,1}$ and $\mathbf{h}_{i,n_i}$ are among the eigenvectors of $\text{Arr}\,(y_i)$. The dimension of the matrix $\hat{\mathbf{H}}_i$ is equal to $n_i \times (n_i - 2)$, all its columns are orthogonal to the vectors $\mathbf{h}_{i,1}$ and $\mathbf{h}_{i,n_i}$ and have the unit length. Therefore,

$$y_i = \mathbf{H}_i \theta_i, \quad \theta_i = [\theta_{i,1}; 0; \ldots; 0], \qquad i \in J_F^r. \tag{8}$$

Denote by $\mathbf{H}_{i,\mathrm{R}}$ the right $n_i \times (n_i - 1)$ sub-matrix of the matrix $\mathbf{H}_i$, i.e. the matrix $\mathbf{H}_i$, from which the first column $\mathbf{h}_{i,1}$ is removed. It follows from (8) that

$$y_i = [y_{i,1}; \; y_{i,\mathrm{R}}], \quad y_{i,1} = \theta_{i,1} \mathbf{h}_{i,1}, \quad y_{i,\mathrm{R}} = \mathbf{H}_{i,\mathrm{R}} \theta_{i,\mathrm{R}}, \quad \theta_{i,\mathrm{R}} = 0_{n_i - 1}$$

for $i \in J_F^r$. Denote also by $A_i^{\mathbf{H_R}}$ the matrices $A_i^{\mathbf{H_R}} = A_i \mathbf{H}_{i,\mathbf{R}}$, $i \in J_F^r$, and compose the matrix

$$\mathcal{A}_F^{\mathbf{H_R}} = \left[A_1^{\mathbf{H_R}}, \ \ldots, \ A_{r_F}^{\mathbf{H_R}} \right].$$

Below, we will need in the matrix $\mathcal{A}_{FI}^{\mathbf{H_R}} = \left[\mathcal{A}_F^{\mathbf{H_R}}, \ \mathcal{A}_I \right]$. The dimension of this matrix is equal to $m \times (n_{FI} - r_F)$, where

$$n_{FI} = n_F + n_I, \quad n_F = \sum_{i \in J_F^r} n_i, \quad n_I = \sum_{i \in J_I^r} n_i.$$

There is the following criterion of the extreme point $[u, y] \in \mathcal{F}_D$ in the dual problem (4).

Proposition 1. [7] *The point $[u, y] \in \mathcal{F}_D$ is extreme if and only if rows of the matrix $\mathcal{A}_{FI}^{\mathbf{H_R}}$ are linear independent.*

According to Proposition 1 at any extreme point $[u, y] \in \mathcal{F}_D$ the inequality

$$n_{FI} - r_F \geq m. \tag{9}$$

must hold. Below, the extreme point $[u, y] \in \mathcal{F}_D$ is called *regular*, if $m = n_{FI} - r_F$. Otherwise, when the inequality in (9) is strict, the extreme point $[u, y] \in \mathcal{F}_D$ is called *irregular*.

Let us give the definition of a non-degenerate point $[u, y] \in \mathcal{F}_D$ [1].

Definition 1. *The point $[u, y] \in \mathcal{F}_D$ is called non-degenerate, if $\mathcal{T}(y) + \mathcal{R}(A^T) = \mathbb{R}^n$, where $\mathcal{T}(y)$ is the tangent space to the cone $\mathcal{K}$ at y and $\mathcal{R}(A^T)$ is the image of the matrix A^T.*

Proposition 2. [1] *Let the point $[u, y] \in \mathcal{F}_D$ be such, that for y the decomposition (6) take place with $y_i = y_{i,1} = \theta_{i,1} \mathbf{h}_{i,1}$, $i \in J_F^r$. Let also $\mathcal{A}_F^{\mathbf{h_R}}$ denote the matrix $\mathcal{A}_F^{\mathbf{h_R}} = \left[A_1 \mathbf{h}_{1,n_1}, \ \ldots, \ A_{r_F} \mathbf{h}_{r_F,n_{r_F}} \right]$. Then the point $[u, y]$ is non-degenerate if and only if columns of the matrix $[\mathcal{A}_F^{\mathbf{h_R}}, A_I]$ are linear independent.*

It follows from Proposition 2 that at non-degenerate point $[u, y] \in \mathcal{F}_D$ the inequality $m \geq r_F + n_I$ holds. The regular extreme point is a non-degenerate point without fail.

3   Pivoting at the Regular Extreme Point

Suppose, that the point $[u, y] \in \mathcal{F}_D$ is extreme, and $[u, y]$ is not the solution of the dual problem (4). In this case it is possible to make a passage from $[u, y]$ to another extreme point of $\mathcal{F}_D$ with more value of the objective function. Let us describe such passage assuming for simplicity that the point $[u, y]$ is regular.

Since $[u, y] \in \mathcal{F}_D$, we have $y = y(u)$. Suppose that for y the partition (6) is valid. This partition of $y \in \mathcal{K}$ onto three blocks of components induces the partition of the vector $x \in \mathbb{R}^n$ on the same three blocks of components, namely, $x = [x_F; x_I; x_O]$, where

$$x_F = [x_1; \ldots; x_{r_F}], \quad x_I = [x_{r_F+1}; \ldots; x_{r_F+r_I}], \quad x_O = [x_{r_F+r_I+1}; \ldots; x_r].$$

Introduce into consideration vectors $\xi_i \in \mathbb{R}^{n_i}$, $i \in J_F^r$. Let $x_i = \mathbf{H}_i \xi_i$, $i \in J_F^r$. Moreover, among all coordinates of ξ_i, $i \in J_F^r$, we mark out the first coordinate $\xi_{i,1}$ and all rest coordinates. In other words, we present the vector ξ_i as $\xi_i = [\xi_{i,1}; \xi_{i,\mathrm{R}}]$. Then $x_i = [x_{i,1}, x_{i,\mathrm{R}}]$, where $x_{i,1} = \xi_{i,1} \mathbf{h}_{i,1}$, $x_{i,\mathrm{R}} = \mathbf{H}_{i,\mathrm{R}} \xi_{i,\mathrm{R}}$. If set $\xi_{i,1} = 0$, then under such choice of x_i the equality $\langle x_i, y_i \rangle = \langle x_{i,\mathrm{R}}, y_{i,\mathrm{R}} \rangle = 0$ for $i \in J_F^r$ is preserved.

The rest coordinates $\xi_{i,\mathrm{R}}$, $i \in J_F^r$, we get from the second equality (5). We demand that $\xi_{i,\mathrm{R}}$, $i \in J_F^r$, and also x_i, $i \in J_I^r$, satisfy the following system of linear equations

$$\sum_{i \in J_F^r} A_i^{\mathbf{H}_\mathbf{R}} \xi_{i,\mathrm{R}} + \sum_{i \in J_I^r} A_i x_i = b. \tag{10}$$

The matrix of this system at the regular extreme point $[u, y] \in \mathcal{F}_D$ is square. It is of order m, and has the form

$$\mathcal{A}_{FI}^{\mathbf{H}_\mathbf{R}} = \left[A_1^{\mathbf{H}_\mathbf{R}}, \ldots, A_{r_F}^{\mathbf{H}_\mathbf{R}}, A_{r_F+1}, \ldots, A_{r_F+r_I} \right]. \tag{11}$$

By Proposition 1 rows of the matrix $\mathcal{A}_{FI}^{\mathbf{H}_\mathbf{R}}$ are linear independent, hence (11) is a nonsingular matrix. Solving the system (10), we obtain

$$p_{FI} = [\xi_{F,\mathrm{R}}; x_I] = \left(\mathcal{A}_{FI}^{\mathbf{H}_\mathbf{R}} \right)^{-1} b, \tag{12}$$

where $\xi_{F,\mathrm{R}} = [\xi_{1,\mathrm{R}}; \ldots; \xi_{r_F,\mathrm{R}}]$. It is evident, that for any $i \in J_F^r$ the vector $x_{i,\mathrm{R}} = \mathbf{H}_{i,\mathrm{R}} \xi_i$ is orthogonal to the vector y_i.

Below, we will use the notation $J_{FI}^r = J_F^r \cup J_I^r$. Denote also

$$x_{FI} = [x_{1,\mathrm{R}}; \ldots; x_{r_F,\mathrm{R}}; x_{r_F+1}; \ldots; x_{r_F+r_I}]. \tag{13}$$

We say that x_{FI} is the *non-degenerate basic solution* of the equation $Ax = b$, if rows of the matrix $\mathcal{A}_{FI}^{\mathbf{H}_\mathbf{R}}$ are linear independent. It follows from Proposition 1 that x_{FI} is a non-degenerate basic solution, when $[u, y]$ is an extreme point of $\mathcal{F}_D$.

Proposition 3. *Let* $x_i = \mathbf{H}_i \xi_i$, $\xi_i = [0; \xi_{i,\mathrm{R}}]$ *for* $i \in J_F^r$. *Then* $x_i \in \mathbb{K}^{n_i}$ *if and only if* $\xi_i = [0; \ldots; 0; \xi_{i,n_i}]$ *and* $\xi_{i,n_i} \geq 0$.

Proposition 4. *Let* $\xi_i = [0; \xi_{i,\mathrm{R}}]$ *and* $x_i = \mathbf{H}_i \xi_i$, *where* $i \in J_F^r$. *Moreover, let al.l* x_i, $i \in J_{FI}^r$, *be such that* $x_i \in \mathbb{K}^{n_i}$. *Then the point* $[u, y]$ *is a solution of the dual problem* (4). *The point* $x = [x_F; x_I; x_O]$, *where the block* x_O *consists of zero vectors, is the solution of the primal problem* (3).

Proof. This assertion follows from (5) and inclusions $x \in \mathcal{K}$, $y \in \mathcal{K}$. $\quad\square$

In what follows, we assume that there exists the index $k \in J^r_{FI}$ such that the inclusion $x_k \in \mathbb{K}^{n_k}$ is violated.

For any nonzero x_i, $i \in J^r_{FI}$, the following decomposition

$$x_i = \eta_{i,1} \mathbf{g}_{i,1} + \eta_{i,n_i} \mathbf{g}_{i,n_i}, \tag{14}$$

takes place, where by analogy with (7)

$$\mathbf{g}_{i,1} = \frac{1}{\sqrt{2}} \left[1; \frac{\bar{x}_i}{\|\bar{x}_i\|} \right], \qquad \mathbf{g}_{i,n_i} = \frac{1}{\sqrt{2}} \left[1; -\frac{\bar{x}_i}{\|\bar{x}_i\|} \right].$$

The coefficients $\eta_{i,1}$ and η_{i,n_i} in (14) are following:

$$\eta_{i,1} = \frac{1}{\sqrt{2}} \left(x^1_i + \|\bar{x}_i\| \right), \qquad \eta_{i,n_i} = \frac{1}{\sqrt{2}} \left(x^1_i - \|\bar{x}_i\| \right).$$

Denote by $\Gamma_{i,\mathrm{R}}$, $i \in J^r_F$, the hyperplane

$$\Gamma_{i,\mathrm{R}} = \left\{ y \in \mathbb{R}^{n_i} : y = \mathbf{H}_{i,\mathrm{R}} \nu, \ \nu \in \mathbb{R}^{n_i - 1} \right\}$$

with $\mathbf{h}_{i,1}$ as a directing vector of this hyperplane.

Proposition 5. *Let $k \in J^r_F$ and let*

$$x_k = x_{k,\mathrm{R}} = \eta_{k,1} \mathbf{g}_{k,1} + \eta_{k,n_k} \mathbf{g}_{k,n_k} \notin \mathbb{K}^{n_k}.$$

Then one of the following two possibilities is realized:

1. $\mathbf{g}_{k,1} = \mathbf{h}_{k,n_k}$ *and* $x_{k,\mathrm{R}} = \eta_{k,n_k} \mathbf{h}_{k,n_k}$ *with* $\eta_{k,n_k} < 0$;
2. *the inequality* $\eta_{k,1} \eta_{k,n_k} < 0$ *holds, i.e. one of two coefficients* $\eta_{k,1}$, $\eta_{k,1}$ *is positive and another is negative.*

Proof. The vector x_k is nonzero and belongs to the hyperplane $\Gamma_{k,\mathrm{R}}$, which is a support hyperplane to the cone $\mathbb{K}^{n_k}$. The ray

$$l_{k,n_k} = \left\{ z \in \mathbb{R}^{n_k} : z = \sigma \mathbf{h}_{k,n_k}, \ \sigma \geq 0 \right\}$$

belongs to $\Gamma_{k,\mathrm{R}}$, and also it belongs to $\mathbb{K}^{n_k}$. Moreover, points from this ray are unique points from $\mathbb{K}^{n_k}$ belonging to $\Gamma_{k,\mathrm{R}}$. Thus, the first possibility realizes, when $\mathbf{g}_{k,1} = \mathbf{h}_{k,n_k}$ and $\eta_{k,1} < 0$.

If the first possibility is not realized, then because of $x_{k,1} = 0$ and $x_k \notin \mathbb{K}^{n_k}$, the vectors $\mathbf{g}_{k,1}$ and $\mathbf{g}_{k,n_k} \in \mathbb{K}^{n_k}$ without fail are such, that $\mathbf{g}^1_{k,1} > 0$ and $\mathbf{g}^1_{k,n_k} < 0$. Here by $\mathbf{g}^1_{k,1}$ and $\mathbf{g}^1_{k,n_k}$ are denoted first coordinates of the vectors $\mathbf{g}_{k,1}$ and $\mathbf{g}_{k,n_k}$, respectively. We derive from here that $\eta_{k,1} \eta_{k,n_k} < 0$. $\square$

Now we take x_k, where $k \in J^r_{FI}$. Since $x_k \notin \mathbb{K}^{n_k}$, then at least one coefficient from $\eta_{k,1}$ and η_{k,n_k} is negative. Let it be the first coefficient $\eta_{k,1} < 0$. Compose the vector $z_k \in \mathbb{R}^{n_{FI} - r_F}$ by setting

$$z_k = [z_{k,1}; \ \dots \ ; z_{k,r_F + r_I}], \tag{15}$$

where $z_{k,i} \in \mathbb{R}^{n_i - 1}$, $i \in J_F^r$, and $z_{k,i} \in \mathbb{R}^{n_i}$, $i \in J_I^r$. If $k \in J_F^r$, then we set

$$
z_{k,i} = \begin{cases} \mathbf{H}_{k,R}^T \mathbf{g}_{k,1}, & i = k, \\ 0_{n_i - 1}, & i \in J_F^r, \ i \neq k, \\ 0_{n_i}, & i \in J_I^r. \end{cases} \tag{16}
$$

In the case, where $k \in J_I^r$, we take

$$
z_{k,i} = \begin{cases} \mathbf{g}_{k,1}, & i = k, \\ 0_{n_i}, & i \in J_I^r, \ j \neq k, \\ 0_{n_i - 1}, & i \in J_F^r. \end{cases} \tag{17}
$$

Let make the passage from $u \in \mathcal{F}_{D,u}$ to the new point

$$
\bar{u} = u - \alpha \Delta u, \tag{18}
$$

where $\alpha > 0$. We demand that the vector Δu satisfy to the system of linear equations

$$
\left(\mathcal{A}_{FI}^{H_R} \right)^T \Delta u = z_k. \tag{19}
$$

Since the matrix $\mathcal{A}_{FI}^{H_R}$ is nonsingular, we get

$$
\Delta u = \left(\mathcal{A}_{FI}^{H_R} \right)^{-T} z_k \tag{20}
$$

where $\left(\mathcal{A}_{FI}^{H_R} \right)^{-T}$ denotes the matrix $\left(\left(\mathcal{A}_{FI}^{H_R} \right)^T \right)^{-1}$.

Proposition 6. *Let Δu be defined by* (20). *Then*

$$
\langle b, \Delta u \rangle = \eta_{k,1} < 0. \tag{21}
$$

Proof. According to (12) and (20)

$$
\langle b, \Delta u \rangle = \langle b, \left(\mathcal{A}_{FI}^{H_R} \right)^{-T} z_k \rangle = \langle \left(\mathcal{A}_{FI}^{H_R} \right)^{-1} b, \ z_k \rangle = \langle p_{FI}, z_k \rangle.
$$

Therefore, in the case, where $k \in J_I^r$, we have by (15), (17)

$$
\langle b, \Delta u \rangle = \langle x_k, \ \mathbf{g}_{k,1} \rangle = \langle \eta_{k,1} \mathbf{g}_{k,1} + \eta_{k,n_k} \mathbf{g}_{k,n_k}, \ \mathbf{g}_{k,1} \rangle = \eta_{k,1} \| \mathbf{g}_{k,1} \|^2 = \eta_{k,1} < 0.
$$

Similarly, if $k \in J_F^r$, then by (16)

$$
\begin{aligned}
\langle b, \Delta u \rangle &= \langle b, \left(\mathcal{A}_{FI}^{H_R} \right)^{-T} \mathbf{H}_{k,R}^T \mathbf{g}_{k,1} \rangle = \langle \left(\mathcal{A}_{FI}^{H_R} \right)^{-1} b, \ \mathbf{H}_{k,R}^T \mathbf{g}_{k,1} \rangle = \langle \xi_k, \ \mathbf{H}_{k,R}^T \mathbf{g}_{k,1} \rangle \\
&= \langle \mathbf{H}_{k,R} \xi_k, \ \mathbf{g}_{k,1} \rangle = \langle x_k, \ \mathbf{g}_{k,1} \rangle = \langle \eta_{k,1} \mathbf{g}_{k,1} + \eta_{k,n_k} \mathbf{g}_{k,n_k}, \ \mathbf{g}_{k,1} \rangle = \eta_{k,1} < 0.
\end{aligned}
$$

Thus, in any case the inequality (21) holds. $\qquad\square$

Compute $\Delta y = \mathcal{A}^T \Delta u$. We have $\Delta y = \mathcal{A}^T \left(\mathcal{A}_{FI}^{\mathbf{H_R}}\right)^{-T} z_k$. Introduce the block-diagonal matrix

$$\mathbf{H}_{F,R} = \begin{bmatrix} \mathbf{H}_{1,R} & 0 & \ldots\ldots & 0 \\ 0 & \mathbf{H}_{2,R} & 0 & \ldots & 0 \\ \ldots & \ldots & \ldots\ldots & \ldots \\ 0 & \ldots & \ldots & 0 & \mathbf{H}_{r_F,R} \end{bmatrix}. \tag{22}$$

Then $\mathcal{A}_{FI}^{\mathbf{H_R}} = [\mathcal{A}_F \mathbf{H}_{F,R},\ \mathcal{A}_I]$. The matrix $\mathcal{A}_{FI}^{\mathbf{H_R}}$ is of dimension $n_{FI} \times (n_{FI} - r_F)$. After transposition we obtain $\left(\mathcal{A}_{FI}^{\mathbf{H_R}}\right)^T = \left[\mathbf{H}_{F,R}^T \mathcal{A}_F^T;\ \mathcal{A}_I^T\right]$. Therefore,

$$\Delta y_{FI} = \begin{bmatrix} \mathcal{A}_F^T \\ \mathcal{A}_I^T \end{bmatrix} \Delta u = \begin{bmatrix} \mathcal{A}_F^T \\ \mathcal{A}_I^T \end{bmatrix} \left(\mathcal{A}_{FI}^{\mathbf{H_R}}\right)^{-T} z_k = \begin{bmatrix} \mathcal{A}_F^T \\ \mathcal{A}_I^T \end{bmatrix} \begin{bmatrix} \mathbf{H}_{F,R}^T \mathcal{A}_F^T \\ \mathcal{A}_I^T \end{bmatrix}^{-1} z_k. \tag{23}$$

Multiplying the equality (23) from the left by the block-diagonal matrix,

$$M_{FI} = \begin{bmatrix} \mathbf{H}_{F,R}^T & 0 \\ 0_{n_I} & I_{n_I} \end{bmatrix},$$

we amount to

$$M_{FI} \Delta y_{FI} = \begin{bmatrix} \mathbf{H}_{F,R}^T \mathcal{A}_F^T \\ \mathcal{A}_I^T \end{bmatrix} \begin{bmatrix} \mathbf{H}_{F,R}^T \mathcal{A}_F^T \\ \mathcal{A}_I^T \end{bmatrix}^{-1} z_k = z_k.$$

Assume that $k \in J_I^r$. In this case

$$\Delta y_i = \begin{cases} \mathbf{g}_{k,1}, & i = k \\ 0_{n_i} & i \neq k \end{cases}, \qquad i \in J_I^r. \tag{24}$$

For $i \in J_F^r$ we have, respectfully,

$$\mathbf{H}_{i,R}^T \Delta y_i = 0_{n_i - 1}, \quad i \in J_F^r. \tag{25}$$

The equality (25) means, that the vector Δy_i, $i \in J_F^r$ is orthogonal to the hyperplane $\Gamma_{i,R}$. Hence, $\Delta y_i = \sigma \mathbf{h}_{i,1}$, where $\sigma \in \mathbb{R}$.

Now, let $k \in J_F^r$. Then $\Delta y_i = 0_{n_i}$, if $i \in J_I^r$. For $i \in J_F^r$, $i \neq k$, the equality (25) takes place, that is $\Delta y_i = \sigma \mathbf{h}_{i,1}$. If $i = k$, then by (16)

$$\mathbf{H}_{k,R}^T \Delta y_k = \mathbf{H}_{k,R}^T \mathbf{g}_{k,1}.$$

Proposition 7. *Let the point $[u, y] \in \mathcal{F}_D$ be such, that the index $k \in J_I^r$. Moreover, let the vector Δu from (20), satisfy*

$$\Delta y_i = A_i^T \Delta u \in \mathbb{K}^{n_i}, \qquad i \in J_O^r, \tag{26}$$

and

$$\Delta y_i = A_i^T \Delta u = \sigma_i \mathbf{h}_{i,1}, \qquad i \in J_F^r, \qquad \sigma_i > 0. \tag{27}$$

Then the feasible set $\mathcal{F}_{D,u}$ is unbounded. Moreover,

$$\langle b, \bar{u} \rangle = \langle b, u \rangle - \alpha \langle b, \Delta u \rangle \to +\infty, \tag{28}$$

when $\alpha \to +\infty$.

Proof. First of all, remark that by (24), (25) and (27) the following inclusions $\Delta y_i \in \mathbb{K}^{n_i}$, $i \in J_{FI}^r$ take place. Hence, if the inclusion (26) also holds, then for $\alpha > 0$ we get $y(\bar{u}) = y(u - \alpha \Delta u) = y - A^T \Delta u \in \mathcal{K}$. Thus, $\bar{u} \in \mathcal{F}_{D,u}$ for all $\alpha > 0$. The limit equality (28) follows from (21).   □

The similar assertion is valid in the case, where $k \in J_F^r$.

Proposition 8. *Let the point $[u, y] \in \mathcal{F}_D$ be such, that the index $k \in J_F^r$. Besides, let for the vector Δu, defining by (20), the equalities (26) and the equalities (27), where $i \neq k$, hold. Then, if*

$$\mathbf{h}_{k,1}^T \Delta y_k = -\mathbf{h}_{k,1}^T A_k^T \Delta u \geq \eta_{k,1}, \tag{29}$$

the feasible set $\mathcal{F}_{D,u}$ is unbounded and the limit equality (28) takes place.

Proof. The proof almost word by word repeats the proof of Proposition 7. Remark only, that when the inequality (29) holds, the inclusion $\Delta y_k = -A_k^T \Delta u \in \mathbb{K}^{n_k}$ takes place. Therefore $\bar{y}_k = y_k + \alpha \Delta y_k \in \mathbb{K}^{n_k}$ for all $\alpha \geq 0$. □

Further, consider the case, when conditions of Propositions 7 and 8 do not fulfilled. In this situation it is possible to make passage from $u \in \mathcal{F}_{D,u}$ to another point of the set $\mathcal{F}_{D,u}$ with the greater value of the objective function in the problem (4).

Proposition 9. *Let $k \in J_I^r$, and let at least one condition (26) or (27) does not satisfied. Then there exists $0 < \bar{\alpha} < +\infty$, such that $\bar{y} = y + \alpha \Delta y \in \mathcal{K}$ for all $0 < \alpha \leq \bar{\alpha}$. Moreover, $\bar{y} \notin \mathcal{K}$, when $\alpha > \bar{\alpha}$.*

Proof. Denote by $\mathcal{C}_\mathcal{K}(y)$ the cone of feasible directions with respect to the cone $\mathcal{K}$ at the point $y \in \mathcal{K}$. This cone is the direct product of cones of feasible directions $\mathcal{C}_{\mathbb{K}^{n_i}}(y_i)$ with respect to $\mathbb{K}^{n_i}$ at the points $y_i \in \mathbb{K}^{n_i}$, $1 \leq i \leq r$, i.e.

$$\mathcal{C}_\mathcal{K}(y) = \mathcal{C}_{\mathbb{K}^{n_1}}(y_1) \times \cdots \times \mathcal{C}_{\mathbb{K}^{n_r}}(y_r).$$

The direction $s \in \mathbb{R}^{n_i}$ belongs to the cone $\mathcal{C}_{\mathbb{K}^{n_i}}(y_i)$ if and only if $s = s_1 + s_2$, where $s_1 \in \text{lin}(\mathbf{F}_{\min}(y_i | \mathbb{K}^{n_i}))$ and $s_2 \in \mathbb{K}^{n_i}$. Here $\text{lin}(\mathbf{F}_{\min}(y_i | K^{n_i}))$ is a linear hull of the minimal face $\mathbb{K}^{n_i}$, containing the point y_i.

It follows from (25) that $\Delta y_i \in \text{lin}(\mathbf{F}_{\min}(y_i | \mathbb{K}^{n_i}))$ for all $i \in J_F^r$. Moreover, $\Delta y_i = 0_{n_i}$ for $i \in J_I^r$ and $i \neq k$. We have also $\Delta y_k = \mathbf{g}_{k,1} \in \mathbb{K}^{n_k}$. For $i \in J_O^r$ it is evident that $\mathcal{C}_{\mathbb{K}^{n_i}}(y_i) = \mathbb{R}^{n_i}$. We derive from here that $\Delta y \in \mathcal{C}_\mathcal{K}(y)$. Hence $y + \alpha \Delta y \in \mathcal{K}$ for $\alpha > 0$ sufficiently small. Since both conditions of Proposition 7 are not satisfied simultaneously, there exists $\bar{\alpha}$, such that $\bar{y} = y + \bar{\alpha} \Delta y \in \partial \mathcal{K}$. □

Let $J_{FO}^r = J_F^r \cup J_O^r$. Denote also by $v(y)$ the value $v(y) = n - \dim \Gamma_{\min}(y | \mathcal{K})$, where $\dim \Gamma_{\min}(y | \mathcal{K})$ is the dimension of the minimal face of the cone $\mathcal{K}$, containing the point $y \in \mathcal{K}$. This dimension is equal to

$$\dim \Gamma_{\min}(y | \mathcal{K}) = \sum_{i=1}^r \dim \Gamma_{\min}(y_i | \mathbb{K}^{n_i}) = \sum_{i \in J_{FO}^r} \dim \Gamma_{\min}(y_i | \mathbb{K}^{n_i}) = r_F + n_O.$$

$$\tag{30}$$

Here $n_O = \sum_{i \in J_O^r} n_i$. Therefore, $v(y) = n_{FI} - r_F$. The value in the right part of (30) coincides with the left part in the inequality (9).

Proposition 10. *Let conditions of Proposition 9 hold, and let $\bar{\alpha}$ be the maximal possible step-size, under which $\bar{y} \in \mathcal{K}$. Assume also that the index $j \in J^r$ is such, that $\bar{y}_j \in \partial \mathbb{K}^{n_j}$, when $\alpha = \{\bar{\alpha}$, and this inclusion is violated, when $\alpha > \bar{\alpha}$. Then $v(\bar{y}) \geq v(y)$. Moreover, $v(\bar{y}) = v(y)$, if the index j is unique and $j \in J_F^r$.*

Proof. Recall, that the index j with mentioned properties can't belong to the index set J_I^r, since $\Delta y_i \in \mathbb{K}^{n_i}$ for all indices i from this set.

There are two possibilities, when the assumptions of the proposition are fulfilled.

1) The index $j \in J_F^r$ and $\bar{y}_j = 0_{n_j}$. In this case the index j from the set $J_F^r = J_F^r(y)$ passes to the set $J_I^r = J_I^r(\bar{y})$.

2) The index $j \in J_O^r$ and $\bar{y}_j \in \partial \mathbb{K}^{n_j}$. Now, in contrast to the first case j from the set $J_O^r = J_O^r(y)$ passes to the index set $J_F^r = J_F^r(\bar{y})$. Theoretically, the situation is possible when $\bar{y} = 0_{n_j}$. Then instead of $J_F^r(\bar{y})$ we get $J_I^r(\bar{y})$.

Supposing that the index j is unique, we obtain in the first case

$$J_F^r(\bar{y}) = J_F^r(y) \cup \{k\} \setminus \{j\}, \qquad J_O^r(\bar{y}) = J_O^r(y).$$

Hence, the number of indices in the set J_F^r decreases at one, but the number of indices in the set J_I^r increases at same value one. As a result, we have $v(\bar{y}) = v(y)$.

In the second case

$$J_F^r(\bar{y}) = J_F^r(y) \cup \{k\} \cup \{j\}, \qquad J_O^r(\bar{y}) = J_O^r(y) \setminus \{j\}.$$

Therefore $v(\bar{y}) = v(y) + n_j - 2$. Since by the assumption $n_i \geq 2$ for all $i \in J^r$, we have $v(\bar{y}) \geq v(y)$. Again, the inequality (9) is preserved, but it turned out that the updated extreme point $\lfloor \bar{u}, \bar{y} \rfloor$ may be irregular. □

Thus, updating of the regular extreme point $[u, y] \in \mathcal{F}_D$ consists of the following three steps:

1) Determine the index set J_{FI}^r.
2) Determine the vectors p_{FI} and x_{FI} by formulas (12) and (13). Choose the index k.
3) Verify the conditions of Propositions 7 or 8. If these conditions are not fulfilled, compute the maximal possible step-size $\alpha = \bar{\alpha}$. Make the passage from u to the new point $\bar{u} \in \mathcal{F}_{D,u}$ according to formula (18) with the maximal possible step-size $\alpha = \bar{\alpha}$.

4   The Dual Simplex-Type Algorithm

Using the described passage from the regular extreme point to another one it is possible to construct the dual simplex-type algorithm for solving the SOCP. This algorithm can be described in more usual for simplex-type methods terms, namely, in terms of basic and nonbasic variables and in updating of basis. For this purpose introduce the notions of the basic system of cones and of the basic cone variables.

At first, consider the problem (3), in which $x \in \mathcal{F}_P$. Let $A_i(\mathbb{K}_i^n)$ denote the image of the cone $\mathbb{K}^{n_i}$ under the linear mapping $f(y_i) = A_i y_i$, i.e.

$$A_i(\mathbb{K}^{n_i}) = A_i \mathbb{K}^{n_i} = \{p \in \mathbb{R}^{n_i} : \ p = A_i z, \ z \in \mathbb{K}^{n_i}\}. \tag{31}$$

The set $A_i(\mathbb{K}_i^n)$ is also a cone. Define additionally to (31) a one-dimensional cone. Let $\mathbf{g}_{i,1} \in \mathbb{K}^{n_i}$ and $\mathbf{g}_{i,1} \neq 0_{n_i}$. We have, respectfully,

$$A_i(\mathbf{g}_{i,1}) = \{q \in \mathbb{R}^{n_i} : \ q = A_i z, \ z = \eta \mathbf{g}_{i,}, \ \eta \geq 0\}.$$

Assume that $x \in \mathcal{F}_P$ and $x = [x_F; x_I; x_O]$. Assume also that $x_i = \eta_{i,1} \mathbf{g}_{i,1}$, when $i \in J_F^r$, and $x_B = [x_F; x_I]$.

Compose the matrix

$$\mathcal{A}_B = \left[A_1, \ldots, A_{r_F}, A_{r_F+1}, \ldots, A_{r_F+r_I} \right]. \tag{32}$$

Then the sub-matrices forming (32) may be interpreted as a matrix basis of the point x, since $\mathcal{A}_B x_B = b$. Together with (32) consider the following set of cones

$$\left\{ A_1(\mathbf{g}_{1,1}), \ldots, A_{r_F}(\mathbf{g}_{r_F,1}), A_{r_F+1}(\mathbb{K}^{n_{r_F}+1}), \ldots, A_{r_F+r_I}(\mathbb{K}^{n_{r_F}+r_I}) \right\}. \tag{33}$$

we will call the set (33) by a *conic basis* of the point x, if these cones are linear independent. It means that any nonzero vectors $p_i \in A_i(\mathbb{K}^{n_i})$, $i \in J_I^r$, and any nonzero vectors $q_i \in A_i(\mathbf{g}_i)$, $i \in J_F^r$, are linear independent in common. Due to Assertion 3.2 from [7] the set (33) may be a conic basis, if and only if x is an extreme point of $\mathcal{F}_P$. The following inclusion

$$b \in \sum_{i \in J_F^r} A_i(\mathbf{g}_{i,1}) + \sum_{i \in J_I^r} A_i(\mathbb{K}^{n_i})$$

takes place. Cones $A_1(\mathbf{g}_{1,1}), \ldots, A_{r_F}(\mathbf{g}_{r_F,1})$ we will call by *facet basic cones.* Cones $A_{r_F+1}(\mathbb{K}^{n_{r_F}+1}), \ldots, A_{r_F+r_I}(\mathbb{K}^{n_{r_F}+r_I})$ we will call by *interior basic cones.* Remark, that the facet basic cones are depended from the points x_i, $i \in J_F^r(x)$. The matrix $\mathcal{A}_B$ can be a matrix basis of the point $x \in \mathcal{F}_P$, if and only if the set of cones (33) is a conic basis of x.

In similar way, it is possible to define the matrix and conic bases in the dual problem (4). Let $[u, y] \in \mathcal{F}_D$ and let $y = [y_F; y_I; y_O]$, $y_i = \theta_i \mathbf{h}_{i,1}$, $i \in J_F^r(y)$. Consider the set of cones

$$\left\{ A_1(\mathbf{h}_{1,1}), \ldots, A_{r_F}(\mathbf{h}_{r_F,1}), A_{r_F+r_I+1}(\mathbb{K}^{n_{r_F}+r_I+1}), \ldots, A_r(\mathbb{K}^{n_r}) \right\}. \tag{34}$$

By *dual conic basis* of the point $[u, y] \in \mathcal{F}_D$ we will understand a "supplement" of the (34) to the total set of cones. Namely,

$$\left\{ A_1(\mathbf{H}_i'), \ldots, A_{r_F}(\mathbf{H}_{r_F}'), A_{r_F+1}(\mathbb{K}^{n_{r_F}+1}), \ldots, A_{r_F+r_I}(\mathbb{K}^{n_{r_F}+r_I}) \right\}. \tag{35}$$

where $\mathbf{H}'_i = \text{cone}(\pm \mathbf{h}_{i,2}, \ldots, \pm \mathbf{h}_{i,n_i-1}, \mathbf{h}_{i,n_i})$ is a conic hull of the vector $\mathbf{h}_{i,n_i}$ and of the vectors $\pm \mathbf{h}_{i,2}, \ldots, \pm \mathbf{h}_{i,n_i-1}$. By *dual basic variables* we will understand $y_{i,\text{R}} = 0_{n_i-1}$, $i \in J_F^r$ and $y_i = 0_{n_i}$, $i \in J_I^r$. Correspondingly, by *dual non-basic variables* we will understand $y_i = \theta_{i,1}\mathbf{h}_{i,1}$, $i \in J_F^r$ and $y_i \in \text{int }\mathbb{K}^{n_i}$, $i \in J_O^r$. By the dual matrix basis of $[u,y] \in \mathcal{F}_D$ we will understand the set of corresponding matrices consisting the matrix (32). As in the case of primal problem, the matrix (32) forms the dual matrix basis if and only if the set of cones (35) is a dual conic basis of $[u,y] \in \mathcal{F}_D$.

Let us describe the dual simplex-type algorithm. Assume that at k^{th} iteration there are the regular extreme point $u \in \mathcal{F}_{D,u}$ and the corresponding dual slack variable $y = y(u)$ with dual basic variables y_{FI} defining by the index set J_{FI}^r.

STEP 1. Using the index set J_{FI}^r, compute the primal basic solution x_{FI} of the equation $Ax = b$ and define x_i, $i \in J_{FI}^r$.

STEP 2. If $x_i \in K^{n_i}$ for all $i \in J_{FI}^r$, then STOP.

STEP 3. Let $k \in J_{FI}^r$ and $x_k \notin K^{n_i}$. Form the vector z_k and compute the displacement vector Δu.

STEP 4. Compute the maximal step-size $\bar{\alpha}$ and define the index $j \in J_F^r \cup J_O^r$ such that $y_j - \bar{\alpha} A_j^T \Delta u \in \partial K^{n_j}$. Make the passage from the point u to the point $\bar{u} = u - \bar{\alpha} u$ with the updated slack variable $\bar{y} = y(\bar{u})$.

As it follows from the proposed algorithm, the passage from the point $[u,y] \subset \mathcal{F}_D$ to the updated point $[\bar{u}, \bar{y}] \in \mathcal{F}_D$ can be interpreted as removing one cone from the dual conic basis and replacing it by another cone from the set (34) defined by the index j. Independently, is the index k belongs to the set J_F^r or to the set J_I^r, the newly introduced cone does not belong to the previous basis.

5   Convergence of the Method

Let the starting extreme point $[u_0, y_0] \in \mathcal{F}_D$ be given. Assume that this point is regular, and all consequent points generated by the method are regular too. Such sequence of extreme points $\{[u_t, y_t]\}$ we will call *regular*.

Theorem 1. *Let the point $[u_*, y_*]$ be regular unique solution of the dual problem* (4). *Let also the starting extreme point $[u_0, y_0] \in \mathcal{F}_D$ be such that the sequence $\{[u_t, y_t]\}$ is regular. If $\{[u_t, y_t]\}$ is finite, then the last point of the sequence coincides with $[u_*, y_*]$.*

Further, consider the case when the sequence $\{[u_t, y_t]\}$ is infinite. Suppose that the problem (4) is *non-degenerate* in the sense that all extreme points of the feasible set $\mathcal{F}_D$ are non-degenerate. We say also that the problem (3) is *strongly non-degenerate*, if all vectors x_{FI} of the form (13) are non-degenerate basic solutions of the equation $\mathcal{A}x = b$.

Theorem 2. *Let the problem* (4) *be non-degenerate with the extreme regular point $[u_*, y_*] \in \mathcal{F}_D$ being its unique solution. Let additionally the problem* (3) *be strongly non-degenerate. Assume also that the starting point* $[u_0, y_0] \in \mathcal{F}_D$ *is such that the set*

$$\mathcal{F}_{D,u}(u_0) = \{u \in \mathcal{F}_{D,u} : \langle b, u \rangle \geq \langle b, u_0 \rangle\} \tag{36}$$

is bounded, and the dual simplex-method generates the infinite sequence $\{[u_t, y_t]\}$. *Then the sequence* $\{[u_t, y_t]\}$ *converges to* $[u_*, y_*]$.

Proof. Since the set (36) is bounded, the sequence $\{[u_t, y_t]\}$ is bounded too. Therefore, there exist the limit points of this sequence. Let the subsequence $\{[u_{t_l}, y_{t_l}]\}$ be such that $[u_{t_l}, y_{t_l}] \to [\bar{u}, \bar{y}]$, when $l \to \infty$. It is clear that $[\bar{u}, \bar{y}] \in \mathcal{F}_D$.

The number of possible dual matrix bases in the problem (4) is finite. Therefore, we may assume without loss of generality that the points $[u_{t_l}, y_{t_l}]$ are such, that the corresponding sub-vectors y_{FI, t_l} belong to one and the same dual matrix basis. This basis is defined by the index set $\overline{J_{FI}^r}$. We have by aforesaid that $J_{FI}^r(y_{t_l}) = \overline{J_{FI}^r}$ for all $l \geq 1$. Hence, $\overline{J_{FI}^r} \subseteq J_{FI}^r(\bar{y})$, and the inequality (9) at the point $\bar{y}$ takes place.

Let $\overline{J_F^r}$ be a corresponding subset of the set $\overline{J_{FI}^r}$. Consider the matrices $\mathbf{H}_{i,R}$, $i \in \overline{J_F^r}$, which are the right $n_i \times (n_i - 1)$ sub-matrices of the orthogonal matrices $\mathbf{H}_i$. These sub-matrices are restricted by the norm in total. Hence, it is possible to extract from sequences $\{\mathbf{H}_{i,R}(y_{t_l})\}$, $i \in \overline{J_F^r}$, the converged subsequences. Assume without loss of generality that the sequence $\{[u_{t_l}, y_{t_l}]\}$ itself is chosen in such a way that the sequences $\{\mathbf{H}_{i,R}(y_{t_l})\}$ converge to the corresponding matrices $\overline{\mathbf{H}}_{i,R}$ for all $i \in \overline{J_F^r}$.

Since $[u_{t_l}, y_{t_l}]$ are regular extreme points, the matrices $\mathcal{A}_{FI}^{\mathbf{H}_R}$, defined at the points y_{t_l}, are nonsingular. Moreover, all x_{FI, t_l} are non-degenerate basic solutions of the equation $\mathcal{A}x = b$. If we take the solution $\bar{x}_{FI}$ of the equation $\mathcal{A}x = b$ with the matrix $\mathcal{A}_{FI}^{\bar{\mathbf{H}}_R}$, then by assumption of strong non-degeneracy of the primal problem, the solution $\bar{x}_{FI}$ is basic and non-degenerate. Therefore, the rows of the matrix $\mathcal{A}_{FI}^{\bar{\mathbf{H}}_R}$ are linear independent. We obtain that the limit point $[\bar{u}, \bar{y}]$ is an extreme point of the set $\mathcal{F}_D$.

Suppose that the limit point $[\bar{u}, \bar{y}]$ is not a solution of the problem (4). By the proposed algorithm in this case for the vectors $\bar{y}_i$, $i \in J_{FI}^r(\bar{y})$, we have for some $k \in J_{FI}^r(\bar{y})$ that $\bar{\eta}_{k,1} < 0$. Here the bar over $\eta_{1,k}$ indicates that this coefficient is related to the decomposition (14) just for the point $\bar{y}_k$.

We obtain by Proposition 6 that $\langle b, u_{t_l+1} \rangle = \langle b, u_{t_l} \rangle - \alpha_{t_l} \eta_{k t_l, 1} > \langle b, u_{t_l} \rangle$. As Δu_{t_l} are bounded, there exists the constant $c > 0$ such that $\alpha_{t_l} \geq c$ for l sufficiently large. Thus, $\langle b, u_{t_l+1} \rangle > \langle b, \bar{u} \rangle$ at some iteration. This inequality contradicts to convergence of the sequence $\{u_{t_l}\}$ to $\bar{u}$, because of the objective function in the dual problem must increase monotonically. Hence, $[\bar{u}, \bar{y}]$ may be only the optimal solution $[u_*, y_*]$ of (4). □

Consider the partial case of the problem (3), in which the dimension of the all cones $\mathbb{K}^{n_i}$ coincides between themselves and is equal to two. Then any extreme point $[u, y] \in \mathcal{F}_D$ is such that the number of columns in the matrix $\mathcal{A}_{FI}^{\mathbf{H}_R}$ is equal to $d_{FI} = r_F + 2r_I$. The extreme point will be regular if $d_{FI} = m$.

Proposition 11. *Let the problem* (4) *with* $n_i = 2$, $i \in J^r$, *be non-degenerate and have the solution. Moreover, let* $[u_t, y_t] \in \mathcal{F}_D$ *be a regular extreme point. Then the consequent point* $[u_{t+1}, y_{t+1}]$ *is a regular extreme point too.*

Proof. Assume that index $k \in J^r_{FI}$ belongs to the set J^r_I. Then it is turned out that the index j belongs either to the set $J^r_F(y_{t+1})$ or to the set $J^r_I(y_{t+1})$. The first case appears, when $j \in J^r_O(y_t)$, and the second case, when $j \in J^r_F(y_t)$. At any case the index k passes from the set $J^r_I(y_t)$ to the set $J^r_F(y_{t+1})$. If $j \in J^r_F(y_t)$, then we obtain that in each set J^r_F and J^r_I the amount of indices does not change, since

$$J^r_I(y_{t+1}) = J^r_I(y_t) \cup \{j\} \setminus \{k\}, \quad J^r_F(y_t) = J^r_F(y_{t+1}) \cup \{k\} \setminus \{j\}.$$

Thus, in both cases $d_{FI}(y_{t+1}) = d_{FI}(y_t) = m$. The equality $d_{FI}(y_{t+1}) = d_{FI}(y_t) = m$ is valid also, when $k \in J^r_F(y_t)$.

Since the point $[u_{t+1}, y_{t+1}]$ is non-degenerate, it is an extreme point too. $\square$

According to Proposition 11 the sequence $\{[u_t, y_t]\}$, generated by the algorithm of the dual simplex-method is regular, if the starting point $[u_0, y_0]$ is also regular. Therefore, solving the problem (3) with $n_i = 2$, $i \in J^r$, by theorem 2 we get the convergent sequence $\{[u_t, y_t]\}$. This sequence converges to the solution of the problem (4).

6 Conclusion

We presented the dual simplex method for the SOCP. The main attention have been given to updating of regular extreme points. In principle, it is possible to develop the approach for updating irregular extreme points. However, this approach is more complicated in compare with the regular case.

References

1. Alizadeh, F., Goldfarb, D.: Second-order cone programming. Math. Program. Ser. B. **95**, 3–51 (2003)
2. Lobo, M.S., Vandenberghe, L., Boyd, S., Lebret, H.: Applications of second order cone programming. Linear Algebra Appl. **284**, 193–228 (1998)
3. Wolkowicz, H., Saigal, R., Vandenberghe, L. (eds.): Handbook of Semidefinite Programming. Kluwer Acad. Publ, Dordrecht (2000)
4. Anjos, M.F., Lasserre, J.B. (eds.): Handbook of Semidefinite, Cone and Polynomial Optimizatin: Theory, Algorithms, Software and Applications. Springer, New York (2011)
5. Nesterov, Y.E., Todd, M.J.: Primal-dual Interior-Point Methods for Self-Scaled Cones. SIAM J. Optimization **8**, 324–364 (1998)
6. Monteiro, R.D.C., Tsuchiya, T.: Polynomial convergence of primal-dual algorithms for second-order cone program based on the MZ-family of directions. Math. Program. **88**(1), 61–83 (2000)

7. Pataki, G.: Cone-LP's and semidefinite programs: Geometry and a simplex-type method. In: Cunningham, W.H., McCormick, S.T., Queyranne, M. (eds.) IPCO 1996. LNCS, vol. 1084, pp. 162–174. Springer, Heidelberg (1996). https://doi.org/10.1007/3-540-61310-2_13

8. Muramatsu, M.: A pivoting procedure for a class of second-order cone programming. Optim. Methods Softw. **21**(2), 295–314 (2006)

9. Hayashi, S., Okuno, T., Ito, Y.: Simplex-type algorithm for second-order cone programming via semi-infinite programming reformulation. Optim. Methods Softw. **31**(6), 1272–1297 (2016)

10. Goberna, M.A., Lopez, M.A.: Linear Semi-Infinite OPtimization. John Wiley and Sons Ltd., New York (1998)

11. Zhadan, V.: A variant of the simplex method for second-order cone programming. In: Khachay, M., Kochetov, Y., Pardalos, P. (eds.) MOTOR 2019. LNCS, vol. 11548, pp. 115–129. Springer, Cham (2019). https://doi.org/10.1007/978-3-030-22629-9_9

About Difference Schemes for Solving Inverse Coefficient Problems

Vladimir Zubov$^{(\boxtimes)}$ⓘ and Alla Albuⓘ

Dorodnicyn Computing Centre, Federal Research Center "Computer Science
and Control" of Russian Academy of Sciences, Moscow, Russia
vladimir.zubov@mail.ru

Abstract. The investigation deals with the choice of a finite-difference
scheme for approximating the heat diffusion equation when solving the
inverse coefficient problem in a three-dimensional formulation. Using the
examples of a number of nonlinear problems for a three-dimensional
heat equation whose coefficients depend on temperature, a comparative
analysis of several schemes of alternating directions was performed. The
following schemes were examined: a locally one-dimensional scheme, a
Douglas-Reckford scheme, and a Pisman-Reckford scheme. Each numerical method was used to obtain the temperature distribution inside the
parallelepiped. When comparing methods, the accuracy of the obtained
solution and the computer time to achieve the required accuracy were
taken into account. The inverse coefficient problem was reduced to the
variational problem. Based on the carried out research, recommendations were made regarding the choice of a finite-difference scheme for the
discretization of the primal problem when solving the inverse coefficient
problem.

Keywords: Three-dimensional heat equation · Numerical methods ·
Nonlinear problems · Alternating directions schemes

1 Introduction

In [1–7] the inverse coefficient problem of identification of the temperature-
dependent thermal conductivity coefficient of a substance was studied. It was
based on the Dirichlet boundary value problem for the one-dimensional and two-
dimensional nonstationary heat equations. The studied objects were a layer of
material and a rectangular plate. The inverse coefficient problem was reduced to
a variational one. In the above-mentioned works an effective algorithm for the
numerical solution of the inverse coefficient problem based on the modern Fast
Automatic Differentiation technique (see [8–10]) was used.

The one- and two-dimensional variants of the above considered inverse problem are of great mathematical interest. However in practice experimental data is
collected for three-dimensional objects. Therefore it is desirable to solve the problem of identifying the thermal conductivity coefficient in the three-dimensional

N. Olenev et al. (Eds.): OPTIMA 2020, LNCS 12422, pp. 317–330, 2020.
https://doi.org/10.1007/978-3-030-62867-3_23

formulation. A major part of the algorithm proposed for solving the inverse coefficient problem is the solution of the primal problem (determination of the temperature field at any point of the object at any time). The effectiveness of solving the inverse problem as a whole depends on the effectiveness of its solution.

When the variational problem is solved numerically using the Fast Automatic Differentiation technique, the approximation of the conjugate problem is uniquely determined by the choice of the primal problem approximation and, as shown by the study of the inverse coefficient problem in a two-dimensional formulation, in the transition from the one-dimensional case to the multidimensional case the influence of this choice becomes more significant. So, if the initial boundary value problem in the two-dimensional case is approximated by an implicit scheme with weights (see [3]), the approximation of the conjugate problem will be a linear system of algebraic equations, for the solution of which (due to two-dimensionality in space) it is also necessary to construct an iterative process. In this regard, to solve the primal problem in the multidimensional case, it was recommended to use some economical finite-difference scheme (see [11]).

In [4–6] to approximate the thermal conductivity equation the Pisman-Reckford scheme of alternating directions was chosen, which is unconditionally stable when considering the problem in two-dimensional space (see [12]). In this case to solve the conjugate problem it is not necessary to organize an iterative process. The resulting linear system of algebraic equations is split into a number of subsystems and each is solved by applying the tridiagonal matrix algorithm.

The algorithm for solving a direct problem in the three-dimensional case is even more complicated and requires a lot of computation time. One of the ways to reduce the computation time is to choose a good approximation of the primal problem. In the literature there are a large number of numerical methods that can be used to solve the heat diffusion equation in the three-dimensional case. Therefore a natural question arises: which method to choose for solving a concrete problem?

This question was investigated in the work of Jules Thibault [13]. There, a practical comparison of the nine more common three-dimensional methods is presented. Each numerical method was used to obtain the temperature distribution inside the parallelepiped. The difference schemes studied in [13] are divided into three main classes: explicit schemes, implicit schemes and schemes of alternating directions. When comparing schemes, the accuracy of the resulting solution, ease of programming, calculation time and required computer memory were taken into account. The methods of alternating directions were considered the most effective for solving three-dimensional problems of thermal conductivity. These methods also proved their effectiveness in the study of many practical problems that were solved over a long period of time after the publication of the above-mentioned work.

Comparison of numerical methods in [13] was carried out for a rather simple problem in which all coefficients of the heat equation were constant. When solving inverse coefficient problems for a non-stationary heat equation, we deal with nonlinear problems, since we consider the temperature-dependent coefficients of the thermal conductivity of a substance. However in the literature we have not found any recommendations on which method to give preference to when solving this type of problem, thus we decided to investigate this question. In addition, we would like to note the fact that the comparisons made in [13] have lost their significance, since there the approximation of the equation in space was carried out on a template that contained too few points in each direction (5 or 10 points). The computer technology currently used allows us to solve practical problems on much more detailed grids.

In the present work, on the example of several nonlinear problems for the three-dimensional heat equation with coefficients dependent on temperature, a comparative analysis of the difference schemes of alternating directions is carried out, which in our opinion should be used in solving the problem of identifying the thermal conductivity coefficient of a substance in the three-dimensional case. Locally-one-dimensional schemes, the Douglas-Reckford scheme, and the Pisman-Reckford scheme were used (see [14–16]). When comparing the methods, the accuracy of the resulting solution and the time to achieve the required accuracy on the computer were taken into account.

The problem of identifying the thermal conductivity coefficient of a substance is related to the study of the characteristics of newly created materials. When experimental studies are being carrying out, as a rule, samples of material of a simple form are used (usually this is a parallelepiped). It is reasonable to consider the inverse coefficient problem arising in the three-dimensional case also for a parallelepiped object. Therefore, all the numerical methods used in this work were applied to obtain the temperature distribution inside a parallelepiped based on the first boundary value problem for a three-dimensional non-stationary heat equation.

2   Statement of the Problem

Suppose that the specimen under study is a parallelepiped of length R, width L and height H. The initial temperature T of the parallelepiped is known. The law of temperature variation on the surfaces of the parallelepiped is also known. For a mathematical description of the heat conduction process in a parallelepiped we use the Cartesian coordinates x, y and z. The points $s = (x, y, z)$ of the parallelepiped form a domain $Q = \{[0, R] \times [0, L] \times [0, H]\}$ with a boundary $\Gamma = \partial Q$. The temperature field at each time is described by the initial-boundary value (mixed) problem:

$$C(s)\frac{\partial T(s,t)}{\partial t} = div_s(K(T(s,t))\nabla_s T(s,t)), \qquad (s,t)\{\in Q \times (0,\Theta]\}, \quad (1)$$

$$T(s,0) = w_0(s), \qquad\qquad s \in \overline{Q}, \qquad\qquad (2)$$

$$T(s,t) = w_\Gamma(s,t), \qquad\qquad s \in \Gamma, \quad 0 \le t \le \Theta. \qquad (3)$$

Here t is time; $T(s,t) \equiv T(x,y,z,t)$ is the temperature of the material at the point s with the coordinates (x,y,z) at time t; $C(s)$ is the volumetric heat capacity of the material; $K(T)$ is the thermal conductivity coefficient; $w_0(s)$ is the given temperature at the initial time $t = 0$; $w_\Gamma(s,t)$ is the given temperature on the boundary of the object. The volumetric heat capacity of a substance $C(s)$ is considered a known function of the coordinates.

If the dependence of the thermal conductivity coefficient $K(T)$ on the temperature T is known, then we can solve the mixed problem (1)–(3) to find the temperature distribution $T(s,t)$ in $Q \times [0, \Theta]$. Problem (1)–(3) described above we will call the primal problem.

3   Algorithm for Solving a Direct Problem

To solve the primal problem numerically, we introduce grids (generally nonuniform) in time and space.

The time grid was constructed by a set of nodal values $\{t^j\}_{j=0}^J$, $t^0 = 0$, $t^J = \Theta$. The steps τ^j of this grid were determined by the relations $\tau^j = t^{j+1} - t^j$, $j = \overline{0, J-1}$.

Additionally we introduce two spatial grids: basic and auxiliary. To construct the basic spatial grid on the segment $[0, R]$, a system of nodal points $\{x_n\}_{n=0}^N$ was chosen so that $x_0 = 0$, $x_N = L$ and $x_n < x_{n+1}$ for all $0 \le n < N$. The distance between x_n and x_{n+1} was denoted as h_n^x, i.e. $h_n^x = x_{n+1} - x_n$, $n = \overline{0, N-1}$. Similarly, on the intervals $[0, L]$ and $[0, H]$ we choose sets of reference points $\{y_i\}_{i=0}^I$ and $\{z_l\}_{l=0}^L$ such that $y_0 = 0$, $y_I = R$ and $y_i < y_{i+1}$ for all $0 \le i < I$ and $z_0 = 0$, $z_L = H$ and $z_l < z_{l+1}$ for all $0 \le l < L$, respectively. In this case $h_i^y = y_{i+1} - y_i$, $i = \overline{0, I-1}$ and $h_l^z = z_{l+1} - z_l$, $l = \overline{0, L-1}$. The points of the basic grid are the set of points $\{x_n, y_i, z_l\}$ where $n = \overline{0, N}$, $i = \overline{0, I}$ and $i = \overline{0, I}$.

Auxiliary grid $\{\tilde{x}_n, \tilde{y}_i, \tilde{z}_l\}$, $n = \overline{0, N+1}$, $i = \overline{0, I+1}$ and $l = \overline{0, L+1}$ was built in a similar way:

$$\tilde{x}_0 = x_0, \quad \tilde{x}_{N+1} = x_N, \quad \tilde{x}_n = x_{n-1} + h_{n-1}^x/2, \quad n = \overline{1, N},$$

$$\tilde{y}_0 = y_0, \quad \tilde{y}_{I+1} = y_I, \quad \tilde{y}_i = y_{i-1} + h_{i-1}^y/2, \quad i = \overline{1, I}.$$

$$\tilde{z}_0 = z_0, \quad \tilde{z}_{L+1} = z_L, \quad \tilde{z}_l = z_{l-1} + h_{l-1}^z/2, \quad l = \overline{1, L}.$$

The lines $\{\tilde{x}_n, \tilde{y}_i, \tilde{z}_l\}$ divide the domain Q into computational cells. An elementary cell is assigned the indices (n, i, l) if the cell vertex nearest to the point (x_0, y_0, z_0) coincides with the grid point $\{\tilde{x}_n, \tilde{y}_i, \tilde{z}_l\}$. The volume of this cell is denoted by V_{nil}, i.e.,

$$V_{nil} = \frac{h_n^x + h_{n-1}^x}{2} \cdot \frac{h_i^y + h_{i-1}^y}{2} \cdot \frac{h_l^z + h_{l-1}^z}{2}, \quad n = \overline{0, N-1}, i = \overline{0, I-1}, l = \overline{0, L-1}.$$

The numerical algorithm for solving the primal problem is based on the heat balance equation in the unit cell, which means that the variation in the heat content within a volume V_{nil} over a fixed time interval $\tau^j = t^{j+1} - t^j$ is equal to the amount of heat transferred through the surface S_{nil} of V_{nil} over the same time interval:

$$\iiint\limits_{V_{nil}} [E(x,y,z,t^{j+1}) - E(x,y,z,t^j)]dV = \int\limits_{t^j}^{t^{j+1}} \iint\limits_{S_{nil}} K(T)T_{\overline{n}}dSdt,$$

where $E(x,y,z,t)$ is the density of the heat content.

The average temperature in the cell with indices (n,i,l) is denoted by $T_{nil}(t)$, and its value at the time t^j by T_{nil}^j. Then the last relation can be approximately written as:

$$C_{nil} \cdot V_{nil} \cdot (T_{nil}^{j+1} - T_{nil}^j) \cong \int\limits_{t^j}^{t^{j+1}} \left\{ \iint\limits_{S_{nil}^{x+}} K(T) \left.\frac{\partial T}{\partial x}\right|_{x=\tilde{x}_{n+1}} dydz \right.$$

$$-\iint\limits_{S_{nil}^{x-}} K(T) \left.\frac{\partial T}{\partial x}\right|_{x=\tilde{x}_n} dydz + \iint\limits_{S_{nil}^{y+}} K(T) \left.\frac{\partial T}{\partial y}\right|_{y=\tilde{y}_{i+1}} dxdz - \iint\limits_{S_{nil}^{y-}} K(T) \left.\frac{\partial T}{\partial y}\right|_{y=\tilde{y}_i} dxdz$$

$$\left. + \iint\limits_{S_{nil}^{z+}} K(T) \left.\frac{\partial T}{\partial z}\right|_{z=\tilde{z}_{l+1}} dxdy - \iint\limits_{S_{nil}^{z-}} K(T) \left.\frac{\partial T}{\partial z}\right|_{z=\tilde{z}_l} dxdy \right\} dt. \quad (4)$$

Here, S_{nil}^{x+} denotes the part of S_{nil} that belongs to the plane $x = \tilde{x}_{n+1}$ and S_{nil}^{x-} denotes the part of S_{nil} that belongs to the plane $x = \tilde{x}_n$; S_{nil}^{y+}, S_{nil}^{y-}, S_{nil}^{z+} and S_{nil}^{z-} are defined in a similar fashion. The heat flow on the surface $x = \tilde{x}_{n+1}$ of the cell at a time t is approximated by the formula:

$$C_{nil} \cdot V_{nil} \cdot (T_{nil}^{j+1} - T_{nil}^j) \cong \int\limits_{t^j}^{t^{j+1}} [\Lambda_x T_{nil}(t) + \Lambda_y T_{nil}(t) + \Lambda_z T_{nil}(t)] \, dt, \quad (5)$$

$$\Lambda_x T_{nil}(t) = \left[\frac{K(T_{n+1,il}(t)) + K(T_{nil}(t))}{2} \cdot \frac{T_{n+1,il}(t) - T_{nil}(t)}{h_n^x} \right.$$

$$\left. - \frac{K(T_{n-1,il}(t)) + K(T_{nil}(t))}{2} \cdot \frac{T_{nil}(t) - T_{n-1,il}(t)}{h_{n-1}^x} \right] \cdot S_{il}^{yz},$$

$$\Lambda_y T_{nil}(t) = \left[\frac{K(T_{n,i+1,l}(t)) + K(T_{nil}(t))}{2} \cdot \frac{T_{n,i+1,l}(t) - T_{nil}(t)}{h_i^y} \right.$$

$$\left. - \frac{K(T_{n,i-1,l}(t)) + K(T_{nil}(t))}{2} \cdot \frac{T_{nil}(t) - T_{n,i-1,l}(t)}{h_{i-1}^y} \right] \cdot S_{nl}^{xz},$$

$$\Lambda_z T_{nil}(t) = \left[\frac{K(T_{ni,l+1}(t)) + K(T_{nil}(t))}{2} \cdot \frac{T_{ni,l+1}(t) - T_{nil}(t)}{h_l^z} \right.$$

$$-\frac{K(T_{ni,l-1}(t)) + K(T_{nil}(t))}{2} \cdot \frac{T_{nil}(t) - T_{ni,l-1}(t)}{h_{l-1}^z}\right] \cdot S_{ni}^{xy},$$

$$S_{il}^{yz} = \frac{h_i^y + h_{i+1}^y}{2} \cdot \frac{h_l^z + h_{l+1}^z}{2}, \qquad S_{nl}^{xz} = \frac{h_n^x + h_{n+1}^x}{2} \cdot \frac{h_l^z + h_{l+1}^z}{2},$$

$$S_{ni}^{xy} = \frac{h_n^x + h_{n+1}^x}{2} \cdot \frac{h_i^y + h_{i+1}^y}{2}.$$

Further, depending on how Eq. (5) is approximated in time, we obtain one or another difference scheme with which the primal problem (1)–(3) is solved.

The experience of studying the problem of identifying the thermal conductivity coefficient in a two-dimensional formulation allowed us to conclude that it is desirable to use one of the schemes of alternating directions when solving multidimensional inverse coefficient problems (see [17]). The most popular among them are locally-one-dimensional scheme, the Douglas-Reckford scheme, and the Pisman-Reckford scheme. Along with the main values T_{nil}^j of the desired grid function, the above schemes use two intermediate values $T_{nil}^{j+\frac{1}{3}}$ and $T_{nil}^{j+\frac{2}{3}}$ of the temperature function, which can be formally considered as the values of the temperature at the moment of time $t^j + \tau^j/3$ and $t^j + 2\tau^j/3$ (see [11]).

For a more compact representation of difference schemes, we introduce the following notation

$$\Lambda_x^p(T_{nil}^j) = \left[\frac{K(T_{n+1,il}^p) + K(T_{nil}^p)}{2} \cdot \frac{T_{n+1,il}^j - T_{nil}^j}{h_n^x}\right.$$
$$\left. -\frac{K(T_{n-1,il}^p) + K(T_{nil}^p)}{2} \cdot \frac{T_{nil}^j - T_{n-1,il}^j}{h_{n-1}^x}\right] \cdot S_{il}^{yz},$$

which we will use for the time discretization of the operator $\Lambda_x T_{nil}(t)$. Similar notations $\Lambda_y^p(T_{nil}^j)$ and $\Lambda_z^p(T_{nil}^j)$ are introduced for the time discretization of the operators $\Lambda_y T_{nil}(t)$ and $\Lambda_z T_{nil}(t)$.

The local one-dimensional scheme has the form:

$$x - direction$$
$$\frac{C_{nil} \cdot V_{nil} \cdot (T_{nil}^{j+1/3} - T_{nil}^j)}{\tau^j/3} = \Lambda_x^p(T_{nil}^{j+1/3});$$
$$y - direction$$
$$\frac{C_{nil} \cdot V_{nil} \cdot (T_{nil}^{j+2/3} - T_{nil}^{j+1/3})}{\tau^j/3} = \Lambda_y^q(T_{nil}^{j+2/3});$$
$$z - direction$$
$$\frac{C_{nil} \cdot V_{nil} \cdot (T_{nil}^{j+1} - T_{nil}^{j+2/3})}{\tau^j/3} = \Lambda_z^r(T_{nil}^{j+1});$$
$$(n = \overline{1, N-1}, \ i = \overline{1, I-1}, \ l = \overline{1, L-1}, \ j = \overline{0, J-1}).$$

As noted in [15] this scheme can be considered in two variants:

i) $p = j + 1/3$, $q = j + 2/3$, $r = j + 1$;
ii) $p = j$, $q = j + 1/3$, $r = j + 2/3$.

In case (i), to determine the unknown function at each stage, we obtain a nonlinear equation, which is solved by an iterative method and at every iteration step, the unknown function is determined using the tridiagonal matrix algorithm. In case (ii), at each stage we obtain a linear equation, which is solved by applying the tridiagonal matrix algorithm.

The Douglas-Reckford scheme

$$x - direction$$
$$\frac{C_{nil}\cdot V_{nil}\cdot(T_{nil}^{j+1/3}-T_{nil}^{j})}{\tau^{j}/3} = \Lambda_{x}^{j+1/3}(T_{nil}^{j+1/3}) + \Lambda_{y}^{j}(T_{nil}^{j}) + \Lambda_{z}^{j}(T_{nil}^{j});$$
$$y - direction$$
$$\frac{C_{nil}\cdot V_{nil}\cdot(T_{nil}^{j+2/3}-T_{nil}^{j+1/3})}{\tau^{j}/3} = \Lambda_{y}^{j+2/3}(T_{nil}^{j+2/3}) - \Lambda_{y}^{j}(T_{nil}^{j});$$
$$z - direction$$
$$\frac{C_{nil}\cdot V_{nil}\cdot(T_{nil}^{j+1}-T_{nil}^{j+2/3})}{\tau^{j}/3} = \Lambda_{z}^{j+1}(T_{nil}^{j+1}) - \Lambda_{z}^{j}(T_{nil}^{j});$$
$$(n = \overline{1,N-1},\ i = \overline{1,I-1},\ l - \overline{1,L-1},\ j - \overline{0,J-1}).$$

The above schemes are stable and are most often used by researchers in practice.

The following Pisman-Reckford scheme for the three-dimensional heat conduction equation is conditionally stable:

$$x - direction$$
$$\frac{C_{nil}\cdot V_{nil}\cdot(T_{nil}^{j+1/3}-T_{nil}^{j})}{\tau^{j}/3} = \Lambda_{x}^{j}(T_{nil}^{j+1/3}) + \Lambda_{y}^{j}(T_{nil}^{j}) + \Lambda_{z}^{j}(T_{nil}^{j});$$
$$y - direction$$
$$\frac{C_{nil}\cdot V_{nil}\cdot(T_{nil}^{j+2/3}-T_{nil}^{j+1/3})}{\tau^{j}/3} = \Lambda_{x}^{j}(T_{nil}^{j+1/3}) + \Lambda_{y}^{j}(T_{nil}^{j+2/3}) + \Lambda_{z}^{j}(T_{nil}^{j+1/3});$$
$$z - direction$$
$$\frac{C_{nil}\cdot V_{nil}\cdot(T_{nil}^{j+1}-T_{nil}^{j+2/3})}{\tau^{j}/3} = \Lambda_{x}^{j}(T_{nil}^{j+2/3}) + \Lambda_{y}^{j}(T_{nil}^{j+2/3}) + \Lambda_{z}^{j}(T_{nil}^{j+1});$$
$$(n = \overline{1,N-1},\ i = \overline{1,I-1},\ l = \overline{1,L-1},\ j = \overline{0,J-1}).$$

In [12] the stability of the Pisman-Reckford difference scheme was analyzed using the von Neumann criteria for the n-dimensional heat equation and it was shown that for $n \geq 3$ it is conditionally stable even if the coefficients of the equation do not depend on temperature. It requires working with a much smaller time step than the first two schemes. But this fact allows us to use a known temperature value on the previous time layer to determine the heat-conductivity coefficient on the current sublayer, i.e. to work with the above-mentioned non-iterative version of this scheme.

Which of these schemes and in which cases to give preference, we will try to clarify on the basis of numerical calculations, the results of which are presented below.

4   Results of Numerical Calculations

A huge number of numerical experiments were conducted regarding the solution of the first boundary value problem for a three-dimensional nonlinear heat-conductivity equation with different input data. The most interesting of them are given in this section in the form of three series of calculations.

The results of numerical calculations showed that both versions of the locally one-dimensional scheme give almost the same calculation accuracy. Therefore, here we present the results of calculations obtained only when using a non-iterative scheme, since it is more efficient.

The temperature interval $[a, b]$ on which the function $K(T)$ will be restored is defined as the set of values of the given functions $w_0(x, y, z)$ and $w_\Gamma(x, y, z, t)$, i.e. the boundaries of the segment $[a, b]$ were assigned as the minimum and maximum values of the indicated functions. This interval is partitioned by the points $\widetilde{T}_0 = a, \widetilde{T}_1, \widetilde{T}_2, ..., \widetilde{T}_M = b$ into $M = 100$ parts. The function $K(T)$ to be found was approximated by a continuous piecewise linear function in the same way as shown in [4]. If the temperature at the point fell outside the boundaries of the interval $[a, b]$, then the linear extrapolation was used to determine the function $K(T)$.

All calculations were performed on four uniform spatial grids with the same step in each spatial direction equal to $1/N$ ($N = 25, 50, 100, 200$). In cases where a stable scheme was used to solve a primal problem, the time step value varied within fairly large (acceptable) limits: for each spatial grid, calculations were performed at $\tau = 0.1, 0.04, 0.02, 0.01, 0.002, 0.001$. When using the Pisman-Reckford scheme for time discretization, the time step value was selected for each spatial grid independently, taking into account the stability criterion. For the chosen space-time grid, the numerical solution of the mixed problem was determined using the finite-difference schemes described above, and it was compared with the analytical solution. The norm $C[a, b]$ was used as a measure of deviation of the calculated solution from the analytical one.

4.1   First Series of Calculations

The example that was considered in this series of calculations is the simplest of all considered. The purpose of the first series of calculations is not only to compare the difference schemes under study, but also to verify the correctness of the generated machine codes that implement the algorithms in question.

Numerical experiments were based on the function

$$T(x, y, z, t) = 2x + 2y + 2z + 12t + 0.05. \tag{6}$$

If the trace of function (6) on the parabolic boundary of the domain $Q \times (0, \Theta) = (0, 1) \times (0, 1) \times (0, 1) \times (0, 1)$ is chosen as the initial function $w_0(x, y, z)$ and boundary function $w_\Gamma(x, y, z, t)$, then, as it is easy to verify, function (6) is a solution of the mixed problem (1)–(3) for $C(s) = 1$ and $K(T) = T$. The temperature at the parabolic boundary of the region under consideration varied from $a = 0.05$ to $b = 18.05$.

It should be noted that the average temperature of the object at the initial moment of time $t = 0$ is approximately 3, and at the last moment of time $t = 1$ it is approximately 15. The temperature field in the object under consideration changes 5 times during the time interval.

As expected, in this series of experiments the accuracy of numerical solutions obtained using all three difference schemes on all spatial grids was very high. It varies from 10^{-14} to 10^{-13} in norm C and depends mainly on the time step value.

4.2   Second Series of Calculations

In the second series of calculations, we considered the option when the volumetric heat capacity $C(s) = 1$ and the thermal conductivity coefficient $K(T) = 1/T$. Numerical experiments were based here on the function

$$T(x, y, z, t) = -3.94(1.2x + 1.2y + 1.2z + 2.5t - 6.3). \tag{7}$$

If the trace of function (7) on the parabolic boundary of the domain $Q \times (0, \Theta) = (0, 1) \times (0, 1) \times (0, 1) \times (0, 1)$ is chosen as the initial function $w_0(x, y, z)$ and boundary function $w_\Gamma(x, y, z, t)$, then, as it is easy to verify, function (7) is a solution of the mixed problem (1)–(3) for $C(s) = 1$ and $K(T) = 1/T$. The temperature at the parabolic boundary of the region under consideration varied from $a = 0.2743$ to $b = 8.62$.

It should be noted that the average temperature of the object at the initial moment of time $t = 0$ is approximately 0.39, and at the last moment of time $t = 1$ it is approximately 0.99. The temperature field in the object under consideration changes 2.5 times during the time interval.

Tables 1, 2, 3 and 4 present the results of numerical calculations obtained using the local one-dimensional scheme (LOS) and the Douglas-Reckford scheme. Here, and in the following tables, the denotation $max - dev$ is the maximum relative deviation of the temperature field obtained as a result of numerical solution of the problem with the specified parameters from the analytical solution of the problem at the corresponding points. The denotation $time$ is the machine time spent on getting a solution for the specified calculation case.

Table 1. Spatial grid with $N = 25$

τ	LOS $max - dev$	LOS $time$	Douglas-Reckford $max - dev$	Douglas-Reckford $time$
0.1	$4.641593 \cdot 10^{-2}$	0.061	$1.109968 \cdot 10^{-1}$	0.228
0.04	$1.315722 \cdot 10^{-2}$	0.112	$1.643127 \cdot 10^{-2}$	0.468
0.02	$5.315369 \cdot 10^{-3}$	0.216	$5.503808 \cdot 10^{-2}$	0.934
0.01	$2.377212 \cdot 10^{-3}$	0.446	$7.785936 \cdot 10^{-3}$	2.241
0.002	$7.044234 \cdot 10^{-4}$	2.028	$5.408524 \cdot 10^{-4}$	7.058
0.001	$5.193939 \cdot 10^{-4}$	3.958	$1.109968 \cdot 10^{-4}$	12.499

Table 2. Spatial grid with $N = 50$

τ	LOS $max - dev$	LOS $time$	Douglas-Reckford $max - dev$	Douglas-Reckford $time$
0.1	$5.849756 \cdot 10^{-2}$	0.276	$1.317934 \cdot 10^{-1}$	2.889
0.04	$1.555370 \cdot 10^{-2}$	0.638	$6.115955 \cdot 10^{-2}$	3.782
0.02	$5.638431 \cdot 10^{-3}$	1.244	$2.630058 \cdot 10^{-2}$	6.661
0.01	$2.185763 \cdot 10^{-3}$	2.481	$8.740507 \cdot 10^{-3}$	11.762
0.002	$4.746511 \cdot 10^{-4}$	12.262	$5.550070 \cdot 10^{-4}$	55.079
0.001	$3.377296 \cdot 10^{-4}$	24.584	$3.347942 \cdot 10^{-4}$	97.884

Table 3. Spatial grid with $N = 100$

τ	LOS $max - dev$	LOS $time$	Douglas-Reckford $max - dev$	Douglas-Reckford $time$
0.1	$6.726045 \cdot 10^{-2}$	1.971	$1.459949 \cdot 10^{-1}$	87.532
0.04	$1.743157 \cdot 10^{-2}$	4.657	$7.368512 \cdot 10^{-2}$	40.556
0.02	$5.990507 \cdot 10^{-3}$	9.665	$3.705856 \cdot 10^{-2}$	67.348
0.01	$2.118525 \cdot 10^{-3}$	18.590	$1.439084 \cdot 10^{-2}$	118.880
0.002	$4.174050 \cdot 10^{-4}$	91.899	$6.466021 \cdot 10^{-4}$	526.100
0.001	$3.382760 \cdot 10^{-4}$	184.552	$3.356402 \cdot 10^{-4}$	906.239

Table 4. Spatial grid with $N = 200$

τ	LOS $max - dev$	LOS $time$	Douglas-Reckford $max - dev$	Douglas-Reckford $time$
0.1	$7.262894 \cdot 10^{-2}$	20.466		
0.04	$1.864297 \cdot 10^{-2}$	49.764		
0.02	$6.270611 \cdot 10^{-3}$	101.229		
0.01	$2.115959 \cdot 10^{-3}$	203.675	$2.123247 \cdot 10^{-2}$	2169.999
0.002	$4.010316 \cdot 10^{-4}$	987.788	$1.205677 \cdot 10^{-3}$	4031.018
0.001	$3.454224 \cdot 10^{-4}$	2012.961	$5.978218 \cdot 10^{-4}$	7268.598

Calculations using the Douglas-Reckford scheme at $N = 200$ and $\tau \geq 0.02$ give a solution that is far from analytical.

Using the Peaceman-Rachford scheme calculations were performed on spatial grids $N = 25$ and $N = 50$. For a spatial grid with $N = 25$ the Peaceman-Rachford scheme is stable when the time step is $\tau \leq 1/1300$. For $\tau = 1/1330$, we have $max - dev = 3.814857 \cdot 10^{-4}$, and $time = 8.496$. For the grid $N = 50$ and $\tau = 1/5500$, we have $max - dev = 3.396681 \cdot 10^{-4}$ and $time = 259.030$.

For smaller spatial grids, it is not advisable to perform calculations using this scheme, since it requires working with a very small time step. This, in turn, significantly increases the calculation time and increases the requirements for the used computer memory.

4.3   Third Series of Calculations

Function

$$T(x, y, z, t) = \sqrt{\frac{x^2 + (y + 1)^2 + (z + 2)^2}{7.5(10 - 97}} \tag{8}$$

is the solution of equation (1), in which the volumetric heat capacity $C(s) = 1$ and the thermal conductivity coefficient $K(T) = 1.5 \cdot T$.

As in the previous examples, in the formulation of the mixed problem (1)–(3) the trace of the function (8) on the parabolic boundary of the domain $Q \times (0, \Theta) = (0, 1) \times (0, 1) \times (0, 1) \times (0, 1)$ was chosen as the initial function $w_0(x, y, z)$ and boundary function $w_\Gamma(x, y, z, t)$. The temperature at the parabolic boundary of the region under consideration varied from $a = 0.6124$ to $b = 3.24$.

The average temperature of the object at the initial moment of time $t = 0$ is approximately 0.818, and at the last moment of time $t = 1$ is approximately 2.588. The temperature field in the object under consideration changes 3.2 times during the time interval.

Tables 5, 6, 7 and 8 present the results of numerical calculations obtained using the local one-dimensional scheme (LOS) and the Douglas-Reckford scheme.

Table 5. Spatial grid with $N = 25$

τ	LOS	LOS	Douglas-Reckford	Douglas-Reckford
	$max - dev$	$time$	$max - dev$	$time$
0.1	$1.164413 \cdot 10^{-2}$	0.150	$2.627338 \cdot 10^{-2}$	0.917
0.04	$5.996658 \cdot 10^{-3}$	0.148	$1.968669 \cdot 10^{-2}$	1.045
0.02	$4.974296 \cdot 10^{-3}$	0.292	$1.277706 \cdot 10^{-2}$	1.680
0.01	$3.583451 \cdot 10^{-3}$	0.556	$6.308627 \cdot 10^{-3}$	3.098
0.002	$1.020272 \cdot 10^{-3}$	2.697	$4.296597 \cdot 10^{-4}$	12.849
0.001	$5.287457 \cdot 10^{-4}$	5.416	$1.953000 \cdot 10^{-4}$	23.813

Calculations using the Douglas-Reckford scheme at $N = 100$ and $\tau > 0.01$ as well as when $N = 200$ and $\tau > 0.002$ give a solution that is far from analytical.

Using the Peaceman-Rachford scheme, calculations were performed on spatial grids $N = 25$ and $N = 50$. For a spatial grid with $N = 25$ the Peaceman-Rachford scheme is stable when the time step is $\tau \leq 1/5000$. For $\tau = 1/5000$, we have $max - dev = 1.403993 \cdot 10^{-5}$, and $time = 76.171$. In this case, using

Table 6. Spatial grid with $N = 50$

τ	LOS $max - dev$	LOS $time$	Douglas-Reckford $max - dev$	Douglas-Reckford $time$
0.1	$6.462635 \cdot 10^{-3}$	0.377	$1.316652 \cdot 10^{-1}$	18.236
0.04	$6.149037 \cdot 10^{-3}$	0.895	$7.181004 \cdot 10^{-2}$	21.844
0.02	$5.114160 \cdot 10^{-3}$	1.752	$1.386934 \cdot 10^{-2}$	19.180
0.01	$3.703145 \cdot 10^{-3}$	3.517	$7.613120 \cdot 10^{-3}$	25.716
0.002	$1.071107 \cdot 10^{-3}$	17.712	$7.052866 \cdot 10^{-4}$	97.943
0.001	$5.561920 \cdot 10^{-4}$	35.537	$1.949762 \cdot 10^{-4}$	182.303

Table 7. Spatial grid with $N = 100$

τ	LOS $max - dev$	LOS $time$	Douglas-Reckford $max - dev$	Douglas-Reckford $time$
0.1	$6.540254 \cdot 10^{-3}$	3.075		
0.04	$6.225773 \cdot 10^{-3}$	6.670		
0.02	$5.185260 \cdot 10^{-3}$	13.419		
0.01	$3.763439 \cdot 10^{-3}$	27.105	$9.566647 \cdot 10^{-3}$	238.834
0.002	$1.099449 \cdot 10^{-3}$	134.075	$1.152257 \cdot 10^{-3}$	819.354
0.001	$5.716632 \cdot 10^{-4}$	267.889	$3.400706 \cdot 10^{-4}$	1560.066

Table 8. Spatial grid with $N = 200$

τ	LOS $max - dev$	LOS $time$	Douglas-Reckford $max - dev$	Douglas-Reckford $time$
0.1	$6.578837 \cdot 10^{-3}$	26.168		
0.04	$6.263228 \cdot 10^{-3}$	66.390		
0.02	$5.220001 \cdot 10^{-3}$	132.344		
0.01	$3.793479 \cdot 10^{-3}$	249.948		
0.002	$1.114479 \cdot 10^{-3}$	1353.468	$1.607011 \cdot 10^{-3}$	10763.078
0.001	$5.801992 \cdot 10^{-4}$	2625.000	$5.755174 \cdot 10^{-4}$	13947.215

the Peasman-Reckford scheme on the spatial grid $N = 25$, the problem was solved with an accuracy of 0.0015% in a relatively short time. This high accuracy was not obtained using any other of the above schemes. And for the spatial grid with $N = 50$ and $\tau = 1/22500$ using the Peaceman-Rachford scheme, $max - dev = 2.897161 \cdot 10^{-6}$ and $time = 2648.968$, i.e. the accuracy of solving the problem in this case is even higher – about 0.0003%.

The analysis of the results of numerical experiments shows that the accuracy of solutions obtained using the locally one-dimensional scheme and the

Douglas-Reckford scheme for the same parameters can differ greatly from each other, and this is despite the fact that both these schemes are stable. The Douglas-Reckford scheme proved to be more "capricious".

The Peasman-Reckford scheme is noticeably different from them. This scheme is conditionally stable, which requires the use of a much smaller step in time than the other schemes under consideration. Therefore, it is advisable to use it in cases where calculations are performed on relatively rough spatial grids (20–50 nodes in each direction). If the calculations are carried out with a time step that is only slightly less than the allowed for stable operation of the scheme, the solution of the problem is obtained quickly and with high accuracy.

5   Conclusion

The choice of an effective difference scheme is very important when solving inverse and optimization problems, when the primal problem is solved at different values of parameters a large number of times. The efficiency of a particular scheme depends strongly on the dynamics of the temperature field.

To solve the three-dimensional heat equation, many researchers recommend using either one of the variants of the locally one-dimensional scheme or the Douglas-Reckford scheme, since both of these schemes are stable. However, the results of the research presented in this paper have shown that often the most effective (high accuracy at a small cost of machine time) for solving a primal problem on rough spatial grids is the Peasman-Reckford scheme, despite the fact that it requires calculations with very small time steps. The authors of this article carried out a series of works (see, for example, [18,19]) concerning the optimization of the metal crystallization process in the foundry. As a result of numerous studies, the Peasman-Reckford scheme was chosen to solve the primal problem in these works, which has proved itself well.

Of course, the schemes considered may behave differently on different problems. Therefore, when solving concrete inverse problem one should choose a suitable difference scheme based on the solution of the test problems.

As shown by the numerical calculations, the Peasman-Reckford scheme on rough spatial grids at a time step close to the maximum acceptable allows us to obtain high accuracy of calculations. Therefore, we recommend first solving the inverse problem on a rough spatial grid (20–50 nodes in each direction) using the Peasman-Reckford scheme. Then, having already a good approximation to the solution of the inverse problem, choosing a good grid and a difference scheme based on the required accuracy of the final result.

References

1. Zubov, V.I.: Application of fast automatic differentiation for solving the inverse coefficient problem for the heat equation. Comput. Math. Math. Phys. **56**(10), 1743–1757 (2016)

2. Albu, A.F., Evtushenko, Y.G., Zubov, V.I.: Identification of discontinuous thermal conductivity coefficient using fast automatic differentiation. In: Battiti, R., Kvasov, D.E., Sergeyev, Y.D. (eds.) LION 2017. LNCS, vol. 10556, pp. 295–300. Springer, Cham (2017). https://doi.org/10.1007/978-3-319-69404-7_21
3. Zubov, V.I., Albu, A.F.: The FAD-methodology and recovery the thermal conductivity coefficient in two dimension case. In: Proceedings of the VIII International Conference on Optimization Methods and Applications "Optimization and Applications", pp. 39–44 (2017). https://doi.org/10.1007/s11590-018-1304-4
4. Albu, A.F., Zubov, V.I.: Identification of thermal conductivity coefficient using a given temperature field. Comput. Math. Math. Phys. **58**(10), 1585–1599 (2018)
5. Albu, A.F., Zubov, V.I.: Identification of the thermal conductivity coefficient using a given surface heat flux. Comput. Math. Math. Phys. **58**(12), 2031–2042 (2018)
6. Albu, A., Zubov, V.: Identification of the thermal conductivity coefficient in two dimension case. Optim. Lett. **13**(8), 1727–1743 (2019)
7. Albu, A., Zubov, V.: On the stability of the algorithm of identification of the thermal conductivity coefficient. In: Evtushenko, Y., Jaćimović, M., Khachay, M., Kochetov, Y., Malkova, V., Posypkin, M. (eds.) OPTIMA 2018. CCIS, vol. 974, pp. 247–263. Springer, Cham (2019). https://doi.org/10.1007/978-3-030-10934-9_18
8. Evtushenko, Y.G.: Computation of exact gradients in distributed dynamic systems. Optim. Methods Softw. **9**, 45–75 (1998)
9. Evtushenko, Y.G., Zubov, V.I.: Generalized fast automatic differentiation technique. Comput. Math. Math. Phys. **56**(11), 1819–1833 (2016)
10. Albu, A., Evtushenko, Y., Zubov, V.: On optimization problem arising in computer simulation of crystal structures. In: Jaćimović, M., Khachay, M., Malkova, V., Posypkin, M. (eds.) OPTIMA 2019. CCIS, vol. 1145, pp. 115–126. Springer, Cham (2020). https://doi.org/10.1007/978-3-030-38603-0_9
11. Samarskii, A.A.: Theory of Finite Difference Schemes. Marcel Dekker, New York (2001)
12. Gao, Ch., Wang, Y.: A general formulation of Peaceman and Rachford ADI method for the N-dimensional heat diffusion equation. Int. Commun. Heat Mass Transf. **23**(6), 845–854 (1996)
13. Thibault, J.: Comparison of nine three-dimensional numerical methods for the solution of the heat diffusion equation. Numer. Heat Transf. Fundam. **8**(3), 281–298 (1985)
14. Peaceman, D.W., Rachford, H.H.: The numerical solution of parabolic and elliptic differential equations. J. Soc. Ind. Appl. Math. **3**(1), 28–41 (1955)
15. Samarskii, A.A., Vabishchevich, P.N.: Computational Heat Transfer. Editorial URSS, Moscow (2003). (in Russian)
16. Douglas, J., Rachford, H.H.: On the numerical solution of heat conduction problems in two and three space variables. Trans. Am. Math. Soc. **8**, 421–439 (1956)
17. Albu, A.F., Zubov, V.I.: Application of the fast automatic differentiation technique for solving inverse coefficient problems. Comput. Math. Math. Phys. **60**(1), 18–28 (2020)
18. Albu, A.V., Albu, A.F., Zubov, V.I.: Control of substance solidification in a complex-geometry mold. Comput. Math. Math. Phys. **52**(12), 1612–1623 (2012)
19. Albu, A.F., Zubov, V.I.: Investigation of the optimal control of metal solidification for a complex-geometry object in a new formulation. Comput. Math. Math. Phys. **54**(12), 1804–1816 (2014)

Author Index